CAMBRIDGE GEOGRAPHICAL STUDIES

17 CENTRAL GRANTS TO LOCAL GOVERNMENTS

CAMBRIDGE GEOGRAPHICAL STUDIES

Cambridge Geographical Studies presents new methods of geographical analysis, publishes the results of new research work in all branches of the subject, and explores topics which unite disciplines that were formerly separate. In this way it helps to redefine the extent and concerns of geography. The series is of interdisciplinary interest to a wide range of natural and social scientists, as well as to planners.

CENTRAL GRANTS TO LOCAL GOVERNMENTS

The political and economic impact of
the Rate Support Grant in England and Wales

R. J. BENNETT

CAMBRIDGE UNIVERSITY PRESS

CAMBRIDGE
LONDON   NEW YORK   NEW ROCHELLE
MELBOURNE   SYDNEY

Published by the Press Syndicate of the University of Cambridge
The Pitt Building, Trumpington Street, Cambridge CB2 1RP
32 East 57th Street, New York, NY 10022, USA
296 Beaconsfield Parade, Middle Park, Melbourne 3206, Australia

First published 1982

Printed in Great Britain
at The Pitman Press, Bath

Library of Congress catalogue card number: 82-4378

British Library Cataloguing in Publication Data
Bennett, R. J.
Central Grants to local governments.—
(Cambridge geographical studies; 17)
1. Grants-in-aid—Great Britain
I. Title
336'.185   HJ9424
ISBN 0 521 24908 2

CONTENTS

PREFACE

This book is concerned with one of the major aspects of public policy in western economies, that of central grants to local governments. The grants allocated from national government to local governments are surrounded in all countries by intense controversies and strongly-held political positions. Within the wide set of possible ramifications of such important and complex programmes this book has two primary aims. First, it seeks to evidence the distributional effects of one major grant programme, that of the Rate Support Grant in England and Wales. The second major aim of the book is to relate grant programmes to the overall structure of the tax and expenditure system of an economy, and from this relation to draw a conclusion as to the optimal form of central grants which can achieve a given pattern of fiscal equity.

The major thesis of this book is that central grants can be appraised only within the aggregate system of financial, legal and other relations between central and local government, and hence only within the aggregate fiscal structure of the state as a whole. As such, central grants can be appraised only against clearly specified objectives of central–local relations. The thesis of the book is developed in three parts. In part one of the book the objectives which have been applied to central grant programmes are examined, and the major species of grant formulae which can be used for allocation are defined; namely, grant formulae aimed at need equalisation, resource equalisation, cost equalisation, or combinations of each. International comparisons are used to develop the argument at this stage.

Part two of the book develops, against this background, a detailed analysis of one central grant programme, the Rate Support Grant used in England and Wales since 1967. The distributional effects of this programme are appraised in order to determine how far the objectives of need, resource and cost equalisation have been achieved. In addition, major examples of local authorities which have been relatively favoured or disfavoured by grant allocation are identified.

The thesis of the book then develops to encompass the interrelations between central allocation of grants and local decisions on the setting of levels of taxing and spending. This is first developed in the last chapter of part two of the book (chapter 9) in which the first major simultaneous equation estimates of British local finances are made. These estimates enable the joint supply and demand components of local financial decisions to be studied, whilst simultaneously controlling for central policy in grant allocation. The full interrelation of central grants with local taxes and expenditures is then developed in part three of the book. In this section the main burden of the conclusions of the book is developed in the context of a set of suggested reforms to the whole system of central–local relations in Britain. These suggestions rest upon three objectives: first, to attain full fiscal equity between both local authorities and client groups; second, to differentiate clearly the ranges of central and local responsibility and discretion for services; and third, to provide a more adequate revenue system. The suggestions for reform which are developed are: a modified grant system using reforms to the block grant principle, but resting upon the clear differentiation of discretionary and non-discretionary services; a modest increase in the use of specific grants and user charges; and the introduction of a shared income tax with variable local tax rates.

Within its objectives it is hoped that the analysis and suggestions for reform presented in this book will be found particularly apposite and stimulative at a time in Britain in which there is intense discussion of local financial reform: first in the aftermath of the 1976 Layfield Report and the 1978 proposals on devolution; and second in the light of the Green Paper of 1981 which has sought to examine alternatives to the domestic rates in Britain.

Since this book presents a specific thesis as to the position of central grants within the network of central–local relations, attention is restricted in the discussion to particular aspects of grants; in particular the book deals with the impacts of decisions, rather than with the process by which decisions were reached; and it treats aggregate statistical outcomes, rather than the individual decision-making process in individual local authorities. I hope, therefore, that the book will be read within its objectives: of proposing reforms within the context of aggregate statistical characteristics, developed through historical analysis of past distributional attributes of British grants, and discussed in the context of the international literature on the subject.

In writing this book I have become enormously indebted to a large number of individuals and organisations. I would like particularly to record my thanks to the many people with whom I have discussed the work and who have proved to be an important stimulus over the long period in which this book was written and rewritten, especially to David Smith who commented critically on the manuscript and kindly allowed me the use of his calculations in figure 5.4. Some of the early stages of the work benefited from the data files constructed by Mrs. Joan Rees whilst working on a separate project, supported by SSRC, concerned with the application of optimal control theory to allocation of the Rate Support Grant. I have also benefited, from the help of my Fitzwilliam College students Sally Howes, Virman Man, Bernard Szczech and Carol Wyatt; from the strong support of my departments, first in University College, London, and then in the University of Cambridge; and from the help of the Computer Centres in both institutions. The manuscript was typed by Mrs. J. Ashman, cartography was provided by Roy Versey, Arthur Shelley and Ian Gulley, and photographs were prepared by Bob Coe. To each of these I offer my grateful thanks. Figures 10.1 and 11.1 are reproduced by kind permission of Methuen and Co.

CHAPTER 1

CENTRAL GRANTS TO LOCAL GOVERNMENTS

1.1 Introduction

Central grants to local government have attracted major controversy and debate in recent years. They have been attacked as inequitable, wasteful, encouraging inefficient government and expansion of public expenditure, and as too heavily influenced by the political 'pork-barrel' in the competition for governmental funds. As a result, programmes such as the British Rate Support Grant, US Revenue-Sharing and Community Development Block Grant, Australian Income Tax Sharing, the French VRTS, and the German income tax sharing have increasingly become political footballs in election contests and debate. Central grants form a very significant proportion of public expenditure in all western countries and it is not surprising, therefore, that they should attract the attention alike of 'tax-cutters' and 'social reformers'. This book is concerned in various ways with these important social, political and economic issues. It employs detailed discussion of the Rate Support Grant in England and Wales as a means of highlighting many general issues of central grants common to most western countries.

Central grants to local governments are one method of sharing responsibility for taxation and expenditure between levels of government within a country. The constitutional organisation of most western countries involves some division of political power between central and local governments, with many countries possessing, in addition, a highly developed set of intermediate governments, such as the States of USA and Australia, the Provinces of Canada, the German Länder or the Swiss Cantons. As a result a major aspect of the organisation of the public sector in the modern state involves *intergovernmental* co-operation. Most important in such co-operation is the manner in which the local and intermediate levels of government are financed. Although methods of revenue-sharing, intergovernmental tax credits and deductions, loans, and even horizontal transfers between local governments, are employed to an important extent in many countries, it is grant transfers from central to local government which usually form the major component of intergovernmental financial co-operation. Certainly these are the most important transfers in both the USA and UK.

The aims of central grants to local governments have been variously ones of improving social equity, increasing economic efficiency, or stimulating economic growth. However, the issue of equity has most often been the overriding criterion adopted, and it is certainly the criterion which has occupied most attention in the literature. The distribution of grants has profound implications for both the level of public services available to individuals in different locations, and the external framework within which industrial and commercial enterprises must make decisions. Hence grants have a significant effect on both the 'social wage' and commercial 'externalities'. Moreover, the distribution of grants, and the behaviour of the local governments to which they are directed, are fundamentally geographical phenomena. They have profound impacts on the differential costs of location decisions and on the level of benefits received from the state by residents in different political subdivisions. Hence, grants are a major determinant of the degree of equity or inequity in public finance and, hence are a fundamental component in the economies of the space economy, determining what some commentators have termed: 'who gets what, where, at what cost'.[1]

The issues surrounding the procedures used to allocate central grants to local governments have received increasing debate in recent years in both academic and governmental circles. The main reasons for this growth in concern have, perhaps, been twofold. First, in most countries the last ten years has seen increasing criticism of the overall size of the public sector. As a result, central grants, because they tend to underprice local services, have been seen as a major cause of the expansion of public spending. Second, over the period since the late 1960s the proportion of local expenditure funded by grants has increased more rapidly than most other components of public spending. This has been mainly the result of the necessity for central government to expand local revenue bases which are often inflexible and non-buoyant to rises in incomes and the underlying rate of inflation. This, combined with calls for devolution and a reduction in central control, has given rise to calls for new sources of local finance based on local income taxes, or for the sharing of central revenues with central government. The subject of central grants is, therefore, not only topical but of major recurrent importance. However, grants are only one aspect of a complex interplay between the revenue and service responsibilities assigned to central, local and intermediate levels of government. Hence, in this book it is found necessary not only to discuss grants, which is the book's major purpose, but also to relate questions concerning grants to the related governmental environment. In particular three main aspects of the size, type and distribution of grants are developed at various points below: first, the relation of grants to the service responsibilities at each government level; second, the relation of grants to the distribution of revenues, their size and their buoyancy; and third, the relation of grants to the total tax and expenditure system as a question of total fiscal incidence as between different income groups, and between people and industry. Throughout the discussion it is the aim to use the grants of England and Wales to highlight general issues which are common to most western countries.

1.2 The Rate Support Grant (RSG)

The Rate Support Grant (RSG) has existed in Great Britain since 1967. Its form differs considerably between England and Wales on the one hand, and Scotland on the other. Most of this book is concerned solely with the RSG in England and Wales. The Rate Support Grant is an immensely important policy programme. Amounting to over £11 000 m. in 1982/3, it is responsible for the major proportion of transfers from central to local government in England and Wales. At this level of funding it must also rank as one of the most significant intergovernmental grant programmes in the western world.

The Rate Support Grant is a very complicated programme which has been modified continuously since its inception. However, it has possessed a common underlying theme of being concerned primarily with equalisation. Equalisation has been consistently tackled through three main aspects.

(i)   *Domestic rate relief:* a per capita subsidy to local residential taxpayers as opposed to commercial and non-domestic taxpayers.

(ii) *Resource equalisation:* a component aimed at equalising differences in the tax base of different local authorities.

(iii) *Needs equalisation:* a component aimed at equalising differences in the expenditure requirements of different local authorities.

The form of the RSG in Britain is not dissimilar from equalisation grants adopted in other countries. For example, the emphasis on attempting to equalise differences in local tax capacity is an important constituent in many US grant programmes (such as Revenue Sharing and Community Developments Block Grant), and is a major element in the transfer of funds from the federal level to Länder in Germany, Switzerland and Austria, to the Provinces in Canada, and to the States of Australia. In contrast the use of a component aimed at equalising the expenditure need of different local governments constitutes a more significant element of British grant programmes than in most other countries. The domestic element of the Rate Support Grant represents a specific subsidy element which has no relevant counterpart in the intergovernmental grant programmes of other countries.

The RSG has a complex history, which is detailed in later chapters, but over the years it has grown to become an enormously significant component of both UK local finance and of total public finance in Britain. The England and Wales rate support grant in 1981/2 represented about 13% of the total public sector expenditure at all levels of government. As such it is the largest single item of public expenditure excepting only defence. Moreover, it is enormously significant for local governments where, in 1981/2, it constituted about 40% of total local expenditure.

With such large sums of public expenditure involved, it is important that the eligibility criteria used for allocating grants to local governments are based on sound criteria of economic efficiency, social equity, and economic management. However, it has become increasingly clear that the manner in which the RSG has been organised does much to encourage economic inefficiency and although it attempts to be equitable, has been a major cause of inequalities in the social wage and commercial externalities between areas. As a result, a barrage of criticism has been directed at the RSG, and various suggestions have been made for its reform. Perhaps the most significant suggestions have derived from the 1971 Green Paper *The Future Shape of Local Government Finance*, the Layfield Report produced from the *Committee of Enquiry into Local Government Finance* in 1976, and the 1977 Green Paper *Local Government Finances* (see G.B. Government 1971, 1976, 1977). The *Local Government, Planning and Land Act* (1980) and the 1981 Green Paper (G.B. Government, 1981b) have attempted to overcome some of these criticisms but, as later chapters will demonstrate, many difficulties still remain.

1.3 Economic and political hypotheses

The literature on intergovernmental grants falls into *four* broad classes. First, there is the major body of works from economics and public finance. This emphasises the effects of grants on the price of goods and hence upon economic efficiency in the allocation of resources.[2] Second, and springing from the economic literature, is a subset of economic theory concerned with fiscal federalism.[3] This places emphasis on the division of public functions between levels of government and concludes that grants are appropriate for attacking problems of inefficient pricing, and perhaps even for local stimulation of economic growth, but are inappropriate for income redistribution or stabilisation of the economic cycle. The literature on fiscal federalism is particularly important in suggesting which grants are most appropriate for achieving a given social or economic goal. A third body of literature derives from political science.[4] This emphasises the interplay of power at each level of government and the political behaviour of actors in both the allocation of funds and in their subsequent use. A major component of this literature emphasises the role of grants in purchasing support for 'log-rolling' or 'pork-barrelling' votes; and this model has been applied particularly to US House and Senate Committees, and to the German Bund, but somewhat less successfully to the Westminster Parliament.[5] An additional component of the political science literature has emphasised the effect of grants on local financial behaviour. This analyses local behaviour as selfish political action and seeks to assess whether new programmes have been implemented, if new programmes have been substituted for old, whether tax cuts made, or whether salary rises and other forms of economic inefficiency have been fostered.[6] A fourth group of literature derives from economic geography. This has placed most emphasis on the expenditure side of local finance: the spatial distribution of social and economic benefits deriving from grant programmes.[7] Recently, however, a new emphasis has grown on the revenue side of the local sector examining the effect of grants on local tax bills and the distribution of costs for local programmes.[8] Each of these groups of literature has been mutually overlapping, and some flavour of each approach will be found in the following chapters.

Previous research on the Rate Support Grant has flowed from each of these groups of literature. For example, there has been a considerable economic literature concerned with the price and equity aspects of the RSG.[9] In addition, both the Layfield report and the debate on devolution in Great Britain stimulated a large range of research studies concerned with amendments to the RSG to integrate it with national economic management, to improve its aspects of equalisation, and to assess its effects on fiscal incidence.[10] The political science literature has also attacked these questions, but in addition has encompassed aspects of political party, and managerial or organisational effects.[11] In economic geography a diverse range of literature has been concerned with assessment of local expenditure need, measurement of access to local benefits, and comparison of revenue and expenditure positions of local authorities.[12]

This literature has brought us to the position of possessing a number of well-understood economic and political hypotheses of the Rate Support Grant and, in addition, it has allowed formulation of a series of partially confirmed, and often controversial, hypotheses of the economic and political impacts of RSG. As a consequence there are, perhaps, now four fairly well-accepted features of the Rate Support Grant which should be emphasised. First, there has been the effect of the dominance of the grant. Because it is so large and because it constitutes such a large proportion of local revenue it has often been claimed to have adversely affected economic efficiency by underpricing of local services, markedly reduced local autonomy, and greatly confused the lines of accountability for governmental services.[13]

A second feature of the Rate Support Grant has been the extent to which it has been allocated on purely objective principles and the extent to which it has been affected by political and other influences. Despite many criticisms, the RSG is, by international standards, a programme which has allowed a high degree of equalisation of expenditure and tax burdens between local authorities. However, it has often been claimed that there have also been significant political influences operated by central government, in general to favour urban areas under Labour administrations and rural areas under Conservative ones.[14]

A third feature of the Rate Support Grant, which is common to all grant programmes, has been the difficulty of measuring expenditure need. Considerable research[15] has been devoted to this issue from various points of view: analysing past expenditure patterns, quantifying service inputs, assessing service outputs, and measuring client groups. The method of need assessment used since 1981/2 is some improvement on the previous complex set of methods, but the issue is still far from resolved.

A fourth feature of the Rate Support Grant has been the difficulty of coping simultaneously with important area and service differences within a single grant programme. Allowance for service differences between rural and urban areas, Welsh and English counties, and London and non-London areas has been incorporated into the distribution formula for the RSG in various ways, but the procedures used have been essentially ad hoc, have been claimed to allow political influence, and certainly leave considerable room for improvement.

The literature on central grants to local governments, and the particular results of analysis of the Rate Support Grant, lead, therefore, to a number of outstanding questions as to the economic and political impact of the RSG. It is the intention of this book to resolve some of these questions. Particular attention is directed, in the following chapters, to the question of economic efficiency, means of improving autonomy and accountability, the extent of political influence on local spending through the level of grant allocations, the problems of needs assessment, and the question of the appropriate weighting for different types of jurisdiction within the same grant programme.

1.4 Research approach

The analysis reported in this book concerns the whole period of the existence of the Rate Support Grant in England and Wales, from 1967/8 up to the present. However, because of the reorganisation of local government in 1974, the reorganisation of London government in 1963, and a number of other factors, comparisons between years are not always reliable. For this reason detailed analysis is concentrated mainly on the years since 1974/5 for which reliable and comparable data are available. However, where possible comparisons are drawn from the period 1967/8 to 1973/4, and in addition some analysis also includes the later years of the General Grant and Rate Deficiency Grants for 1962/3 to 1965/6.

The analysis for the years 1962/3 to 1973/4 involves 163 local authorities in England and Wales: the Counties, County Boroughs and London Boroughs (some omissions of data result in a reduction from the total of

168 authorities which were extant over this period). The way in which the data for these years were handled is described in Appendix 1. For the period since 1974/5 the analysis uses the 116 local authorities: the non-metropolitan Counties (18 in Wales, 39 in England), 36 metropolitan Districts, and London Boroughs (12 Inner London and 20 Outer London, plus the City). The Isles of Scilly is omitted from the analysis in all cases. These units represent the so-called 'need authorities' used for initial distribution of the Rate Support Grant needs element up to 1980/81. The use of these units allows a simplified and more general analysis, but does require aggregating non-metropolitan Districts to give the County totals. In the case of most variables the County total is given as the sum of the representative District variables. For tax rates (rate poundages), the average rate poundage of the County is given as the rateable value of each District multiplied by its rate poundage, summed for all Districts, and then divided by the County rateable value. The calculation of the County averages in this way maintains the veracity of the original data and eliminates any distortion of the results. The post-1974 local authorities are shown in figure 1.1. The data sources used for this period are listed in Appendix 1.

For various purposes in the subsequent analysis classifications of local authorities are used. The definition of most of these is made clear in the text, but for two groups it is useful here to describe the definitions employed. The definition of party colour of the local council was derived by finding the controlling party and then grouping councils into four groups, as follows:

Party colour		*No. of authorities 1974–8*
A.	Conservative-controlled Councils	57
B.	Labour-controlled Councils	40
C.	Other party-controlled Councils	8
D.	Councils which change party control or which have no overall majority	11

This party control was averaged over periods corresponding to different party control at Westminster, i.e.

1.	1962–64	Conservative party at Westminster
2.	1964–70	Labour party at Westminster
3.	1970–74	Conservative party at Westminster
4.	1974–79	Labour party at Westminster
5.	1979–	Conservative party at Westminster

Use of this averaging process overcame two difficulties in subsequent analysis. First, it allows inclusion of a party continuity effect (cf. Alt, 1971). Second, this approach allows for the well-established feature of local government elections in Britain, that local party control tends to swing in the reverse direction of that at Westminster. However, as noted in Bennett (1982a), this classification has five main difficulties: first, it ignores

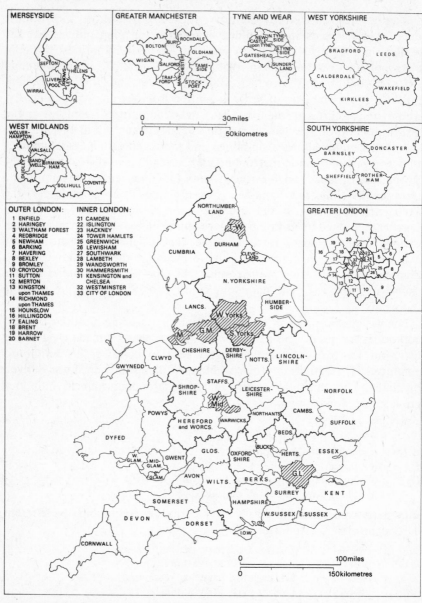

Fig. 1.1 Administrative areas in England and Wales. The map shows the 'needs authorities' as used for distribution of the rate support grant 1974–80 and are as constituted after reorganisation of local government in 1974. The insets show the metropolitan Districts within each metropolitan County and the London Boroughs within the Greater London Council.

differences in opposition parties; second, it ignores major differences in 'other party' attributes (group three in the classification); third, it ignores major differences in the majority party holding power; fourth, it ignores coalitions of non-partisan councils which often act as Conservative authorities under another name over this period; and fifth, it ignores the effect that party changes before or during the study period take time to have a significant effect on expenditure or grant patterns. Nevertheless, the classification employed is a simple and workable one, but the definitional difficulties, especially with respect to group three of 'other party' control, must be carefully borne in mind in interpreting the results.

This definition of party colour was one of the most difficult of the classifications to resolve. There have been a variety of approaches to measurement of local political party colour in expenditure and grant studies and these have followed four main approaches

(1)   Dummy variables; e.g. code an independent variable as 1 for Labour areas and 0 for Conservative or other areas.[16]
(2)   Absolute majority of a given party; usually, absolute Labour majority, in Britain.[17]
(3)   Degree of marginality of a Council's party control; e.g. ratio of controlling party to total seats.[18]
(4)   Political interaction variables; multiplicative relation of party majority or marginality with grants or other variables.[19]

These approaches differ in that methods (2) and (3) attempt to measure party effects as an interval or ratio variable, and this approach has also characterised method (4), whilst method (1) uses dummy variables to measure party effects at a purely nominal level. The interval data approach has obtained some important results. For example it allows the marginal shift in degree of party control to be assessed. Thus Oliver and Stanyer (1969) were able to estimate that a 10% increase in the number of Labour councillors in County Boroughs in 1964/5 produced a 2.31% increase in local rate poundage ($r^2 = 0.168$). This result was subsequently confirmed by Jackman and Sellars (1977*b*) who found a 10% increase in Labour councillors in metropolitan Districts in 1977/8 to produce a 1.57% increase in rate poundages ($r^2 = 0.111$). Similarly Gibson (1980) has found a multiplicative variable (using method (4)) of Labour councillors as a proportion of total members multiplied by needs elements of the RSG to be statistically significant and positively related to expenditure levels.

Despite these encouraging findings, however, it is the contention here that party control is fundamentally a categorical variable. A council is either controlled by one party or it is ruled by a coalition, and recent history of British local finance demonstrates that in many cases councils are willing to enact policies strongly in line with the ideology of the controlling party, even when this has the slenderest majority. Moreover, the evidence to support increasing enactment of party politics in relation to size of majority is not strong. It seems therefore, that party is best measured by the presence or absence of a particular party control, i.e. as a nominal variable. In the research that follows four categories are employed reflecting Labour, Conservative, other party, and no overall majority or party change. Since measurements of party control are derived from time averages this last category includes councils which have had particular party control but which have changed control over the period in which an average is taken. However, one feature of the interval scale methods of political variable measurement is used at many points below; this is the concept of multiplicative interactions with other variables.

The main consequences of adopting a nominal scale of measurement for party effects is that subsequent analysis must adopt particular statistical methods. Hence, for initial statistical hypothesis testing, analysis of variance (ANOVA) methods are frequently employed below; for estimation of the marginal changes resulting from given changes in grants, needs or other variables, categorical regression methods must be employed. Both methods mark a major departure from most previous approaches to the estimation of party effects used both in Britain and other countries. Using this classification, the distribution of party control over the 1974–78 period is shown in figure 1.2. For this period it is clear that the Conservative party had dominant control of the Counties and many metropolitan Districts, whilst for 1979–83, when central government control switched to the Conservatives, the Labour party took control of most of the metropolitan councils and many County councils.

A second grouping used extensively below is that of areas of differing urban stress. Various definitions of this variable could be employed: for example, in the USA considerable research has been devoted to obtaining an adequate index of urban stress.[20] However, for the purpose of the present analysis, the urban areas defined in the 1978 *Act* (G.B. Government, 1978) are employed. These define two classes of urban policy areas: partnership areas, and areas with urban programmes or special powers. As noted by Bentham (1980), these areas do not correspond very satisfactorily with normal indications of urban stress such as housing quality, social class, unemployment, and so forth. However, these areas do define local authorities which themselves recognised special urban problems and which were also accepted as urban problem areas by the Labour government of the mid-1970s. Whilst some political bias has undoubtedly entered these definitions, they nonetheless provide a useful basis for classification, and one which has been followed to a large extent by the Conservative government since 1978 in defining Enterprise Zones and Urban Development Corporations. The four classification groups which result are shown in figure 1.3 and are defined as follows

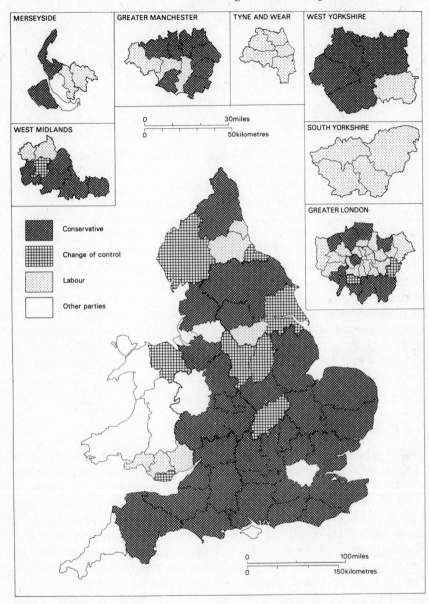

Fig. 1.2 Areas of average control by major parties 1974–78.

		No. of local
Urban stress areas		*authorities*
A.	Major urban stress areas (local authorities which have entered into partnership agreements)	14
B.	Urban stress areas (local authorities containing urban programme areas or special powers)	23
C.	Other urban areas (other metropolitan Districts and London Boroughs)	32
D.	Non-metropolitan Counties	47

A third variable used for subsequent analysis is that of local earnings. Unfortunately reliable data on earnings do not exist prior to 1974. Since that date, however, the *New Earnings Survey* provides information on average weekly wages in Counties and London Boroughs, but not metropolitan Districts.[21] These data show that distribution of earnings between local authorities gives a concentration of high earners in London and the south east as well as some in other areas (Cheshire, Cleveland, South Yorkshire and South Glamorgan). This distribution of earnings, of course, in large part reflects the differing costs of living between areas, but is nevertheless a useful index against which to judge domestic ratebills. One important feature should be noted: that with the exception of Lewisham, all of the inner area of London has a high number of large wage earners and a higher than average wage. Other metropolitan Districts have relatively low average wages, but for London it is clear that inner city problems are not associated as strongly as they are in the USA, with the incidence of personal poverty.

1.5 Plan of book

In the light of the various hypotheses of economic and political impacts of The Rate Support Grant, and following the heated debate which has marked the form of RSG used since 1981/2, this book is directed at two purposes. First, there is an examination of the patterns of inequality and inefficiency that have been induced by the form of RSG allocations used in the period since 1967/8. However, the second, and primary aim of this book is to argue the case for an improved practical form of grant allocations that might be implemented. Towards this aim, two primary sources of material are used: the analysis of present inadequacies in the grant procedure, and international comparisons. On the basis of this analysis it is concluded that, within the constraints of the existing information bases, and the political and constitutional structure of England and Wales, the best form of grant allocation that can be achieved must be based on a modified form of the unitary grant adopted since 1981/2. The modifications suggested relate to the methods of needs assessment, the way in which standard expenditures and standard rate poundages are determined, the treatment of different administrative areas, and the extents of continuity and stability that can be

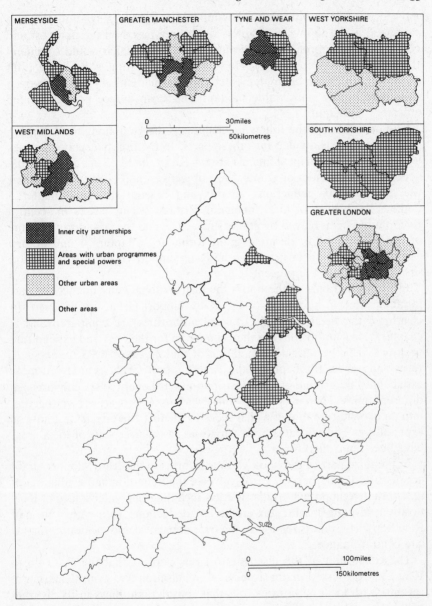

Fig. 1.3 Local authorities with various forms of status under the 1978 *Inner Urban Areas Act*. In many cases the designated zone is a small area contained within the local authority shown.

introduced. Other alternative possibilities for grant allocation are also examined in the following chapters, and those based on the unit cost and control theory approaches have particular virtues which would commend their use in an ideal world. However, in suggesting a *practical* grant procedure which can be implemented the modified unitary grant approach is seen to be the only viable method which can induce both economic efficiency and equity.

The discussion of the following chapters is directed along the following path. The book is divided into three parts. In the first part general issues underpinning British local finance are raised. In chapter 2 are discussed the general characteristics of grants, their objectives, and alternatives to their use, such as tax deductions, credits and revenue-sharing. Chapter 3 outlines the history of local finance in Britain. It emphasises the central role that has been played by grants since 1929, and the continuity of many of the ideas used in defining grant formulae. Chapters 4 and 5 then describe in detail the structure of the Rate Support Grant over the period since 1967/8 up to the present.

In the second part of the book, the distributional consequences of past formulae used in allocation of the Rate Support Grant are assessed. In chapter 6 the need positions of local authorities are compared and in chapter 7 local authorities are then compared for their tax and expenditure positions. This permits determination of the effects of RSG on expenditures and needs. In chapter 8 the Rate Support Grant is analysed more explicitly to determine how its distribution has benefited specific areas at different times. Then in chapter 9 these various aspects are drawn together into an expenditure model which seeks to identify the effects of grants on expenditures and tax rates once variations in other aspects of local areas have been controlled for.

In the third part of the book various reforms to the Rate Support Grant are discussed. Chapters 10 and 11 involve the definition and evaluation of an alternative approach to distribution including the wider issues of local taxation, revenue-sharing, tax-credits and deductions. Hence the reform of Rate Support Grant is related to the overall balance and incidence of local and central finance.

The conclusion of the discussion of this book is that the RSG up to 1980/1 has possessed major aspects of equalisation and central sharing of revenue with local authorities, but that considerable flaws in its distribution led to arbitrary (often incomprehensibly arbitrary) distribution patterns. Hence, although the RSG can be accorded an important role in limiting the development of financial difficulties of local authorities in Britain, it was nevertheless open to considerable improvements. After 1981 the form of RSG was considerably modified. Some of the flaws in the previous grant system were overcome, others were created. As far as the arguments of this book are concerned, the new RSG made an important step in defining standard service expenditure and in implementing a unitary block grant. However, it has tackled inadequately the problem of differentiating discretionary from non-discretionary services and has left untouched the problem of providing local authorities with a new and more buoyant revenue source.

The conclusions of the book also relate to possible local financial systems which could be used to replace the present RSG. These proposals are stimulated by the threee aims of (1) differentiating discretionary and non-discretionary services; (2) achieving full fiscal equity of individual taxes and benefits for non-discretionary service irrespective of location; and (3) providing a more adequate revenue system. The proposals which are made, and their distributional consequences evaluated are not all novel; indeed many date back to suggestions by Goschen in 1872. However, three points will be found fairly novel: first the suggestion for two sets of grants, one for discretionary and one for non-discretionary spending; second, for the implementation of a shared income tax; and third, the integration of central and local taxes *and* benefits through a form of negative income tax. Although full details for practical implementation cannot be given, it is hoped that this book will stimulate the evaluation of these suggestions; and indeed that it will lead to wider debate of central grants and local financial systems in Britain and in other countries.

PART 1

CENTRAL GRANTS: THEORY AND PRACTICE

CHAPTER 2

CATEGORIES OF CENTRAL GRANTS

2.1 Grant categories

Various types of central grants have evolved to satisfy different objectives and purposes and this chapter introduces the terminology generally applied in these different categories. A general typology of grants and revenue sharing can be constructed as shown in table 2.1, depending upon their generality, matching requirements and relation to need or capacity indicators. *General grants* (or block grants) are not tied to any specific spending programme and can be channelled by local governments (usually with relatively minor limitations and legal constraints) to whatever programmes they judge to be of greatest local concern or importance. General revenue-sharing in the United States and Canada, and the British Rate Support Grant are examples of this grant category. *Specific grants* (also termed selective, or categorical grants) are tied to specific expenditure programmes, e.g. education, health, public infrastructure[1] and so forth. They are also often very constrained even with these categories. Special revenue-sharing and categorical grants in the USA and the British specific grants for police, transport, national parks and the urban programme are examples of this category.

It is normally argued that general grants are consistent with 'dual federalism' in which revenues are re-assigned between levels of government, but that specific grants are consistent with 'co-operative federalism' in which revenue assignment is coupled with sharing of responsibilities between government levels: a balance of national funding for merit goods, local funding for local public goods, and so forth (Musgrave 1959; Oates, 1972, 1977).

Both general and selective grants can be adjusted by matching requirements. *Matching grants* in the USA, termed conditional grants in Canada, are not extensively used in Europe. They require local government to contribute a sum equal or proportional to the grant sum allocated by central government. *Non-matching grants* are unconditional and do not possess such requirements. The aim of matching is to induce a degree of local involvement, commitment, accountability and responsibility for that particular expenditure programme. Non-matching and general, uncon-

19

Table 2.1 *Typology of grants (after Musgrave and Musgrave, 1980)*

Type of grant		Unrelated to need	Need-related
General	non-matching	1	5
	matching	2	6
Specific	non-matching	3	7
	matching	4	7

ditional grants are more favoured in countries with a high degree of heterogeneity such as the USA, Canada and Australia. Non-matching grants are also often preferred on grounds of equalisation since it is easier for higher income areas to generate sufficient matching funds. On the other hand, matching and specific grants are often more applicable in smaller and unitary states with a high degree of local homogeneity in which the demands, standards and preferences for services may be more similar between jurisdictions.

Grants can also be differentiated by the degree to which they are need-related. *Need-related grants* are allocated on the basis of measured geographical differences in needs, or costs of servicing those needs, or a combination of these elements. Examples of such grants are included in the British Rate Support Grant, Canadian Equalisation Grants and many US State Education Finance Grants. *Non-need-related grants* do not take geographical variations in unit costs of need into account.

Grants can also be distinguished by whether they are *open-* or *closed-ended*, i.e. whether central government will finance all eligible requirements, or whether an upper limit of grant transfer is set. A major problem of open-ended grants is that they give a subsidy to local areas in direct proportion to their spending and hence give an incentive to increase public expenditure. Closed-ended grants, in contrast, have the disadvantage that they require a level of cut-off in aid to be defined, with subsidies being paid only to those Local Authorities below this level. This leads to a non-proportional level of subsidy between areas which in turn gives unequal pricing and induces inequity between local authorities.

In addition to the various types of central grants displayed in table 2.1, a number of other means of transferring resources between levels of government are also available. The most important of these are revenue-sharing, tax credits and deductions, reassignments of tax revenues, direct expenditure, and spatially differentiated central taxes. This book is primarily concerned with central grants to local governments and hence the discussion of alternative means of inducing intergovernmental co-operation is fairly limited. However, when in part III of the book we evaluate a more rational basis for local finance in Britain, it is necessary to return to this issue, and in particular to appraise the potential for forms of both revenue-sharing and reassignment of tax sources.

2.2 Objectives of central grants

Different categories of grants will be appropriate in different situations. The following are twelve of the major explicit and implicit objectives which are normally envisaged in choosing between grants which are general or specific, matched and unmatched, need-related and non-need-related or are open- or closed-ended.

(i) *Equalisation*

This is the major aim of RSG and concerns how far the formula for grant distribution achieves equalisation between jurisdictions, in terms of revenues, needs, costs, other criteria, or a combination of several criteria. The primary grants used for equalisation are general block grants (such as the Rate Support Grant) and general revenue-sharing allocated on the basis of need. Matching is often employed when it is sought to affect the balance of public and private choice in the local area, or where it is sought to control the direction of funds, e.g. when equalisation is aimed at only single sectors, e.g. the specific need-related matched grants used in the USA HEW categorical grants.

(ii) *Neutrality to income distribution*

A major aim of grants to local governments is to ensure a similar pattern of services to each local authority. Some of these services are orientated specifically to client groups of need (the old, the young, disabled, those needing public transport, and so forth). Such services are clearly redistributional in intent. Hence, the expenditure side of local finance has an intentional income effect. However, on the revenue side, it has usually been taken as an explicit aim of local taxation and grant programmes that they should be relatively neutral in their income effects. This is important for two reasons. First, it is usually deemed preferable to solve the income distribution issue of taxation at a national level; people should not bear different tax burdens merely because they live in different areas (see Musgrave and Musgrave, 1980; Oates, 1972, 1977; Hunter, 1977; Bennett, 1980). Second, marked differences in tax burdens for income groups between areas provides an important stimulus to migration, generating so-called fiscal migration (Tiebout, 1956; Oates, 1969). Such migration presents important difficulties since it frequently allows the rich to escape their 'just' tax burden and hence induces regressivity into the tax system.

Hence a major aim of RSG, like grant programmes in most countries, has been to reduce, as far as possible, differences in tax rates and burdens on *similar* groups living in *different* areas.

(iii) *Fiscal balance and adequacy of local resources*

Deriving from the objectives of equalisation and neutrality to income distribution, a third main objective has been to maintain fiscal balance between levels of government. Many public services can be efficiently organised only at local levels, but most of the high-yielding revenue sources are usually best reserved to central governments (especially personal and corporate income tax). In addition most revenue sources which can be used at local level are relatively regressive, e.g. property tax and many forms of sales tax. Hence, it is necessary to employ some mechanism for reallocating revenues between levels of government in order to achieve fiscal balance: both to ensure the adequacy of local revenues and to limit the extent to which regressive tax bases are used. This is a primary motive in the RSG, but methods of revenue-sharing and systems of tax credits and deductions can also be used to achieve fiscal balance as in the USA, Germany, Australia, Canada and Switzerland (see Bennett, 1980). It is also a major motivation for need-related matching grants which are general or selective depending on the extent of revenue mismatches.

(iv) *Neutrality to local behaviour*

A primary aim of most grants is that, after achieving equalisation and fiscal balance, the grant distribution should be free from effects of manipulation by local governments unfairly to increase their allocations. Hence, the grants should be neutral to local authority choice to provide a particular form or quality of service. This requires the distinction of discretionary from non-discretionary spending, and this in turn usually requires the definition of both 'standard' or 'minimum' service levels and 'standard' tax rates. This in turn usually leads to a choice of closed-ended rather than open-ended grants.

(v) *Feasibility*

Grant programmes can be implemented only with data which it is feasible to collect and the accuracy of which is beyond question, i.e. it must be possible to implement distribution in practice. This often leads to rather cruder formulae for grant distribution than would be desired in an ideal world. In particular it is often impossible to implement grants requiring detailed costing of local services or, in Britain, grants which use detailed data on personal incomes for local areas.

(vi) *Facilitating local discretion*

Related to the previous aims, grants may seek to encourage or suppress local discretion in providing given services. It is usually considered that local governments are important agents of democracy. As a result, grant programmes should aim to facilitate that degree of local discretion in taxing and service provision which is consistent with encouraging the democratic process at local level. This aim requires the differentiation of responsibilities for different expenditure functions between levels of government: central government being responsible for national 'merit' goods, local governments for 'local' goods, intermediate governments for 'regional' or State goods, and so forth. For example, to achieve 'minimum standards' in national goods at local level, matched non-need-related forms of grants are usually employed and these may be general or specific depending upon the spending sector involved (see Musgrave and Musgrave, 1980). With large variations in local fiscal capacity, however, the local matching requirement cannot be employed to achieve 'minimum standards' and this is the case for the Rate Support Grant.

(vii) *Accountability*

Related to the issue of differentiating local from central (and intermediate level) government decisions on taxing and spending is the aim of making local decisions clearly accountable to the local electorate. This also relates to maintaining the democratic process. Accountability requires clear lines of responsibility for local financial decisions to be visible to the local voter. This requires that local taxation and other charges (the local 'costs' of local services) represent significant proportions of local spending and clearly reflect local expenditure decisions: an issue emphasised in the Layfield Report (G.B. Government Committee of Enquiry, 1976). This in turn requires that grants should be kept below a certain level as a proportion of local spending or else local services will become markedly underpriced relative to their costs, and this will affect demand, efficiency (q.v. below) and hence accountability.

(viii) *Intelligibility*

A major aim of grant programmes should be that they are readily comprehensible both to the voters and to the elected representatives at each level of government concerned. This aim relates to the accountability objective and ensures that the role of political and professional judgements is laid bare. This often encourages preference for specific grants with clearly specified aims rather than general grants.

(ix) *Encouragement to efficiency/productivity*

Grants should not subsidise profligate local spending or unjustifiably high labour/output ratios. This requires that grants allow as close as possible a relation between services provided and their 'cost' through taxation. Clearly this aim is often severely limited by that of equalisation such that the question is usually one of the appropriate fiscal balance. Efficiency also requires that grants should aim to have a minimal impact on the price rules in the private sector, thus maintaining efficient allocation of resources in the economy as a whole. This leads to favouring grants which incorporate a measure of cost comparison between local authorities and 'standard' costs for given levels of service provision. A common form of such grants is the 'unit cost' approach (q.v. pp. 135, 138–9).

(x) *Compensation for cost and benefit spillover*

Grants can also provide a primary means for overcoming either local cost spillovers (one jurisdiction pays a part of the costs of another jurisdiction and receives no benefits) or benefit spillovers (one jurisdiction receives benefits from another jurisdiction at no cost – the 'free-rider' problem). The use of grants in such cases derives from Pigou (1947) and is usually referred to as the *compensation principle*. Where compensation for benefit spillover is sought, selective and open-ended matching grants can be used. Where it is sought to equalise differences in the local tax rates required to give equal levels of local service benefits, general unmatched grants such as the RSG are employed, but experience with such grant programmes as the US Medicaid and AFDC funding suggests that they are very vulnerable to corrupt local practices.[2]

(xi) *Stability and certainty*

Grants should normally, as far as possible, maintain a stable pattern of distribution from one time period to another, with certainty as to the magnitude of the total sum available. This encourages efficiency in local authorities and aids their expenditure planning. Changes in grant distribution procedures should be introduced only as a result of objective changes in circumstances: either (a) changes in the local need to spend and ability to raise revenues, or (b) changes in national priorities resulting either from national political shifts (such as the party colour of central government) or from the effects of overall economic management criteria which modify the total sums available.

(xii) *Objectivity*

Related to stability and certainty, it is normally envisaged that grants should be allocated, as far as possible, upon an objective basis which is clearly visible to all. This allows the distribution of grants to be justified by and appraised against clear social, economic and political goals. Clearly objectivity can never be completely achieved, particularly in the measurement of expenditure needs. However, what must be guarded against is a grant programme which allows arbitrary or ad hoc administrative or political decisions which are not justified by clearly identified motivation. In Britain, the Rate Support Grant has often fallen far short of achieving a rational and objective distribution.

Each of these criteria plays a role, relatively major or minor, in the design and practical implementation of any grant programme. For many grants, however, the primary aim is one of equalisation. Since this has also been the primary aim of the Rate Support Grant, this book is primarily devoted to developing in more detail the major categories of equalisation grants that can be implemented.

Equalisation grants rest on a background principle of *fiscal equity*. This principle is applied in this book only to those components of public finance which derive from the actions of local government. Hence fiscal equity applies to local authorities as units, although under various assumptions it may sometimes also be deduced to apply to particular groups of individuals living within those units. Under this definition fiscal equity is attained when each local authority is able to provide services at a given level at the same tax rate. Fiscal equity is not concerned with achieving equal tax rates, equal service levels, or equal levels of expenditure per head between local authorities, except in so far as there may be concern that certain minimum service levels are achieved. Rather, fiscal equity is concerned with ensuring that different local authorities have the same capacity to provide a given standard of service at the same overall cost. This requires the analysis of tax bases, service needs, service costs, productivity and efficiency of service provision, variation in the choice of services provided, and a number of other factors. Hence, in an ideal world, equalisation grants must incorporate measures of tax base, need, costs, productivity, efficiency, and preferences. However, as we shall see in the following discussion, this is by no means a simple task.

Apart from the aims of equalisation, other objectives of grant allocation are also important in different circumstances. For the Rate Support Grant, the issues which are particularly important in addition to that of equalisation are the degree of local discretion, adequacy, neutrality, encouragement of efficiency, stability, certainty and data feasibility. As a result, these criteria emerge at many stages to restrict the form of equalisation grant that might be implemented in an ideal world. Moreover, as will be seen in the following discussion for the RSG the extent to which one objective has been used or publicised rather than another has varied a great deal over time.

2.3 Resource-equalising grants

Resource equalisation seeks to achieve fiscal equity in the revenue base of different local governments. Ignoring differences in services costs and needs for the moment, revenue equalisation will be achieved when each Local Authority can raise the same revenue by exacting the same tax rate. That is, resource equalisation is achieved when each Local Authority makes the same tax *effort*. To implement such grants two measures of the local revenue sector are required: first, a measure of local government tax bases, and second, a measure of local government *tax rate* on efforts.

Differences in the tax resources between local authorities can be measured, in a meaningful sense, only against the base from which taxes are derived. Thus, for example, in Britain where the only local tax is the property tax of the rates, there is no point in assessing the personal incomes of the people in each area to give an indication of tax capacity, since income at present represents a tax base which cannot be tapped at local level. Hence in Britain the appropriate measure of the local tax base is the rateable value of the total property in the Local Authority. In more complex systems of local finance, such as that obtaining in the USA, where many different local taxes are available, the tax base must be measured with respect to some average of the various sources available, such as the representative tax capacity measures suggested by US ACIR (1962, 1971).

The difference in fiscal effort between local governments measures the extent to which a given local authority is using its tax base: the relative pressure it is placing on its tax payers. Measures of such fiscal effort will vary in complexity depending upon the diversity of the local tax system. In Britain with only one local tax, relative tax effort is measured simply as the tax rate exacted in each local authority (the rate in the pound, or rate poundage). With diverse revenue sources such as those characterising USA State and local government, tax effort must be measured by an average of the tax rates weighted by the extent to which each revenue source is used. The problem of inequity that arises from unequal local authority tax bases can be understood from figure 2.1. Four local authorities A, B, C, and D with differing tax bases (rateable value per head) can raise radically different amounts of local tax at the same tax rate. For authorities B and C, for example, the tax yield at a tax rate of 30 p in the £ will be £30 and £45 per head, respectively. Conversely, different local authorities can raise the same total revenue at differing tax rates. Again,

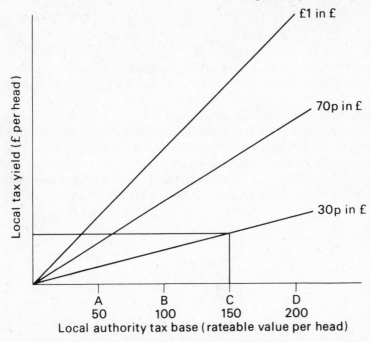

Fig. 2.1 Different levels of tax yield depending upon size of tax base and local tax rate for four local authorities and three specimen tax rates.

for local authorities B and C, the £45 per head of tax yield raised by C at 30 p in the £ can only be obtained by raising taxation at the much higher rate of 45 p in the £ (tax rate = tax yield/tax base).

Revenue equalisation grants seek to overcome this problem of inequities in the tax base by distributing grants preferentially to those local authorities with small tax base and high tax effort as against those local authorities with large tax base and low tax effort. The precise magnitude of these grants which will achieve the stated objective of fiscal equity of the revenues (that each local authority can raise the same revenue by exacting the same tax rate) can be defined only by comparing each local tax base B_i against a 'standard' tax base $\bar{B}$. The difference between this standard and the actual local base then gives the size of the grant

$$G_i = (\bar{B} - B_i)K \tag{2.1}$$

The size of the grant is multiplied by a constant K, which scales the grant to the size of the local area, e.g. its population, expenditure level, etc. However, it is normal to scale the grant by the magnitude of the total local expenditure outlays, E_i, per unit of tax base, i.e.,

$$G_i = (\bar{B} - B_i)(E_i/B_i) \tag{2.2}$$

or

$$G_i = (\bar{B} - B_i)t_i \qquad (2.3)$$

where the second equation derives from recognising that the total local expenditures, divided by the local tax base, gives the local tax rate t_i, or tax effort.

This form of revenue equalisation is achieved by the central government acting as a local taxpayer on that extra quantity of local tax base required to give each local authority a base (rateable value per head) equal to the national standard. Grants employing distribution formulae such as equation (2.3) are essentially those used in the RSG between 1958 and 1980. This form of grant is shown in figure 2.2 using a tax rate of 30 p in the £ as an example. This grant certainly satisfies the criterion of providing each

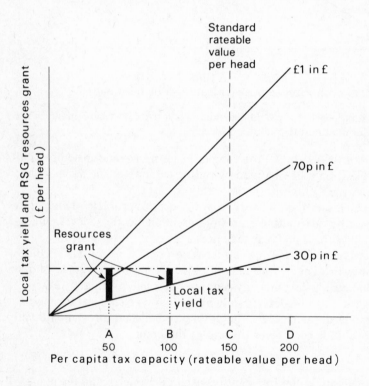

Fig. 2.2 Revenue-equalisation grants based on the deficiency payments principle for local authorities below a specified standard tax base per head at a specified local tax rate.

local authority with the same revenue at the same tax rate and hence equalises fiscal effort. As deficiencies in the local tax base become greater, so proportionately does the grant. Conversely, as surpluses in the local over the standard tax base become greater, so larger negative transfers act to reduce the local 'excess' tax capacity. Hence, formulae such as (2.3) can be seen as a form of fully-equalising, open-ended proportional grant.

Three major difficulties with such grants arise. First, the grant is open-ended and hence the central government cannot determine exactly the level of total grant support it will be required to give, and this creates problems for budget management. A second problem arises because the grant is not independent of (or neutral to) local authority behaviour: the level of grant is determined, in part, by the local authority choice in setting its expenditure level and hence its tax rate: the so-called 'feedback effect'.

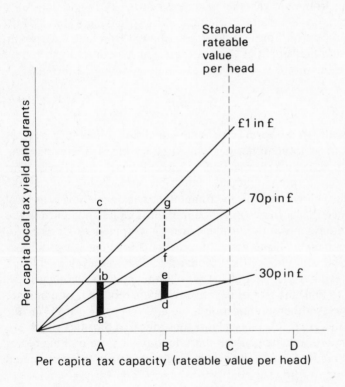

Fig. 2.3 Difference in levels of revenue-equalisation grants on the deficiency payments principle deriving from differences in local tax rate levied.

This problem can be understood from figure 2.3. Since the central government performs as a local ratepayer on the deficit in local property tax base, the actual level of grant payable will vary with the level of local property tax (rate poundage) that is levied and the grant provides an incentive to tax at higher rates. In figure 2.3 two local authorities, A and B, fall below the national standard and receive resource-equalising grants, but the level of grant varies depending on the tax rate levied. For example, at a rate poundage of 30 p in the £ both local authorities receive a lesser grant than at 70 p in the £. (In each case the grant payment *bc* and *fg* for local authorities A and B, respectively at 70 p in the £, exceeds the grant *ab* and *de* at 30 p in the £.) In the British Rate Support Grant the two problems of open-endedness and feedback were overcome up to 1980 by incorporating estimates of local expenditure (and hence tax rate) into the allocation formula and then making proportional ex post adjustments for differences of the estimates from outturns: the 'resources clawback'. In addition, since 1981/2, feedback has been largely eliminated by specifying a standard tax rate $\bar{t}$ for given groups of local authorities (London Boroughs, metropolitan Districts, non-metropolitan Counties, etc.) and by tapering to reduce grant at high expenditure levels. This results in an amended formula

$$G_i = (\bar{B} - B_i)\bar{t} \tag{2.4}$$

A third difficulty in the form of revenue-equalising grants such as (2.3) or (2.4) is that those local authorities with large tax bases greater than the standard are 'taxed' by resources being moved away from them. This procedure is usually both politically infeasible and it also reduces the incentive for those local authorities to maintain reasonable levels of local expenditure since this becomes reflected in their tax rate and hence in the magnitude of their contribution to the central government. As a result of this effect resource-equalising grants in practice are frequently modified. One modification is to choose the 'standard' tax base at a relatively high level, perhaps even at the level of the local authority with the highest tax base, in order to limit the size of negative transfers required, or eliminate them altogether. Another modification is the *deficiency payments principle* which applies the grant formula only to those local authorities which are below the 'standard' tax base.[3] Both these features, choice of a high local authority tax 'standard' and a deficiency payments principle were applied to the Rate Support Grant up to 1980.

2.4 Need-equalising grants

Need equalisation seeks to achieve fiscal equity in the expenditure requirements of different local governments. This will be achieved when each local authority can expend the same proportion of resources on a given service at the same level of service performance. In this case differences in the tax base of local governments are initially ignored. However, such grants can be implemented only after differences in the level of need and service performance between local governments have been measured.

Differences in levels of need arise primarily from different numbers of client-groups (the size of population eligible for each local service) and differences in size, density and area of local authorities. This gives rise to marked differences between local governments in the levels of total expenditures, per capita expenditures, and expenditures per unit of service delivered. However, differences in expenditure need arising from the local choice to provide services at a higher standard must also be allowed for. This is usually accomplished by measuring expenditure need only for those services which are provided on a general basis for all areas; so-called 'beneficial services'. This leads to the definition of a 'standard' of needs, which is frequently termed a 'national minimum standard'. For single-purpose governments, expenditure need is relatively easy to assess. But for multi-purpose local governments such as those in Britain which are involved in delivery of a wide variety of services, a representative needs measure is required which can aggregate the expenditure requirements weighted by the needs in each different service category (see Auten, 1974; Bennett, 1982b, and the discussion of chapters 6 and 7).

Differences in service performance result from different local authorities expending different quantities of resources to produce a given standard of service, i.e. local authorities use different quantities of inputs to produce the same service output. Detailed measurement of performance levels entails the complex procedures of unit cost grants discussed in a following section. A simple procedure for measuring performance levels, which is often used instead of unit costing, however, is to specify a level of outlay per unit of need which is required to produce a given standard of service. This represents the 'on average' or 'standard' expenditure per capita required of a local authority to provide a given standard of service. This ensures that different expenditure needs are compensated for only at that level of services, and for those services, which are provided for all areas at a given standard level of needs. Hence, measurement of the performance level for different services adds to national minimum standards of need equalisation, a level of 'standard' need equalisation. The measurement of spending need is, in fact, an immensely complex and controversial issue which is left for further discussion in later chapters. However, accepting for the moment that differences in local expenditure need of each local government can be defined, central grants which aim at need equalisation will be distributed preferentially to those local authorities with high needs relative to national standards, as against those with relatively low needs.

The magnitude of these grants is given by scaling the difference of local needs N_i from national 'standard' needs $\bar{N}$ as follows

$$G_i = (N_i - \bar{N})K \qquad (2.5)$$

where the size of grant is determined by a constant K. This constant is often chosen so as to reflect the level of minimum performance or outlay M required of each local authority to produce a given standard of service, i.e.

$$G_i = (N_i - \bar{N})M \qquad (2.6)$$

The form of this grant can be understood from figure 2.4. Four local authorities are scaled by their relative expenditure need measured as £ per

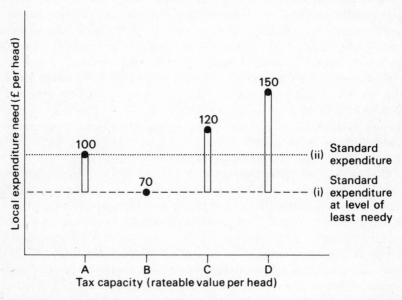

Fig. 2.4 Needs-equalisation grants derived from two settings of standard need: (i) at the level of the least needy – full equalisation with no negative transfers; (ii) at a standard level of need giving full equalisation only if there are negative transfers.

head. Each local authority is then placed on an axis related to its tax capacity (as in figures 2.1 to 2.3). If the standard level of expenditure need (minimum standard) is defined at the level of the lowest need (of for example, local authority B) this yields a form of equalisation fully compensating for need differences above the minimum although in no sense proportional to differences in need. However, a higher level of standard need can also be chosen, such as that of local authority A, in which case only a proportion of need differences is compensated.

   The form of need grant in equation (2.6) has most of the desirable
features of providing a level of grant in proportion to deviations from
standard levels of need. The scale of these transfers becomes greater the
larger the local needs, whilst local authorities with needs below the
standard will suffer large negative 'taxes'. Since, as in the case of the
negative transfers for achieving resources, such 'taxes' give severe disin-
centives for local tax effort, it is normal either to set the standard of needs
very low (usually at the least needy local authority, as in the case of local
authority B in figure 2.4); or to impose a constraint that grants are only
paid to a local authority, not subtracted away from them. Both of these
modifications have been employed in the case of the allocation of the Rate
Support Grant.
   One problem with need-equalising grants such as that given in equation
(2.6) is that they do not compensate for need differences in any propor-
tonal way. In figure 2.4, for example, local authority B using the standard
of least need, receives nothing, whilst authorities A, C or D, with respective expenditure needs of £100, £120 and £150 per head, will receive needs
entitlement of £30, £50, and £80 per head respectively (a total allocation of
£160 per head). Hence, this form of grant places different burdens on the
total rate bill in different local authorities, and hence induces inequities in
overall fiscal equity between areas. Those authorities with high per capita
expenditure need will have a lower rate bill per unit of need than those
areas with low per capita need. Hence, local services will be relatively
underpriced per unit of supply in areas with high need in comparison with
those with low need and this may encourage inefficiency and oversupply of
services. This suggests an alternative *proportional need-equalising* grant
proposed in the Green Paper (G.B. Government, 1977) given by the
equation

$$G_i = \frac{G_{TOT}}{N_{TOT}} N_i \tag{2.7}$$

where G_{TOT} is the total per capita grant available, N_{TOT} is the total level of
per capita need in all local authorities, and N_i is the level of need in
authority i. Under the proportional system of needs allocation the same
proportion of expenditure is met in each local authority in figure 2.4, and
with the same allocation of £160 per head, the entitlements would be £36,
£25, £44 and £55 for authorities A, B, C and D, respectively. This can be
checked since £160 represents 36.9% of the total local authority expenditure needs of £440 per head. The difference between the proportional
needs grants (2.7) and those in equation (2.6) demonstrates that a proportional grant would give more to the lower need areas, and less to the
higher need areas, but that a constant ratio of central government support
to local revenue burden would be maintained.

2.5 Combined resource and need-equalising (Unitary grants)

Unitary grants combine features of each of the two previous approaches of independent resource and need equalisation. The combination of the two forms of equalisation permits each local authority to expend the same proportion of resources on a given service at the same level of service performance, irrespective of differences in absolute levels of service need, whilst simultaneously exacting the same tax rate. The unitary grant approach has the advantage of simultaneously compensating for inequities in tax bases and inequities in the local need to spend. To implement such grants measurements are required of the four elements of: tax base, tax effort, service performance, and level of need.

Using the same measures of these elements as those given in the preceding paragraphs, central grants aimed at simultaneous resource and need equalisation will be distributed preferentially to those local authorities with low tax base, high tax effort and high expenditure need at a given level of service performance. In contrast, smaller grants, or no grants at all, will be paid to local authorities with high tax bases, low tax efforts and low expenditure needs at a given level of service performance. Various forms of unitary grant have been suggested both in the academic literature and in the debate over grants in Britain.[4] Some of the most important grant formulae are discussed below.

Unitary grant (1)

This is the most direct form of combined needs and resource equalisation suggested originally by Musgrave (1961): it has three main forms, the first two of these are given by

$$G_i = (\bar{B} - B_i)t_i + (N_i - \bar{N})M \qquad (2.8)$$

and

$$G_i = (\bar{B} - B_i)t_i + (N_i - \bar{N})\bar{B}t_i \qquad (2.9)$$

In both these equations equalisation is derived by simultaneously equalising tax effort on the standard tax base. Since deficiencies in the tax base are compensated for in the first part of the formula, the second part of the formula compensates only for those differences in the need to spend which will not be provided for in the first part. In the first formula (2.8), there is the simple combination of the two previous equations (2.3) and (2.6) for independent resource and need equalisation. In the second formula (2.9), the need equalisation component is modified by replacing the measure of minimum performance levels M by a term $\bar{B}t_i$ to reflect the yield from local taxes applied to the standard tax base. This second formula recognises that, since differences in the need to spend must be paid for by differences in revenue effort, equalisation of needs can be achieved by scaling grants to the tax effort, but only in relation to the standard tax base. This guarantees equal levels of resources for given expenditure need, i.e. it guarantees equal levels of service performance. However, it has the disadvantage, like the resource-equalisation grant (2.3) of being dependent on the behaviour of local authorities in setting the tax rate t_i: the feedback effect. Hence, this method usually has to be combined with the setting of a standard tax rate $\bar{t}$. This gives a third form of unitary grant

$$G_i = (\bar{B} - B_i)\bar{t} + (N_i - \bar{N})\bar{B}\bar{t} \qquad (2.10)$$

which includes standard tax rates $\bar{t}$ throughout. The magnitude of grants obtained from both formulae (2.8) and (2.10) is shown in figure 2.5. This figure shows two forms of grants. First, figure 2.5A shows the actual pattern of combined need and resource equalisation used in the RSG up to 1975: needs are compensated above a deficit in local tax yield. The second form of grant, figure 2.5B, shows the pattern of equalisation resulting from use of equation (2.10) and used in RSG allocations from 1967 to 1980: needs are compensated only above a standard need (in this case of the least needy). The second method is superior to the first in terms of equalisation effects since it encourages the use of local tax rates to fund the base level of need experienced by all authorities, and which can be supported at equal tax rates because of the effect of the resources element (provided this is fully equalising).

Unitary grant (2)

The unitary grants which follow all derive from rearranging the revenue supply equation given by

$$E_i = B_i t_i + G_i,$$

i.e. that expenditure E_i can be contributed from the two sources of local tax base and grants. The unitary grant (2) derives from setting standard expenditure and tax rate terms and was the original suggestion of the Conservative Government in Britain in 1979.[5] It takes the simple form

$$G_i = \bar{E}_i - KB_i\bar{t}. \qquad (2.11)$$

This equation requires definition of the two new terms $\bar{E}_i$ (the standard expenditure assessed for local authority i) and K, a proportional multiplier used to penalise 'overspending' authorities. It was proposed that K should be set to

$$K = \lambda(E_i/\bar{E}_i) \qquad (2.12)$$

thus scaling or 'tapering' grants by the extent of difference between actual expenditure E_i and assessed expenditure $\bar{E}_i$. It was desired that the

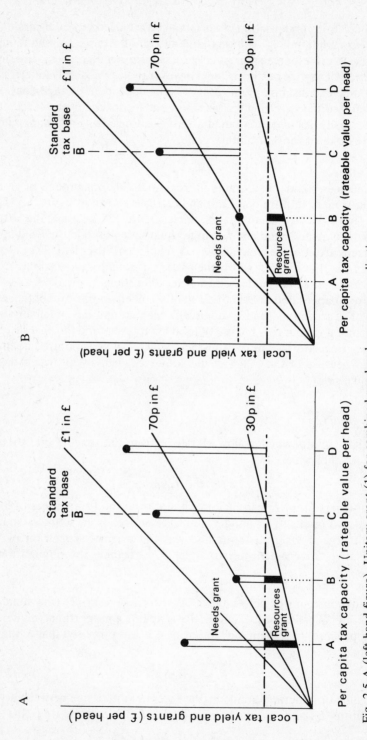

Fig. 2.5 A (left-hand figure). Unitary grant (1) for combined need and resource equalisation.
B (right-hand figure). Unitary grant (1) as actually used for combined need and resource equalisation in the England and Wales Rate Support Grant 1967–80: needs are equalised only above the level of a specified standard.

constant $\lambda > 1$ when $(E_i/\bar{E}_i)$ increased above a certain threshold (i.e. there was 'overspending' by local authority i relative to its assessed standard expenditure).

Equation (2.11) provides a grant equal to the difference between the assessed standard expenditure of an area and the yield from its tax base at a standard tax rate. The determination of the resulting grant can be understood from figure 2.6. The vertical axis shows the estimated expenditure of a given local authority. To find this expenditure from the local tax

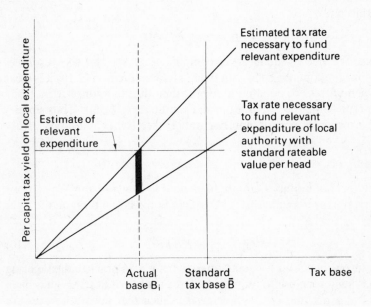

Fig. 2.6 Unitary grant (2). Grants are allocated as the difference between estimated expenditure on the actual and a standard tax base.

base B_i requires a tax rate corresponding to the upper line in the figure. However, to find the same expenditure from a local authority with the standard tax base requires a lower tax rate corresponding to the lower line in the figure. The difference between these two tax rates provides a direct estimate of the size of unitary grant.

Although differing from the Musgrave formulae for unitary grants discussed above, equations (2.11) and (2.10) are clearly related since standard expenditure and standard need are in effect identical. The relationship can be demonstrated as follows. First, substitute equation

(2.12) into (2.11) and then rewrite in terms of the needs-equalising part alone

$$G_i = \bar{E}_i - \lambda(E_i/\bar{E}_i)B_i \bar{t} \qquad (2.13)$$

$$= \bar{E}_i - \lambda(E_i/\bar{E}_i)\bar{B}\bar{t}. \qquad (2.14)$$

Set the resource equalisation component to the standard $\bar{B}\bar{t}$ and then set this equal to unity. On subtracting this result from the full grant equation (2.13) we obtain the resource equalisation component

$$G_i = [\bar{E}_i - \lambda(E_i/\bar{E}_i)B_i \bar{t}] - [\bar{E}_i - \lambda(E_i/\bar{E}_i)\bar{B}\bar{t}]$$

which reduces to

$$G_i = (\bar{B} - B_i)\bar{t}\,\lambda(E_i/\bar{E}_i) \qquad (2.15)$$

Comparing the needs equalisation component (2.14) and the resource-equalisation component (2.15) with the Musgrave equation (2.10) it is clear that if expenditures are equated to needs, then the equations are identical except for the influence of the penalty multiplier $\lambda(E_i/\bar{E}_i)$. Thus unitary grant (2) is a special form of unitary grant (1) (see Heald and Jones, 1980).

Unitary grant (3)

This is the form of unitary grant implemented in England and Wales since 1981/2. It is similar to equation (2.11) except in the initial term on the right hand side[6]:

$$G_i = E_i - KB_i \bar{t} \qquad (2.16)$$

where all terms, including K, are defined as before. In practice the tapering term K has been set close to unity and a large threshold of $(E_i/\bar{E}_i)$ has been set (equal to a tax rate of 155 p in 1980/1) such that few authorities are penalised by 'overspending'. Nevertheless, this power has created great political controversy and has been used. The use of E_i instead of $\bar{E}_i$ in equation (2.11) means that all expenditure up to the threshold, both discretionary and nondiscretionary, is eligible for grant.

Manipulation of equation (2.16) can be accomplished in the same way as with unitary grant (2) to show that they are both special forms of Musgrave's unitary grant (2.10) (see Heald and Jones, 1980).

Unitary grant (4)

This derives from a suggestion by the Department of Environment in evidence to Layfield (G.B. Government Committee of Enquiry, 1976, Appendix 7, pp. 21–50). The three previous unitary grants discussed above each suffer from the defect that authorities above the standard tax base have lower marginal costs of expenditure, i.e. they can raise the same expenditure at lower tax rates. Unitary grant (4) is a form of *proportional grant with pooled marginal expenditure* defined by

$$G_i = (\bar{E}_i - B_i t) + (KE_i - B_i t') \tag{2.17}$$

where

$$K = (E_i - \bar{E}_i)/\bar{E}_i \tag{2.18}$$

is the proportional difference of local expenditure from the standard, and t' is the pooled tax rate, equal to

$$t' = (K \sum_i E_i)/\sum_i B_i \tag{2.19}$$

when set equal to the national average tax rate. This unitary grant allows local authorities to spend more than the standard, but they must charge the pooled tax rate. The excess income for resource-rich authorities deriving from the higher tax rate is then placed in a pool to provide additional finance to other local authorities. Because this form of grant requires *explicit* horizontal transfers between local governments, it is usually deemed to be politically infeasible to implement.

The major advantage of each unitary grant is that it combines resource and need equalisation and thus the two major sources of inequity are simultaneously treated. For example, local authorities with high tax capacity and high needs are penalised with negative transfers from revenue equalisation, but receive high grants from need equalisation, with the combined effect cancelling out to give the local authority small grants and forcing it to fund high needs with its high capacity. Conversely, local authorities with low capacity and low needs are compensated for their small tax base, but penalised for their low needs, again resulting in a small total transfer. A combination of needs and resources grants thus possesses the advantage that it limits the scope of both need and resource equalisation and thus produces smaller total grant payments overall. Separate needs and resource equalisation in the form of equation (2.8) was used in the Rate Support Grant up to 1980/81, but a unitary grant has been implemented since 1981/2 in the form of equation (2.16).

2.6 Cost-equalising grants

Equalisation of unit costs seeks to achieve fiscal equity in needs provision by adding to the normal needs equalisation formula consideration of the differences in the cost of providing a given unit of service at a given level in different local authorities. In the need-equalisation approach discussed above a simpler procedure for scaling the magnitude of grants was chosen by specifying a level of outlay per unit of need required to produce a given

standard of service. Unit costing expands this consideration of performance levels to one which encompasses the costing per unit of the required
level of service outputs in each local authority. To implement such grants
three sets of new measures are required: one of differences in local
authority costs, a second for differences in the standard of local authority
services, and a third for differences in the size of client groups.

Differences in the costs of providing a given standard of service in
different local authorities derive from four main factors. First, the differing
physical environments of local authorities give rise to cost variation such as
those for building and maintenance, road maintenance and repair, and
space heating and insulation. Second, technical factors such as the population density, settlement shape, and economies of scale cause important
cost differences. Third, factor costs vary significantly for different local
authorities, especially for the cost of land, labour and capital, and of
transporting raw materials. This feature was reflected in the RSG needs
element by labour cost differentials. Finally, costs vary due to the elements
of 'sunk costs' deriving from past investments. Especially important are the
age and condition of the capital stock and infrastructure, and the level of
debt arising from past investments which must be serviced.

The measurement of the unit costs of each service at a given level of
service is extremely difficult. Certainly measures such as expenditure per
recipient are not in themselves sufficient since these do not differentiate
service quality. Hence, in Britain the DOE and CIPFA returns by local
authorities cannot be used directly.[7] Instead local costs per unit must be
determined for each quality level, taking account of each variable causing
cost variation. Two practical approaches to this problem are available.
First, regression analysis as suggested by Boyle (1966) and other workers[8]
and used to some extent in the distribution of the Rate Support Grant.
Second, unit costs derived from sectional programme assessments by either
the local authorities themselves, or by central departments.[9] An additional
factor here is the difference in local authority efficiency and productivity.
These should *not* be allowable elements in grant distribution; hence, there
is the need to equalise not 'actual' costs, but those 'potential' costs which
would obtain if all local authorities were equally efficient.

Service standards are particularly difficult to assess because major
differences in such factors as labour-servicing ratios (e.g. the pupil/teacher
ratio in schools) may have to be supported in order to provide a common
standard of service in different local authorities. Moreover, differences in
the standards of local authority services can only be assessed with relation
to local *output* levels i.e. the quality of service benefits available in each
local area.[10] This requires measurement of such indices as the probability
of given crimes being committed as a measure of police protection, given
levels of examination results or university entry as a measure of educative
quality. As emphasised by Le Grand (1975), it is not the input require-

ments which are important for equalisation purposes since, for example, different numbers of policemen and patrol cars will be needed to give the same level of police protection in different areas. Similarly, different pupil/teacher ratios and quantities of equipment aids will be required to produce the same educational standard in different jurisdictions. The problem of defining and measuring these output ratios is particularly difficult and is discussed further by Newton and Sharpe (1977).

The size of client groups is the third factor affecting local expenditure need. The relative number of old or young makes heavy but very different expenditure demands for local public services, and local authorities with relatively more old or young people will have larger revenue requirements than other local authorities, *ceteris paribus*. Again, local authorities with relatively large concentrations of industrial enterprises will have relatively higher expenditure need for road investment and maintenance, refuse disposal, public transport or pollution controls. Hence, a major component of need assessment must be concerned with determining the relative numbers of people and enterprises in differing client groups.

Having defined and measured differences in service costs, service standards (output levels) and client groups, central grants aimed at unit cost equalisation will give high levels of grants to areas with high needs, high costs and low service qualities, and lower grant levels to areas with low needs, low costs and high service qualities. The form of such grants will be scaled by the relative deviation of local costs C_i from 'standard' costs $\bar{C}$, and of local service performance levels (output quality) S_i from 'standard' performance levels $\bar{S}$. This yields the following formula

$$G_i = t_i C_i \left(\frac{\bar{B}}{\bar{C}} - \frac{B_i}{C_i} \right) + t_i C_i \left(\frac{S_i}{t_i} - \frac{\bar{S}}{\bar{t}} \right) \qquad (2.20)$$

with other terms defined as before.

This is a form of combined resource and need equalisation but is modified in three ways. First, costs are taken into account. Second, standard tax rates are introduced. Third, need measures are replaced by output measures S_i which encompass need as a deviation of local service standards from the set national standard. Hence in place of the resource equalisation component, the first bracketed term weights the tax base by the difference in the 'standard' costs $\bar{C}$ and local costs C_i for providing the sum of local service. Similarly, in place of the previous needs-equalisation component, the second bracketed term weights the local service costs S_i and 'standard' service costs $\bar{S}$ by the difference between the respective local and 'standard' tax rates, t_i and $\bar{t}$.

This more complex formula derives from Boyle (1966) and Le Grand (1975). It has a number of advantages, a major one of which is that it differentiates that level of services which is provided in response to a local choice of higher service standards from the 'standard' minimum level of services. In addition it takes account of choices in the setting of local tax rates and differences between local authorities in the unit costs of providing given service levels. Hence equation (2.20) is based on a ratio of the per capita purchasing power for discretionary services after required beneficial services have been provided, i.e. it attempts to achieve fiscal equity between local authorities by giving them the same discretionary purchasing power per unit of local tax effort.

2.7 Conclusion

This chapter has introduced the main categories of grants that are available for central finance of local governments and in particular it has specified a set of clear criteria which should be used in designing grant programmes. Perhaps the chief of these criteria is that of equalisation but the other issues, of encouraging local discretion, neutrality to local behaviour, adequacy, stability, intelligibility and feasibility come to play an increasing role when practical proposals relating to the Rate Support Grant, or any other grant programme, are assessed. With this background discussion in mind, the following chapters provide a detailed assessment of the Rate Support Grant. This assessment is aimed at, first, evaluating how far the Rate Support Grant satisfies criteria of equalisation, and second, at assessing the distribution effects of various alternative forms of equalisation.

CHAPTER 3

THE EVOLUTION OF CENTRAL GRANTS AND LOCAL FINANCES IN BRITAIN

3.1 Introduction

The history of local finances and central grants in Britain presents an interesting picture of continuous and often rapid evolution in which experiments were undertaken with many of the equalisation and other grants discussed in the previous chapter. The main features of this history are threefold. First, the rapid growth of local authority responsibility for expenditure from the late nineteenth century onwards, and especially since 1948, gave the local governments a major role in providing public services. Second, the relative inflexibility of local tax yield, arising from its reliance on the property tax ('the rates') as the main source of local revenue over the whole period from 1888 to the present, has presented local authorities with a continuous problem of raising sufficient revenue to fund their ever-burgeoning expenditures on services. A third feature, which derives from the difficulty of revenue raising in the face of rising expenditure demands, has been the increasing role of central government support for local services through various forms of grants such that now it is the Westminster government, using the tax resources of the nation as a whole, which provides the majority of local revenues.

The various features are clearly shown in figures 3.1 to 3.3. With respect to the sources of income of local authorities, figure 3.1 shows the steady decline in the role of rates since the nineteenth century, and especially since the mid-1960s. Over the period 1930–70 rates, central grants, and other local authority income were each responsible for approximately one third of local income. However, since 1972 the position has changed rapidly to one in which central grants are responsible for about one half of local income whilst rates have declined to a position of providing only about 25%. This has had important implications for the fiscal balance of central and local government in Britain, and has severely eroded both local accountability and independence.

The expenditure pattern of local governments is shown in figure 3.2. The most noticeable feature is the very rapid rise in expenditure levels by local government since 1950 and especially since 1965. Comparing the expenditure pattern of figure 3.2 with the income pattern of figure 3.1, it is

43

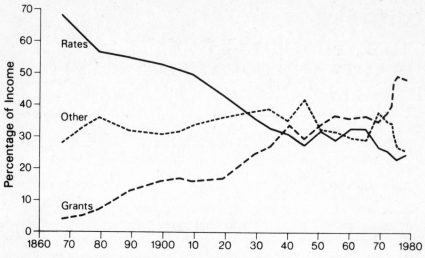

Fig. 3.1 Source of revenue on the current account for local authorities in England and Wales 1867–1980 (SOURCE: *Local Government Financial Statistics*).

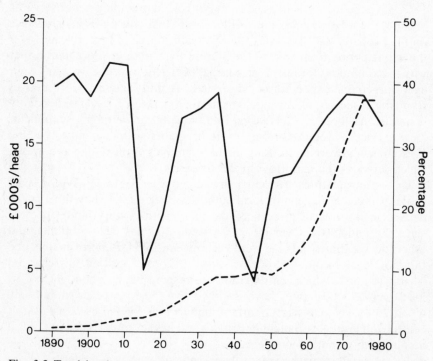

Fig. 3.2 Total local government expenditure in Great Britain 1890–1979. Total expenditure is expressed as a percentage of total expenditure (solid) and expressed as per capita at constant 1978 prices (dashed) (SOURCE: *Local Government Financial Statistics; Statistical Abstract of the United Kingdom*).

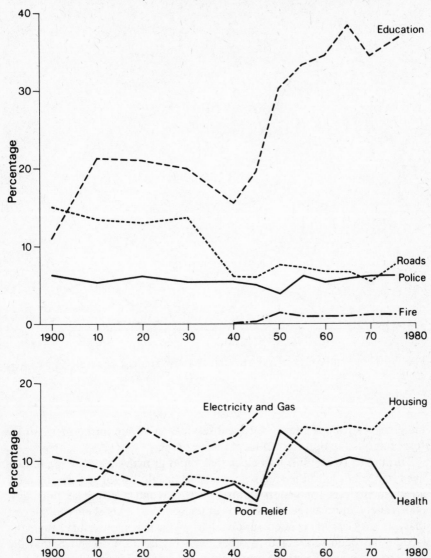

Fig. 3.3 Percentage of local authority expenditure in various service categories 1900–78. Note that these series give only a general indication and are affected by considerable changes of central and local divisions of responsibility (SOURCE: *Local Government Financial Statistics*).

clear that although there has been a rise in locally-borne expenditure, much of the rapid rise in local expenditures has resulted from the greater involvement of central government through central grants. This has important implications for economic efficiency and productivity since it

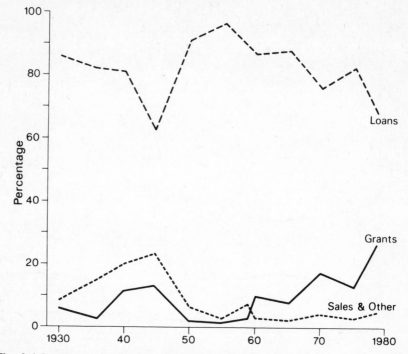

Fig. 3.4 Local authority capital account 1930–79: sources of income in various categories (SOURCE: *Local Government Financial Statistics*).

indicates a strong tendency for local services to be supported increasingly from national as opposed to local tax resources.

Turning to figure 3.3, it is clear that rapid growth of local expenditure has occurred despite the steady loss of responsibilities for services by local government; e.g. for water supply, electricity and gas supply, hospitals, poor relief, and a large number of other services. This seems to suggest that, as services have been removed from local responsibility, then the income has been redirected to other uses. Figure 3.3 also shows that the most spectacular growth of expenditure functions occurs with education and housing in the period immediately following the second World War, and this marks the origin of the modern welfare state as it affects local government finance.

The local capital account also displays important changes. In figure 3.4, showing the sources of income into the capital account, loans in total still play a dominant role, but the importance of central capital grants has increased steadily since 1950 to now constitute 25% of capital income. On the expenditure side, the capital account shown in figure 3.5 displays the strong relation between outstanding debt and total capital expenditure.

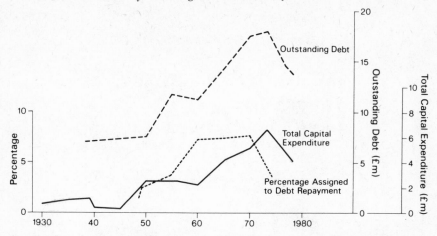

Fig. 3.5 Local authority capital account 1930–78: level of outstanding debt, total capital expenditure, and level of capital expenditure assigned to debt repayment at constant 1978 prices (SOURCE: *Local Government Financial Statistics*).

Outstanding debt has steadily risen especially since 1950, but the proportion of capital expenditure assigned to repayment of debt has been a relatively small proportion of total capital expenditure since the nationalisation of the gas and electricity utilities in 1948.

British public service provision is subject to a high degree of shared responsibility between each level of government. As can be seen in table 3.1, local government is a significant agency for the provision of those services relating to roads and transport, housing, law and order, education, and environmental services. Indeed, for each of these categories, local government is the main agency for providing services. Central government, on the other hand, is the main agency for providing services such as defence, aid, agriculture, trade and industry, health, and social security. Major overlapping of service provision occurs for housing, roads and environmental services. In Britain, therefore, a complex division of expenditure functions has evolved. Complete separation of expenditure functions between levels of government characterises only a small number of services. However, one feature of British local expenditure should be noted in particular (see table 3.1): that there are relatively few local services which relate to the number of poor people living in a local unit. Clearly expenditure on education, police, environment and transport will be related to some extent to the personal incomes of the people living in local units. Also the level of locally-provided personal social services will have a strong income effect, especially as such expenditures relate to facilities for the aged, disabled, and children in care. But there is nowhere near the strength of relation to level of local incomes that occurs, for example, in

Table 3.1. Expenditure levels of central and local government in Great Britain for different service categories

Service category	Total expenditure £m (1979 prices)				% of expenditure centrally undertaken				% expenditure locally undertaken			
	1967/8	70/1	74/5	78/9*	67/8	70/1	74/5	78/9*	67/8	70/1	74/5	78/9*
Defence and overseas aid	9884	8422	9022	9764	100	100	100	100	0	0	0	0
Agriculture, Forestry and Fishing	1335	1101	2353	950	100	98.7	99.4	99.5	0	1.3	0.6	0.5
Trade, Industry and Employment	2964	3520	5540	3665	98.6	98.7	98.7	97.2	1.4	1.3	1.3	2.8
Nationalised industries	N/A	4936	1383	1247	100	100	100	100	0	0	0	0
Roads and Transport	2806	2864	3836	3071	42.4	42.4	46.8	41.3	57.6	57.6	53.2	58.7
Housing	3777	3312	6577	5471	29.3	27.3	29.6	50.0	70.7	72.7	70.4	50.0
Law and Order	1836	1808	2138	2372	25.9	18.5	23.2	23.7	74.1	81.5	76.8	76.3
Education and Arts	7654	7834	9916	9891	13.6	15.7	16.5	15.5	86.4	84.3	83.5	84.5
Health and Personal Social Services	5950	6606	8591	9337	71.0	83.7	85.2	85.3	29.0	16.3	14.8	14.7
Social Security	10403	10681	14024	17420	100	100	100	100	0	0	0	0
Environmental Services	2734	2625	3602	3455	38.5	6.3	11.8	24.2	69.5	93.7	88.2	75.8
Others	18505	9538	1770	1801	N/A	71.8	79.1	78.9	N/A	28.2	20.9	21.1
Total	52196	63532	71591	71394	69.5	71.0	71.2	73.1	30.5	29.0	28.8	26.9

* Estimated.
Source: G.B. Government, Public Expenditure White Papers.

the United States. In Britain welfare payments, medical care, social security, and so forth, are nationally-financed services. As a result, there is a much smaller burden placed on local government in Britain by the presence of a large number of poor people in their jurisdictions.

Against this general background, the present chapter is concerned with dicussing the history of local authority finance and tracing the origins of the present system of central grants. The next chapter discusses in detail the structure of the present system of Rate Support Grants.

3.2 Local finances before 1929

The structure of local government which existed in England and Wales from 1865 until 1974 was established in the *Muncipal Corporations Act* (1835), later amended by the *Local Government Act* (1888) establishing the County Councils, and the *Local Government Act* (1894) establishing the District Councils. These Acts created a general pattern of administration which was to be copied in Australia, Canada and New Zealand and also to a lesser extent in India and Africa.

In the period up to 1929 local authority finance was marked by two main features: first, the rapid evolution of new forms of grants from the Westminster Government; second, enormous growth in the size and range of local expenditures that were undertaken (the rates alone increased from £5.3 m to £16.8 m between 1803 and 1868: Goschen, 1872). Specific grants for local authorities were first instituted in 1825 for the prison service and increased rapidly through the nineteenth century for roads, police, housing, schools and other services. Capital loan schemes between central and local government also expanded rapidly. But the first general block grants, from which the Rate Support Grant was to derive, was not instituted until the 1888 Act. This was in the form of 'assigned revenues'. Assigned revenues were based explicitly on a principle of revenue-sharing. A proportion of yield of central taxes was assigned to local use. Initially it was proposed to distribute this grant between local authorities on the basis of need (measured by the level of indoor pauperism). This was abandoned, however, and instead a combination of existing specific grants and the level of local revenue raised was employed. Education grants remained an important specific grant outside this block grant structure, whilst the inadequacy of assigned revenues to meet new expenditure demands led to a steady rise in new specific grants: for agriculture, probation, tuberculosis eradication, mental treatment, maternity and child welfare, registration of electors, training of midwives, welfare of the blind, treatment of venereal diseases, provision of health visitors, and many others.

An issue separate from that of grants, but having increasing influence on the need for grant transfers, was that of the local rates. Following thirteen and fourteenth century precedents of special levies, the rates were formally instituted in the Elizabethan *Poor Law* of 1597 and the *Poor Relief Act* of 1601. They did not attain great significance, however, until the nineteenth century when the expansion of the service needs of newly growing cities and towns required large quantities of revenue to be raised. This period was marked by the search for a just method of allocating tax burdens but, as with the Elizabethan Poor Law, attempts to relate taxes to ability to pay (*iuxta facultates* – ability and substance) were abandoned in favour of using the rental value of housing. The early history of rental valuation derives mainly from its codification in London in the *Parochial Assessments Act* (1836), the *Metropolitan Management Act* (1856) which instituted the first consolidated rate and replaced a large number of different levies, and the *Valuation (Metropolis) Act* (1869), which instituted the principle of uniform valuation with five-yearly reviews.

Over the period up to 1925 there were many attempts to reform the rates and establish a simple procedure which was uniform for Counties and County Boroughs. But this period was distinguished by rapid city growth with many overlapping jurisdictions, and considerable jealousy and competition between the rural and new urban areas. Particularly influential was the work of Goschen's (1872) *Report on Local Taxation*. This suggested uniform valuation procedures and the rental valuation principle, but was particularly important in emphasising, for perhaps the first time in England, the importance of geographical differentials in tax capacity (rateable values) of different local authorities. Goschen recognised that over the period 1826–68 the local taxes raised had increased in size by only 135% and whilst local rateable values had risen by over 200%. Moreover, the rise in rates had been much higher in the cities than in the rural areas, many of which had actually had declining rate bills. Hence, he concluded that there was a need for better valuation of property, for more consolidated government, and for more central government aid to poorer city areas to relieve the local rate burdens. These proposals on valuation were encompassed in the Chamberlain reforms of the 1920s. For example, the *Rating and Valuation Act* (1925) enforced the final codification of local rates outside London. The main effects were to give a consolidated general rate levy for each area with universal revaluation every five years, to regularise allowances for deductions against rental valuations, and to reduce the total number of local authorities from 15 546 to 1708. The Act was a great improvement but was not entirely successful since it was not until the third five-yearly valuation in 1937/8 that a uniform basis of valuation in England was attempted (and this was postponed until the *Local Government Act*, 1948, and was, in fact, only finally implemented in severely amended form in 1956 under the *Valuation for Rating Act* (1953)). Uniform rating procedures have yet to be introduced in Scotland.

3.3 Local finances 1929–48

The *Local Government Act* of 1929 is significant in the history of the Rate Support Grant since it represents for England and Wales the point at which a major block grant was introduced. The aim of this reform was, first, to relieve the pressure of the rates on 'productive industry' by derating some manufacturers, and second, to reduce the burden of local tax in general by the use of a grant deriving from the central government tax base. For the first time this recognised the role of local authorities as part of the overall government system and formed the basis for a partnership of the two levels of government to achieve minimum standards of services. The new block grant, the *General Exchequer Contribution*, achieved a significant reduction in local rate burdens, but its significance progressively diminished as the grant was kept at the same level until 1948.

The General Exchequer Contribution followed precedents that had been set earlier in Scotland by grants to relieve the burden of local taxation in the Highlands where differences in rateable values were used as a criterion of inequity. In 1920 the *Royal Commission on Local Taxation* had proposed a block grant to equalise local tax rates paid to Scottish local authorities in terms of their relative equality, economy and efficiency. In evidence to the Commission, for example, Lord Balfour suggested a scheme based on 'first, ability to raise local funds, and secondly, necessity for expenditure upon the services assisted' (quoted in Boyle, 1966, p. 30). The grant proposed was, then

$$G_i = C + m + \bar{t}(\bar{B} - B_i) \qquad (3.1)$$

where

C is a constant to scale the total grants to be distributed
m is the minimum local outlay per unit of expenditure need
$\bar{t}$ is standard tax rate
$\bar{B}$ is standard tax base
B_i is per capita tax base in local authority i, and
G_i is per capita grant to local authority i.

Thus both needs and resources differences were incorporated. A more complex procedure for a similar block grant was proposed by the Kempe Committee (House of Commons (1914) Session Vol. *40–41*, Cmnd. 7315) which tried to take better account of differences in expenditure need. The resulting grant formula was given by

$$G_i = C + m(N_i - \bar{N}) + \bar{t}(\bar{B} - B_i) + 0.4 E_i \qquad (3.2)$$

where

N_i is the level of expenditure need in local authority i (measured as proportion of school children to total population)
$\bar{N}$ is a standard level of need
E_i is the total per capita of expenditure in local authority i

and the other terms are the same as in the Balfour formula. This is a form of the unitary grant for simultaneous needs and resource equalisation discussed in chapter 2.

The Kempe Committee formula was instituted by the Labour Government for the education contribution to local authorities in 1917 and continued until 1958, but the Conservative Government General Exchequer Contribution of the 1929 Act followed a different form in which differences in local needs were measured by indicators rather than as differences from a standard; but the measures of resource differences were retained. The formula adopted is summarised well in Chester (1951) and Rhodes (1976) and was given as follows

A. Basic grant (population)

Population for year preceding grant allocation.

B. Supplementary grants (Intermediate weighted population)

(i)   Proportion of basic grant equal to the percentage excess of children under 5 to the level of 50 per 1000 population.

(ii)   A proportion of basic grant equal to the percentage deficit of rateable value below the level of £10 per head.

C. Final weighted population

(iii)   A proportion of intermediate population equal to ten times the percentage excess of male unemployment over the level of $1\frac{1}{2}\%$ of the average population of the Counties and County Boroughs over the three preceding years.

(iv)   For County Councils alone (and excluding the London County Council), proportion of the intermediate weighted population equal to the percentage by which the population per mile of road was less than 200, but only for those local authorities with population per mile less than 100. For those local authorities with 100 or more population per mile of road, weighted population was increased by the percentage relation of 50 to the actual population per mile.

In 1937 the extent of weighting for unemployment and sparsity (indicators (iii) and (iv)) was increased slightly. With this set of need indicators and a separate set of resource grant elements, the foundations for the present Rate Support Grant were laid.

The period following the institution of the General Exchequer Contribution saw continued rises in local expenditure, increasing anomalies in local valuation because of infrequent updating, and a diminishing significance for the grant because it was held at the same level. An additional criticism was that there were no negative needs weights, thus preventing horizontal transfers from resource-rich, low-need areas to resource-poor, high-need

Table 3.2. *Rate burdens by type of town*

RV/Head	Local authority type	Average rate poundage	Average per capita receipts		
			From local rates	From General Exchequer Contribution	Total
High	Seaside resorts	45	575.6	56.4	632.0
	Big spenders	85	639.2	72.8	652.0
	Cathedral towns	50	416.0	71.6	487.6
Medium	Middling stinters	54	341.6	92.8	434.4
	Poor spenders	80	370.0	140.8	510.8
Low	Poor stinters	70	320.8	117.6	438.4
	Unweighted national average	67	431.2	101.6	532.8

Source: Hicks and Hicks, 1944*a*, table 5; converted to New Pence.

local authorities. As a result Hicks and Hicks in a series of three studies (1943, 1944*a*, *b*) castigated the local finance system as one which created enormous anomalies in rate burdens and in local expenditure levels. Their results are summarised in table 3.2.

From this table Hicks and Hicks concluded that the main differences in spending levels and rate percentages resulted from differences in the local resource base measured in terms of rateable value per head. They then recommended, first, the use of a new block grant to act as a really effective equaliser, and second, more up to date valuation. They also recommended that the grant should be paid only to local authorities below a standard income, to bring them up to that standard. The formula proposed was for the central government to act as a ratepayer on that deficiency in local rate percentage which fell below a set national standard $\bar{B}$, i.e. a grant paid according to a deficiency payments principle given by the formula (compare equation 2.3)

$$G_i = t_i(\bar{B} - B_i) \text{ subject to } B_i < \bar{B}. \tag{3.3}$$

This latter suggestion was influential on the form of subsequent resource grants up to 1980.

The period after 1929 was also dominated by the depression years and the attempts of both central and local government to ameliorate distress. However, one of the most iniquitous aspects of local finance still concerned the fragmented nature of local government which resulted in disproportionately high burdens in old city and industrial areas relative to newly-expanded suburbs as noted by Goschen in 1872. Royal Commission Reports in 1935 on Merthyr Tydfil and in 1936–7 on Tyneside, drew particular renewed attention to the coincidence of the highest rate burdens with the local authorities suffering the highest degree of unemployment and economic distress. The Tyneside Report, in particular, drew attention to the need for the expenditure on 'regional' services such as public health, education, public assistance, police, fire and highways to be borne by larger government units which included all of those who benefited from the services and provided sufficient tax base (i.e. the cities and their suburbs).

The evolution of British local government over the late nineteenth and early twentieth centuries thus shared many aspects in common with the local government structure of the present-day N.E. USA or Australia (for discussion, see Bennett, 1980): the division of tax base between, on the one hand, central cities with high expenditure need and low tax base and, on the other hand, newly-expanded suburbs and satellites with lower expenditure need and high tax base; the spillover of benefits from city to suburbs; and the inappropriate assignment of responsibility for some expenditure functions between levels of government. Some of these difficulties were overcome in 1934 when responsibility for poverty relief for the unemployed was vested in the centrally-funded Unemployment Assistance Board, and in 1937 when increased weighting was given to unemployment in the General Exchequer Contribution. But the problems arising from fragmented city administration were not finally reduced until the reorganisation of local government in 1974, just over 100 years after they were first noted by Goschen.

3.4 Local finances 1948–66

The period immediately following 1948 is enormously significant since it sees the establishment of the welfare state in Britain. As such it marks a period of massive rises in both national and local service provision and hence of public expenditure.

The immediate post-war years returned a Labour government (1945–51) which nationalised hospitals, maternity homes, sanitoria and mental hospitals, taking them into centrally-funded Regional Health Authorities. In addition, municipal gas and electricity undertakings were brought under national corporations in 1948–9 and the *New Towns Act* (1946) established a direct flow of funds from the central government Treasury to some local governments. Although local authorities lost the substantial revenue source deriving from gas and electricity charges, this period did see the rearrangement of expenditure functions such that regional services became centrally rather than locally funded. However, the period also saw the shouldering of major increased local expenditure in housing, and some personal social services, and in planning (as a result of the *Town and Country Planning Act*, 1947).

The issue of fragmentation of government was tackled by a Boundary Commission covering the period 1945–51, by a second Commission set up in 1955, and by the Royal Commission on government in Greater London set up in 1957. Little happened as a consequence of these developments however. Many of the anomalies in the valuation of property for the rates were overcome in the *Local Government Act* (1948) by rationalising assessment under the Inland Revenue which produced the first national valuation list in 1956. However, attempts to change from a rental basis of valuation to one of construction cost were not pursued, despite the growing problem of assessing rents following the Rent Restriction Acts of the Second World War. Moreover, the difficulties of changing to the national valuation system for rates in 1956 led to a damping to 1939 rental values for residential ratepayers and the derating of shops and offices to 20% of their assessments. However, derating of industry (introduced in 1929) was reduced from 75 to 50% of assessment.

For the capital account, freer access to the general capital markets was allowed after 1952, but free access to central funds via the PWLB, recommended by the Radcliffe Report on the Working of the Monetary System (G.B. Government, 1959), was rejected. Loan finance burgeoned in the immediate post-war years, especially with respect to the Housing account.

With respect to grants, the period up to and following 1948 was characterised by three features. First, there was concern to provide local government with a larger and more buoyant revenue source. This was stimulated by the continuing problems of the rates and by the loss of substantial income from the electricity and gas utilities. However, no attempt was made to tackle this issue. A second issue concerned the steady increase in specific grants. By 1957/8 these provided 24% of total local expenditure and although popular with central government in allowing direction to and encouragement of specific service categories, were under-mining the independence of local government. This problem was tackled in the 1958 *Local Government Act* by consolidating many specific grants into a new block grant. A third feature of this period was improved attempts to provide more support to the poorer areas by forms of equalisation grants. This stimulated new and enlarged block grant transfers: from 1948 to 1958 the Exchequer Equalisation Grant, and for 1958–66 the Rate Deficiency Grant and General Grant.

The *Exchequer Equalisation Grant* (EEG), deriving from the Labour Government's *Local Government Act* (1948) was paid according to the deficiency payments principle advocated by Hicks and Hicks in 1944. Grants were given only to those local authorities whose rateable value per head fell below a given standard. The standard chosen was the national average, whilst the per capita rateable value was weighted to take account of numbers of children and density features deriving from the 1929 General Exchequer Contribution (given as the actual population, plus the number of children under five, plus, for Counties with less than 70 people per mile of road, one third of the additional population needed to bring the population per mile of road up to a value of 70). For those local authorities receiving grants, the formula used was

$$G_i = \left\{ \frac{\text{Relevant local expenditure}}{\begin{array}{l} \text{Credited} \\ \text{Rateable} + \begin{array}{l} \text{Rate yield at a} \\ \text{rate of £1 in £} \end{array} \\ \text{Value} \end{array}} \times \begin{array}{l} \text{Credited} \\ \text{Rateable} \\ \text{Value} \end{array} \right\} \qquad (3.4)$$

The Credited Rateable Value (CRV) is that additional rate base necessary to bring an authority's rateable value up to the level of the national average. This term is multiplied with the ratio of the relevant local expenditure (which is given as the net expenditure which would have to be met from rates if there were no grants) to the sum of the CRV and the product of a rate of £1 in the £. This last rate product is a measure of actual rateable value, intended to allow for losses and exemptions from rates. Grants were calculated and paid to Counties and County Boroughs, and on average 54 out of 61 Counties, and 55 out of 83 Boroughs received grants.

The County was also treated as the administrative unit for other areas, whilst the allocation of the final grant to districts was made by *Capitation Payments*. For urban Districts and non-county Boroughs this was equal to one half of the taxed EEG paid to counties, excluding London, divided by the unweighted population. For rural Districts, the contribution was one half that of the other Districts. In each case these amounts were deducted from the County total, leaving the County the remainder. A more detailed discussion of these grants is given by Chester (1951), G.B. Government (1953), *Accountancy Research* Research Working Party (1949–53), Woodham (1953), Lees *et al.* (1956), Lees (1957) and Williams (1958, 1959).

From the structure of equation (3.4) it is clear that the Exchequer Equalisation Grant was open to a number of criticisms. First, it was a deficiency payments grant not yielding true equalisation by penalising the richer local authorities. Secondly, the grant was not independent of local authority actions in choosing their tax rates, since the ratio of relevant expenditure to actual rateable value, which forms one part of the equation, was equal to the local tax rate. Third, the grant relied heavily on the accuracy of rateable value assessment, and even by 1958 it was still possible to claim that there was a 'notorious lack of uniformity between authorities in valuation practices' (Williams, 1959, p. 120). Fourth, because the

County level was that used for evaluation, rich rural Districts within poor Counties received large grants, and poor Districts within rich Counties received no grants. Finally, the capitation system for distribution within Counties favoured the densely populated urban Districts and worked in the opposite direction to the principles of the EEG as a whole.

The Rate Deficiency Grant and the General Grant. The Local Government Act (1958), under a Conservative Government, introduced the rerating of industry, which had been partially exempt since 1929; but again alternative revenue sources or revenue-sharing in the form of assigned revenues were rejected (see G.B. Government White Paper, 1957). But, most significant however, it introduced two new grants in place of the Exchequer Equalisation Grant. First, a Rate Deficiency Grant directly replaced the Exchequer Equalisation Grant and gave effect to many of the recommendations of the Edwards Committee (1953) in seeking to remove the criticisms of that grant. Second, a General Grant instituted a new block grant procedure which consolidated a large number of disparate specific grants into a single grant element. This latter also had the effect of terminating the old Education Equalisation Grant that had been working since 1929.

The *Rate Deficiency Grant* (RDG) attempted to overcome the criticisms of the Exchequer Equalisation Grant by paying a proportion of the grant directly to Districts within Counties, broadly in relation to their level of expenditure. In addition, capitation payments were discontinued and a shift was made from rateable values to rate products, thus allowing for noneffective rateable value. The population weighting was removed and the grant became a tax revenue deficit grant only; the attempts at including 'needs' were passed over to the General Grant. The formula for allocation then became as a percentage grant (G.B. Government, 1962)

$$G_i(\%) = \frac{(\text{standard 1d rate product} - \text{actual 1d rate product})\ 100}{\text{standard 1d rate product}} \quad (3.5)$$

where

$$\begin{matrix} \text{standard 1d rate} \\ \text{product} \end{matrix} = \begin{Bmatrix} \text{1d rate product} \\ \text{of all local} \\ \text{authorities in} \\ \text{England \& Wales} \end{Bmatrix} \times \begin{Bmatrix} \dfrac{\text{population of area}}{\text{population of England}} \\ \text{and Wales} \end{Bmatrix} \quad (3.6)$$

This percentage result was then applied to the expenditure eligible for grant. However, in the calculations this grant was modified in two ways. First, for those local authorities with a ratio of population to road mileage of less than 70, an addition to the population was made of two-fifths of the additional population required to bring the population to 70 per mile of road thus favouring rural areas. Second, the total amount of grant that could be paid to any local authority was limited to be within a total percentage rate of increase, but with special allowances for those local authorities with previously low expenditure levels to permit them to catch up.

The RDG was found a generally accepted principle except for the manner in which the expenditure eligible for grant was calculated. G.B. Government (1962) note that there were difficulties in allowing for abnormal needs (such as slum clearance) and in the manner in which the eligible expenditure was calculated. This undoubtedly allowed the incursion of political elements which favoured the Conservative-controlled rural areas. In addition it was felt that the modification of the grant allocation to take account of population density and road mileage (which also favoured rural areas) confused the grant with needs assessment where otherwise it would be purely concerned with resource equalisation.

The *General Grant* was a major effort to consolidate funding of a disparate set of spending programmes and it was also allocated as a block grant allowing local authorities discretion in reallocating between spending sectors. It replaced a series of specific grants for:

> Education (excluding school milk and meals) and Agricultural education
> Health Service under the *NHS Act* (1946)
> Fire Service
> Child care
> Town Planning (excluding war damage)
> Road Safety
> Traffic Patrols
> Registration of electors
> Physical Training and Recreation
> Residential and temporary accommodation under *National Assistance Act* (1949)
> School Crossing Patrols
> Technological and Further Education Grants

The police and housing grants were the principal specific grants remaining outside this system. The new grant aimed to give local authorities a new discretion in local spending while giving them adequate support in terms of their expenditure needs. Thus, as stated by G.B. Government (1957) '. . . successive governments since the war have subscribed to a policy of giving greater freedom to local authorities from central control . . . (but) without the stimulus of specific percentage grants, some local authorities may not spend as much as they otherwise would. But the government . . . are convinced that greater independence and freedom from detailed control is essential to them'.

Whilst the RDG offered support to local authorities with a low tax base, the General Grant offered support to local authorities with large expenditure needs. The initial level of grant to each local authority was derived from the percentage contribution of each of the specific grants to that local authority. But from 1959/60 onwards, the grant was paid according to a group of needs indicators divided into two sets; one to give the basic grant, and a second to give supplementary support for special needs. In 1959/60 these had the following weights

1. *Basic grant*   *(£ per person)*
    (i) Per head of population   5.75
    (ii) For each child under five   0.52
2. *Supplementary grants*
    (iii) Young children (under five) and old people (over 65)   0.421
    (iv) School children per head of population multiplied by number of
         pupils per 1000 population in excess of 110   0.058
    (v) High density; a percentage of basic grant equal to half the
         population in excess of 18 per acre
    (vi) Low density; a percentage of basic grant equal to two and a quarter
         times the road mileage per 1000 population for local authorities with
         mileage exceeding two per 1000 (Maximum of 70%)
    (vii) Declining population; a percentage of basic grant equal to half the
         amount by which population decline exceeded 5% over last 20 years
    (viii) High cost metropolitan districts; 5% of basic grant to local
         authorities in whole or in part in Metropolitan Police District
    (ix) *minus* product of 9d in pound rate poundage.

This structure can be summarised for all the factors except (v) to (viii) as follows (Boyle, 1966)

$$\bar{N}_k = \frac{\sum_{j}^{n} N_{jk} P_j}{P_j} \tag{3.7}$$

where $\bar{N}_k$ is the standardised need of category k averaged over all local authorities j with population P_j. The grant allocated is then given as

$$G_i = C_0 + \sum_{k}^{m} C_k (N_{ik} - \bar{N}_k) + C_R (\bar{B} - B_i) \tag{3.8}$$

where C_k and C_R are constants, N_{ik} is the level of need indicator k of local authority i, and $\bar{B}$ and B_i define the respective standard and local tax bases.

Within the General Grant, then, there were allowances for both needs and tax base differences. Hence, this grant structure contains much of the format of the following RSG structures. Moreover, many of the criticisms of the Rate Support Grant were to derive from this period. The deficiency payments principle prevented true equalisation, whilst the indicators used in the General Grant and the structures for their measurement were often questionable. In addition the use of the average levels of need and rateable value per head as the standard against which to judge local authorities is subject to the criticism that it excludes many local authorities which have problems in raising local revenues.

3.5 The emergence of the Rate Support Grant

The development of local finance since 1966 is one characterised by three periods of evolution. First, from 1966 to 1974 many of the features established in the previous period were carried forward. The most significant changes were successive Rent Acts, which resulted in the almost total disappearance of the market for private rental housing making the rental basis for valuation for rating almost totally arbitrary, and the *Local Government Act* (1966) which established the RSG as a combination of the previous General Grant and Rate Deficiency Grant. Second, 1974 marks the reorganisation of local government in England and Wales (1976 in Scotland) and this caused a transfer of functions between levels of local government, changes of boundaries and other important discontinuities with the previous period. The third period of evolution begins in 1980 with the establishment of a new form of RSG in the *Local Government, Planning and Land Act* (1980) following the election of a Conservative Government in 1979.

In the *Local Government Act* of 1966, introduced by the Labour Government of Harold Wilson, the Rate Support Grant structure was established in a form closely resembling that used up to 1981/2. The previous trend of consolidating specific grants was continued and local relevant expenditure was now measured as a single unit (but excluding mandatory arrangements for higher education students grants, rate rebates, and so forth).

The RSG and the introduction of rate rebates to poor people were intended to provide local authorities with a stable financial base as an interim mechanism to offset lack of reform of the rating system. Again, in 1974, the response to the 1971 (Green Paper G.B. Government, 1971) call for reform of the rates was a further amendment of the RSG. Similarly, the 1980 reform of RSG, following the 1976 Layfield Report (G.B. Government Committee of Enquiry, 1976), again ducked the issue of reforming local taxes. Hence, what the RSG achieved in 1967 was consolidation of most central support to local government into a single unit of payment and negotiation, giving local authorities a single sum of money representing a block grant unhypothecated to specific uses. The population weighting initiated in the Exchequer Equalisation Grant of 1948 was abandoned and a proportion of the grant was paid direct to Districts within Counties and to County Boroughs, as the Rate Deficiency Grant had been. Moreover, the deficiency payments principle and need indicator approach were also retained.

The RSG, although a single sum of money, was divided into three elements up to 1980/1: needs, resources and domestic. First, the *needs element* replaced the General Grant of the 1958 Act and employed a similar formula to that summarised above in equation (2.5). The major addition to the equation was a variable to measure the number of people of pensionable age in a local authority. The major reduction in the equation was the omission of the product of a standard rate poundage which was absorbed into the resources element. As such, the needs element maintained the trend away from specific towards block grants and was still measured against 'standard needs' of local authority expenditure at average levels. The second, *resources element*, contained the principle of the RDG, paying resource deficiency grants to those local authorities with small tax bases but incorporated the recommendation of the Edwards Committee (G.B. Government, 1953), that all needs components such as population density and miles of road were eliminated. The resources grant then becomes a pure resource equalisation formula; but paid on the deficiency payments principle. The standard rateable value per head $\bar{B}$ against which local authorities were measured remained at the national average from 1966 to 1974, but from that time steadily increased so that it included most authorities. The third component, the *domestic element*, was the only really new component of central grants. This was a discount to domestic rate-payers as against non-domestic rate-payers (commercial). It was introduced because of the rapid rise in rate bills during the 1960s, and also to compensate for the greater levels of expenditure local authorities were forced to undertake with the reorganisation of water supply in 1974/5.

Following the 1974 reorganisation of Local Government in England and Wales, the RSG survived with only minor changes. Its detailed form derived in England and Wales from the *Local Government Act* (1972), implemented in 1974; and in Scotland from the *Local Government (Scotland) Act* (1974), implemented in 1975. Under these Acts a new form of more consolidated government replaced the old County Boroughs, Boroughs and County Councils. From the previous structure of government, only the Parish Councils survived unaltered. The new structure of government follows four models. First, in rural areas, the District, although levying the rates, is subordinate to the County which makes precepts on the Districts of 10–20% of the local property tax. Second, in urban metropolitan areas, the metropolitan County is subordinate to the Districts, although it has a precepting power equivalent to 10–20% of the local tax. Third, in London there is a separate arrangement which divides power more equally between the Districts (London Boroughs), and the County level of the Greater London Council (GLC). Finally, in Scotland, Regions replace the County level with the major functions. In addition, three islands have the combined powers of Regions and Districts.

The distribution of the various local authorities is shown in figure 1.1, whilst the relative magnitude of revenue-raising and expenditure at each level of government is shown in table 3.3. This table also lists the major expenditure items for which each type of local authority is responsible. The main contrast which should be noted is the predominant power of the

Table 3.3. *Division of fiscal responsibilities between local authorities in Great Britain, 1976. (Source: G.B. Government Committee of Enquiry into Local Government Finance 1976)*

| Level of government and major functions (1976) | Revenue (%) | | | | Expenditure (% of total local expenditure) | Division of expenditure between tiers (%) |
	General grants	Specific grants	Property tax domestic	Property tax commercial		
ENGLAND AND WALES						
Metropolitan Counties (6) (Housing, transport, fire, police, refuse disposal, consumer protection)	33	23	26	28	4	20
Metropolitan Districts (36) (Education, personal social services, housing, refuse collection, environment)	67	1	13	19	16	80
Non-Metropolitan Counties (47) (Education, personal social services, transport, police, fire, refuse disposal)	61	7	14	18	42.5	85
Non-Metropolitan Districts (333) (Housing, public transport, refuse collection)	41	6	24	29	7.5	15
GLC (1) (Arts, libraries, fire, refuse, transport)	9	16	22	53	9	45
London Boroughs (33) (Education, environment, housing, police, social services)	59	3	11	27	11	55
SCOTLAND						
Regions (9) (Education, transport, roads, police, fire, water, social work)	67	5	12	16	8.5	85
Districts (53) (and Islands) (3) (Youth employment, housing, environment, refuse)	50	3	21	26	1.5	15
Total number of jurisdictions	522					

Districts (and London Boroughs) in metropolitan areas, with the major expenditure functions of education, housing, and social services, whilst in the non-metropolitan areas, the primary powers rest with the County.

As far as the RSG itself is concerned, the major changes following the 1974 reorganisation were four-fold. First, its size increased and has never dropped below 60% of local expenditure from 1974/5 onwards. Second, the number of needs indicators employed was greatly extended. Third, the *a priori* measures of need were abandoned and instead indicators were employed directly for each local authority. Fourth, London, which remained as a separate entity for grant purposes over the whole period up to 1974, slowly became more integrated into the system of RSG used for other local authorities. Indeed, a major controversy in the allocation of RSG has concerned the relative level of support for London in comparison with other areas.

Apart from the reorganisation of local government, the period since 1966 has also been important for the evolution in local finance in another way: it has been characterised by a rapid growth in awareness in deficiencies in the central grant structure, with the result that suggestions have been made for reforming not only RSG, but the local revenue system as a whole. Of course, many of the criticisms of both central grants and the revenue base of the rates were clear in the nineteenth century, and were particularly well highlighted by Goschen, but during the 1970s and early 1980s the rate of inflation in the UK frequently topped 20% and the inadequacy of the local finance system to cope has become increasingly clear. Apart from earlier criticisms, in 1965 the Allen Committee had drawn attention to the regressive nature of the rates, the Seebohm Committee in 1968 had suggested a radical restructuring of local practice in Social Services, the Skeffington Committee in 1969 suggested the need for radical improvements in public participation, and the Green Paper of 1971 (G.B. Government, 1971) had again suggested the use of a local income tax in place of, or in addition to, the rates. In the period between 1971 and 1974 the inflation of public service costs, particularly hard hit because they are either heavily resource-intensive or labour-intensive, was so great that the RSG increased rapidly and became 66% of local expenditure in 1975/6. This greatly undermined local independence and tended to underprice local services. Hence, by 1974, the problems of local finance were so great that reform seemed inevitable. As a consequence the new Labour Government of 1974 set up a Committee of Enquiry with a lawyer, Frank Layfield, as chairman, and this reported in 1976 (G.B. Government Committee of Enquiry, 1976).

The Layfield Committee report has become a benchmark for subsequent studies. It suggested that a clear line of accountability should be established for those services which were centrally funded and those which were locally funded: this would foster both more responsible government and more effective public participation. It required, Layfield suggested, the expansion of local tax resources so that expenditure could be more clearly related to the areas bearing the burdens, and this in turn required a new revenue base which was both more buoyant than the rates, and one which did not possess its undesirable properties of relative regressiveness. The Committee concluded that only a local income tax (LIT) would serve this purpose, and this should be phased in over a period of years to supplement the rates. The Committee also suggested that the basis of valuation for the rates should be modified to one based on capital rather than rental values.

The Layfield report marks a clear datum in studies of local finance in Britain, but its suggestions and the issues it raised have become submerged under four further areas of concern. First, the Labour Government of the 1974–79 period held power at best by only a slender majority, and in its latter years was a minority government surviving only by an informal coalition with various minor parties (the most important being the pact between the Liberal and Labour Parties). This made it extremely unlikely that radically new measures such as LIT could be implemented. (Even though reform of local finance was an issue central to Liberal Party concern, the influence of Liberal MPs was insufficient and they tended to concentrate their attention on the issue of proportional representation.) Second, the UK Treasury, which had always resisted the development of regional administrative bodies, strongly opposed any use of the personal income tax at local level. The arguments used were that it would undermine the capacity of the national government to manage the economy and would increase the rate of income tax overall which was already at very high levels internationally. Both arguments, although of great importance, could have been overcome if there had been a real political desire for LIT.

The LIT issue also became muddled with a third area of concern, that of devolution. Under the 1978 devolution proposals for Scotland and Wales, the proposed form of finance for the regional assemblies was to be a form of the block grant of the RSG, which would then be subsequently divided between the local authorities. The Westminster Government always staunchly denied to Scotland and Wales any separate revenue-raising powers. Hence LIT became muddled with the issue of federalism and separation of Scotland and Wales, and any hope of giving new revenue powers to local authorities was killed, or at least delayed, by this issue.

The fourth impediment to implementation of the Layfield proposals has operated especially since the return of a Conservative Government at Westminster since 1979. Following a policy of strict monetary control and a desire to reduce the total level of public expenditure, local government has been seen by the Westminster government as requiring much greater control and central intervention. This has been evidenced by restrictions on the total level of capital borrowing, reduction in the level of central support in real terms, imposition of spending targets, exhortations to reduce local expenditure and manning levels, and the preparation of improved statistical series of expenditure to permit closer monitoring of local decisions. Most important of all, however, has been the *Local Government, Planning and Land Act* (1980) which introduced a major modification to the RSG to become a unified block grant (combining both resource and need equalisation). Because the allocation of this grant is based on criteria of standard expenditure per head and standard rate poundage, it introduced the possibility of much greater power of central control over local spending. In addition to this RSG block grant the Conservative Government also were reluctant to allow greater local revenue discretion. Hence the need for alternative revenue sources, such as LIT, has not been fully answered. Partly as a result of their aims the Conservative Government were also reluctant to disturb the basis of rating valuation. Only by mid-1982 was a commitment forthcoming to replace nominal rental valuation of domestic property in favour of capital valuation, as recommended by the Layfield Report and endorsed by the previous Labour Government (G.B. Government, 1977).

The 1980s, then, have seen a shift towards greater central control. Major features in this shift in central control have been the issuing of standard expenditures and standard tax rates for each local authority. In addition 1982/3 saw the introduction of legislation to prevent the levying of supplementary rates in England and Wales. This in some ways represents one of the major constitutional changes since the 1929 introduction of block grants since, for the first time, it limited the local freedom of access to the rates as a separate, local source of taxation.

A major aim of this post-1979 period was to achieve better 'value for money' and accountability to local opinion. The activities of local rate-payers associations certainly increased over this period (see Bennett, 1981*a*), but the major aspect of the drive to obtain better monitoring of 'value for money' has been the preparation of two new publications. The first of these, the *Joint Manpower Watch*, was published by the Department of the Environment from 1980 onwards. It has been considerably hampered by some lack of local authority co-operation and has been a very controversial data set attempting to monitor local authority employee levels. The second new publication, *Local Government Comparative Statistics*, is published by CIPFA with Department of Environment support (CIPFA, 1981*a*). This publication was a direct consequence of recommendations by the Layfield Report and was also accepted by the Labour Government (G.B. Government, 1977). It gives about 50 performance indicators which authorities are obliged after 1981 to publish locally in their annual reports. These indicators concern such items as pupil/ teacher ratio; net costs of non-advanced further education, school meals, children in care, maintenance costs of roads; proportion of serious criminal offences per 1000 population; fire risk categories, etc.

'For the first time we now see published a selection of truly comparable figures covering the whole range of local authority services. We also see figures for management set alongside expenditure'. '. . . efficiency and economy will only be secured if (this) essential information about the costs and levels of service is available to everyone concerned' (T. King, Minister for Local Government and Environmental Services, in G.B. DOE, 1981, p. 1).

Despite criticisms of the 'value for money' concept, and the fact that many councils are already highly efficient in use of manpower, resources and capital (see AMA, 1979; CBI, 1979), this shift to better and more visible government can only be welcomed. However, as a result of the 'value for money' movement and other post-Layfield pressures, the development of local finance since 1979 marks the initiation of a period in which central control of local government has strengthened, central support of local spending has steadily declined, and financial support has shifted away from general towards more specific support of local services.

This shift towards greater central control is nowhere more evident than in the 1982 *Local Government Finance Act*. This introduced four new central controls over local authorities. First, and of prime significance in relation to the RSG block grant, was the termination of the right of local authorities to raise supplementary rates; hence, they could not make adjustments by local tax increases in response to penalties exacted through reductions in RSG grant (at least not during the current financial year). A second change was to limit the powers of precepting authorities. This was mainly aimed at preventing metropolitan Counties and the GLC from spending at high levels by placing large rate burdens on the rate levying authorities (the Districts). The third change was the introduction of a more rigorous central auditing of local accounts through an Audit Commission. This had been suggested by an earlier independent enquiry (G.B. Advisory Committee, 1980). Fourth, and most controversial, was the power to penalise grant during a second or even a third round of cuts – so-called 'super-holdback'.

Running parallel to the 1982 *Act* were the Conservative Government's reactions to suggestions for reform of the method of raising local taxation. In December 1981, belatedly fulfilling an election manifesto commitment, they published a consultative Green Paper *Alternatives to Domestic Rates* (G.B. Government, 1981b) which reviewed local revenue sources in the light of the post-1981 RSG block grant and its modifications in the *Local Government Finance Act* (1982). The conclusion of this document was that no alternative to the domestic rates existed which could be implemented in the short term. Of the possible alternatives considered viable, a local sales tax appeared an attractive contender to a Conservative Government dedicated to shifting tax burdens from direct to indirect taxes, with simplifying the overall tax structure, reducing public sector manpower, and easing the burdens on businesses (see G.B. Government, 1981*b*, p. 27). As a result the government seemed to favour sales tax over LIT, the alternative favoured by most previous commentators and that discussed at length in later chapters of this book. However, for LIT or sales tax, the Conservative Government saw no possibility of implementation before 1987/8 because of delays in computerising personal income tax, VAT, and excise administration and even raised the possibility of using a poll tax. Of course, implementation of local tax sources, giving local authorities significant new access to revenues free from strict central controls, was also in conflict with Conservative thinking which suggested that more rigorous control of local spending was what was required. The major consequences of the Green Paper consultations were, therefore, first, a commitment to implement capital valuation of domestic property for levying of domestic rates; and second, to look again at alternative local tax sources at a future date when computerised income and excise tax machinery allowed their implementation.

3.6 Conclusion

The discussion of this chapter has emphasised the continuity of many of the patterns, and the problems, of local government finance over the period since the 1870s to the present day. Some of the problems have been overcome; as, for example, with the consolidation of government since 1974, and the transfer of many national or regional responsibilities (such as health, poverty relief, water, electricity and gas) away from local authorities. Other problems have been diminished but not yet solved: for example, valuation for rating is still too infrequently undertaken and the rental valuation principle is almost totally arbitrary and should be abandoned. Yet other problems still remain to be solved. Especially important among these is the difficulty of balancing the need for local independence, requiring a diversified revenue base, with the need for central control of total public expenditure and concern for equalisation between areas. Also important is the problem of adequately measuring local spending need.

The RSG is now the major central grant to local government and acts as a major key to understanding the whole balance and pattern of local finance. It is described more fully in the next chapter, but from the brief outline given above it should already be clear that the RSG derives much of its form from the developments of previous periods and suffers many of the same criticisms. The measurement of needs derives much from both the spirit of Balfour's minimum standards of 1901 and from the Kempe Committee's suggestions of need equalisation of 1902. The use of a set of need indicators up to 1980 followed down from the General Exchequer Grant of 1929, with the specific indicators used following essentially the same structure as the General Grant of 1958. The pervasive problem of balancing the rural needs of the Counties with the urban needs of the County Boroughs continues through each of these grant proposals by the use of low density weight allowing for population sparsity. For resources equalisation, the Scottish rateable value per head measure of 1890 has survived as the measure of the rate resource base, and the Hicks and Hicks (1944*a*) suggestion of deficiency payments, codified in the 1948 Exchequer Equalisation Grant, has survived until 1980/1. The recent combining of the separate needs and the resources components into a unitary grant has the aim of simultaneously compensating local authorities for high expenditure need and low tax resource base. The RSG has previously fallen far short of achieving this aim but it has a motivation dating back at least to Goschen's (1872) *Report*, and is implicit in the literature of local service provision and tax-raising as early as the Elizabethan Poor Laws.

CHAPTER 4

THE RATE SUPPORT GRANT

4.1 Introduction

The discussion of the previous chapter has emphasised the evolutionary character of the present method of central support for British local authorities through the RSG. Most of the major components of the present method were discussed at least as early as Goschen's *Report* in 1872. As a result of such continuous evolution, many of the characteristic strengths and weaknesses of former grant systems survive to this day. In the present chapter the structure of the RSG is discussed in detail up to 1980/1. The next chapter continues this discussion for the period after 1981.

The allocation of the Rate Support Grant over the period from 1967/8 up to the present has been governed by the *Local Government Act* (1966) which initiated the grant system, the *Local Government Act* (1974) which modified some of the distribution arrangements and re-established the grant for the period following the 1974 reorganisation of local government (G.B. Government; 1966, 1974) and the *Local Governments Planning and Land Act* (1980) which established the present method of allocating the Rate Support Grant. In addition to these Acts, modifications in detail, often very significant in their distributional impacts, are decided on an annual basis by the Secretary of State for the Environment. Up to 1980/1 *Statutory Instruments* were used to specify the Rate Support Grant *Orders* and *Increase Orders* but since 1981/2 House of Commons *Reports* have been employed (G.B. Government, *Statutory Instruments* and House of Commons *Papers*, various dates). These are issued usually in November or December of each year. Further modifications in detail are introduced by the Secretary of State by use of Department of the Environment *Circulars* (G.B. Government: DoE and MHLG, *Circulars*, various dates).

The development of the RSG since 1967/8 has been immensely complicated. The grant has been characterised, however, by a constant set of underlying objectives concerned with achieving equalisation of both local government financial resources and expenditure need. This objective has recently been restated as one of giving to local authorities 'sufficient grant to put them in a position where they can provide similar standards of service for a similar rate in the pound' (G.B. House of Commons *Papers*,

1980, p. 8). Other objectives, as discussed in chapter 2 have also been employed at various stages and particularly important have been the improvement of fiscal balance and improving local discretion. However, the constant underlying feature has been one of equalisation. As such the Rate Support Grant represents perhaps the major example of an equalisation grant used in the western world.

The discussion of this chapter decribes the manner in which the RSG is allocated and details how this method has evolved since 1967/8.[1] This is approached in three stages. In section 4.2 the planning of the aggregate level of RSG is discussed, whilst section 4.3 describes how the grant is distributed between its major elements and how this affected the allocation between local authorities up to 1980/81. Chapter 5 then discusses the method of allocating the grant which has been employed since 1981/2.

4.2 Planning the aggregate level of Rate Support Grant

The size and form of the Rate Support Grant is decided each year by the Secretary of State for the Environment in consultation with the local authority associations.[2] Although this consultation process is always claimed, by both sides, to be constructive and useful there is no doubt, in the context of a centralised state such as Britain, that the central government through the Secretary of State is the dominant voice. As stated by Stanley de Smith (1971, p. 409), the relationship is a partnership, but one 'between the rider and the horse'. Despite this characteristic, however, the local authority associations are influential, most particularly since they present both strong political constraints and strong political forces in the country as a whole.

Up to 1975/6 little attempt was made to integrate local expenditure estimates with central expenditure planning; the estimates, in effect, acted as forecasts which were implemented without very close scrutiny. Since 1975/6, however, the RSG has become steadily more closely linked with central expenditure planning through the Public Expenditure Survey Committee (PESC). This process of central government planning of rate support and the consultation process with local authorities is shown in figure 4.1. The planning process can be approximated as six stages. First, as shown on the right-hand side of the figure, forecasts of local expenditure are produced by the local authorities and co-ordinated by the Chartered Institute of Public Finance[3] (CIPFA). These forecasts are then disaggregated by spending departments then presented to the central government Expenditure Steering Group within the Department of the Environment. This group scrutinises the local expenditure forcasts and makes important decisions as to what is acceptable. It then produces the final forecasts which incorporate those features which the Secretary of State for the

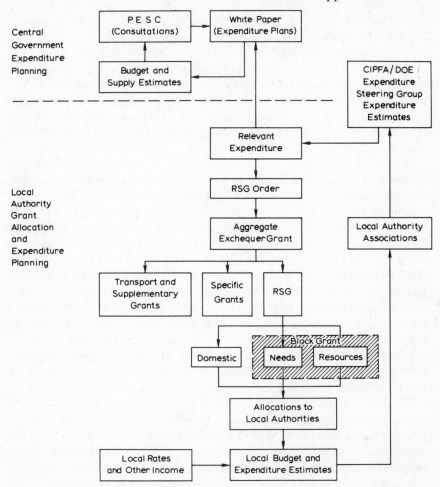

Fig. 4.1 Stages in decision making in the allocation of the Rate Support Grant showing the interactions between central and local government.

Environment is required to take into account in making RSG decisions (G.B. Government, 1966, 1974):

(i)   The latest available information on local relevant expenditure;

(ii)   any probable fluctuation in the demand for relevant services (as far as this can be attributed to national circumstances outside local authority control, e.g. changes in the age structure of the population);

(iii)   the need to develop and increase services in the light of general economic conditions (changes in priorities resulting from legislation in the last year); and

(iv)   the current level of prices, costs and remuneration of wages; and any future variation which is likely to result from decisions already taken which will affect prices, costs and remunerations (derived from Treasury and PESC economic forecasts).

The second stage of planning is the presentation of these agreed expenditure estimates in the RSG *Report* or *Order* in November prior to the start of each financial year. At about the same time as the publication of the *Report*, a series of *Supplementary Reports* and *Increase Orders* is also published which modify the level of grants for the previous one or two years to take account of price and other changes which were not anticipated in the initial expenditure estimates. These usually occur in the November of the current year and in November of the following financial year, respectively.

A third planning stage is the integration of these estimates with central government economic planning in the Treasury. This has involved, since 1969, discussions between central departments through the PESC, which eventually result in firm government plans which are usually published in the Public Expenditure *White Paper* published in December or January (G.B. Government, *White Papers*, various dates). In addition to these firm plans the *Supply Estimates* and *Budget*, debated in Parliament in March, April or May, are also significant in introducing changes and more detailed planning of specific programmes. The expenditure estimates for local authorities in England and Wales are related to the central government's *White Paper* by integrating into the RSG *Report* of November the estimates from the *White Paper* from the preceding December/January. For Scotland, however, the negotiations on RSG are not usually completed early enough and provisional estimates of local expenditure have to be employed instead.

The fourth stage of planning in the RSG procedure is the definition of the Aggregate Exchequer Grant. This includes *all* specific, supplementary, transport, and other central grants in addition to RSG. Up to 1975/6 the expenditure estimate stage, discussed above, directly determined grant totals. However, since 1976/7 increases in grant have been subject to an additional control of *cash limits*. These provide a ceiling amount which will be paid as increased grant (through subsequent *Supplementary Reports* and *Increase Orders*) in respect of increases in local expenditure. Table 4.1 shows the cash limits set for the Aggregate Exchequer Grant since 1976/7 and the degree to which changes in these limits are related to changing economic conditions. As can be seen from the table, the cash limit has not compensated fully for the general rate of price increases. Table 4.1 also shows that cash limits have been used in various ways as instruments of national economic management. In both 1976/7 and 1977/8, for example, final cash limits were reduced as part of general public expenditure policy. However, in 1976/7 final cash limits were not reduced by the £73.1 m originally intended, an additional supplement of £38 m (both in 1978 prices) being eventually allowed (compare *House of Commons Paper* HC-28 (1976) with HC-58 (1977)). Similarly with a change of party of central government in 1979 the cash limits were again used as major instruments of central economic control.

Table 4.1. *The cash limits on relevant expenditure for the Aggregate Exchequer Grant used since 1976/77, and their relation to changes in total costs*

Fiscal year	Initial cash limit (£m)	Final cash limit (£m)	Total allowed increase as % to total relevant expenditure	Changes in total prices %
1976/7	702	655	5.2	15.5
1977/8	706	667	5.4	10.7
1978/9	581	627	4.9	9.6
1979/80	443	673	5.2	8.8
1980/1	1258	1227	9.9	14.9
1981/2	856	—	6.6	—

All figures in 1978 constant prices for England and Wales.
Source: House of Commons Papers.

A fifth stage in RSG planning is a decision as to what proportion of total local estimated expenditure is accepted by central government as 'relevant expenditure' for grant purposes. This decision rests on two criteria, one objective, and one partly political. The first, objective, criterion for defining relevant expenditure derives from estimating that proportion of total local expenditure which derives from local undertaking of services as devolved agents of central government which places a burden directly on local taxpayers as opposed to central taxpayers (see G.B. Government, 1966; G.B. MHLG *Circular* 12/67). As a result of this definition, an attempt is made to differentiate non-discretionary devolved spending from that which is discretionary. Expenditures which are supported directly by user charges are also excluded, i.e. those receipts derived from charges such as local bus fares, legal fees, and public sector housing rents. In addition attempts are made to differentiate expenditures supported by different forms of grant; for example, expenditures deriving from mandatory payments (such as rate rebate and students awards) are excluded from the expenditure accepted as relevant for RSG purposes. The second, political, criterion for defining relevant expenditure derives from the central government's choice of which are and which are not devolved services, the extent to which it will allow increases in wages and other costs to be borne by central grants, the estimate of increases in costs, wages and other inflationary factors which will be accepted at Westminster, the level of resulting local rate bills which can be tolerated at both local and central level, and other factors. For example, from 1971/2 to 1974/5 a term was introduced to reduce relevant expenditure as a result of improved local efficiency but no actuarial or statistical estimate of either the existence or the magnitude of this term was made. In 1974/5, following local government reorganisation, a modified allocation was necessary to take account

Table 4.2. *Local authority total and relevant expenditure in England and Wales at Final Increase Order stage at PES outturn prices*

		Relevant expenditure for RSG purposes (PES basis) £m. (1978 prices)				
Financial year	Total local authority expenditure £m. (1978 prices)	Current	Capital to be met from revenue	Loans	Total	Relevant exp. as % of total
1966/7	10674		N/A		5665	53.1
1967/8	11319		N/A		8485	74.9
1968/9	11728				9049	77.2
1969/70	11959				9618	80.4
1970/1	12483		N/A		10995	88.1
1971/2	12573				11319	90.0
1972/3	13684	9187	256	867	10310	75.3
1973/4	14544	9457	328	1216	11001	75.6
1974/5	16175	10442	305	1299	12046	74.5
1975/6	15979	10910	516	1203	12629	79.0
1976/7	15211	10891	541	1282	12714	83.6
1977/8	14343	10691	564	1145	12400	86.4
1978/9	14650	10982	505	1232	12719	86.8
1979/80	14760	11335	589	1068	12992	88.0
1980/1	13889	9913	508	825	11246	86.7
1981/2[a]	13081	9114	474	734	10322	78.9
1982/3[a]	13267	10625	344	1165	12134	91.5

[a] Planned.
Dashed lines show where data are not directly comparable due to changes in local government responsibility for services and changes in the rendering of statistics.
Sources: Public Expenditure *White Papers* up to Cmnd. 8494 (1982); House of Commons *Papers*; CIPFA *Local Government Trends*, 1979; *Local Government Financial Statistics*.

of the transfer of debts and re-allocation of sewerage and health services; and since 1978/9 interest receipts were introduced to reduce the level of relevant expenditure.

Table 4.2. shows the total level of local expenditure accepted by central governments as relevant. It has risen rapidly in the 1960s and early 1970s but has levelled off since about 1977/8. This pattern is usually accounted for by the influence of three factors. First, the desire to give a greater proportion of central support in a form which is unhypothecated to local use (i.e. as a general rather than specific grants). Second, and related to this first factor, is the question of fiscal balance. The rapid and accelerating rate of inflation in the economy as a whole, particularly marked in the high proportion of labour-intensive local services, outstripped the capacity of the unbuoyant rates to provide sufficient income and this stimulated central support. Third, since the local rates are relatively regressive (up to the introduction of rate rebates in 1967/8) or proportional burdens on local taxpayers, most Westminster governments have thought it undesirable that too great a proportion of local revenue is derived from them. As a result they have sought to achieve fiscal balance by supporting local finances to an increasing extent from the more progressive central tax base. This policy has meant that, with the very rapid inflation in the British economy over the period since 1967, central government has supported an increasing proportion of local service costs up to 1978/9. Since that date there has been a change in philosophy, most clearly represented by the Conservative Government since 1979, that the steady increase in the proportion of public spending deriving from local government should no longer be supported by central government, and that a shift to a greater proportion of local burdens placed on local taxpayers will result in better control of public expenditure as a whole. Also important has been the recent increased desire to make local authorities 'toe the line' in wage controls.

The sixth stage of planning the RSG is the determination of the division of the Aggregate Exchequer Grant between its components of RSG, specific, supplementary, transport and other grants, which are all included within the Aggregate Exchequer Grant. The relative share of these grants since 1966/7 is shown in table 4.3. From this table it is clear that although there has been a steady rise in the total level of Westminster support to local authorities up to 1975/6, followed by a levelling off to the present, the total proportion of RSG has steadily declined (except for 1974/5) since 1971/2. This has been as a result of the growth in significance of specific grants which have increased by 40% since the initiation of the RSG in 1967/8 (mainly as a result of rises in the police and transport grants). However, despite this change of emphasis, the RSG still represents by far the greatest proportion of central government support.

The pattern shown in table 4.3. results from two main features: first, the changing expenditure levels of different local services and the degree to which central government has deemed the expenditure 'relevant'; and second, the variable incidence of inflation on different services and the allowances permitted by central government towards increased costs. For example, the expenditure level supported by the RSG under each of the main service categories is shown in table 4.4. In the period up to 1973/4 expenditure on personal social services, police, local environmental services and planning all expanded relatively rapidly, and only the administration of justice expanded less rapidly than the rate of price increase in the economy as a whole. In this period, therefore, considerable expansion of service quality, manning levels, and service quantities was possible. In the period from 1973/4 to 1978/9, in contrast, only personal social service expenditures have expanded more rapidly than general price increases,

Table 4.3. *Local authority expenditure in England and Wales, current account spending and central government grants 1966–82 at 1978 prices*

(£m)	Aggregate Exchequer Grant final order (£m)	Relevant Expenditure final order (%)	Aggregate Exchequer Grant as % of Relev. Expend. (£m)	RSG (%)	RSG as % of Aggregate Exchequer Grant (%)	Specific grants as % of Aggregate Exchequer Grant (%)	Transport and Supplementary Grants as % of Aggregate Exchequer Grant
General Grant 66/7	4014	4684	80.5	2556	68.1	31.9	—
67/8	4587	8485	54	4162	90.7	9.3	—
68/9	4983	9049	55	4525	90.8	9.2	—
69/70	5396	9618	56	4956	91.8	8.0	—
70/1	6269	10995	57	5755	91.8	8.2	—
71/2	6509	11319	57.5	5977	91.8	8.2	—
72/3	7551	13021	58	6952	92.1	7.9	—
73/4	7736	12914	60	6989	90.3	8.1	—
74/5	9978	15294	60.5	9198	92.1	9.9	0.04
75/6	11914	17877	66.5	10384	87.1	9.9	4.5
76/7	10659	16711	65.5	9203	86.3	8.4	4.7
77/8	9334	15326	60.9	8009	85.7	9.0	3.8
78/9	9186	15157	60.6	7844	85.4	10.5	3.6
79/80	9517	16364	58.1	7966	83.7	11.0	3.7
80/81[a]	9773	16621	58.8	8180	83.7	12.6	3.7
81/82[b]	8263	13866	59.6[d]	6904	83.6	12.7	3.7
82/83[c]	8301	14538	57.1[e]	6797	81.9	14.2	3.9

[a] at first increase order.
[b] before addition of cash limit.
[c] before addition of cash limits.
[d] 59.1% in England and 73.4% in Wales.
[e] 56.1% in England and 72.5% in Wales.
Note the RSG results are modified by outturn figures: up to 1975/76 the grant was paid on outturns, from 1976/77 it was subject to cash limits.
Source: House of Commons Papers.

Table 4.4. *Change in level of Relevant Expenditure allowed in RSG calculations; increase of first Order over final Order*

Service category	Average annual percentage change at 1978 prices		
	1967/8–73/4	1973/4–79/80	1979/80–80/81
Education	10.1	1.5	20.9
Personal social services	20.1	23.3	16.5
Police	13.0	9.3	15.4
Administration of justice	1.0	8.1	16.4
Local transport	11.9	7.5	16.4
Fire	11.1	6.8	11.4
Local environment services	17.5	5.4	18.8
Town planning	13.8	16.6	18.8
Housing (excluding HRA)	—	13.4	—
Total	10.7	3.8	18.8

The 'other services' category is omitted because of lack of comparability arising from changes in local responsibilities. HRA is the Housing Revenue Account.
Source: House of Commons Papers.

and education in particular suffered very marked contraction in cost terms. From 1979/80 to 1980/1 all service expenditures have increased much more rapidly than general price increases. This was in a period when a Conservative Government was trying rigorously to control local spending; since 1980/1 this policy was made more effective through use of the RSG block grant.

Although now a highly developed procedure the integration of local authority expenditure planning with that of central government has been subject to a number of difficulties. First, the estimates of local expenditure derive from discussion between the DoE and the local authority associations and at the early stages do not involve or consider PESC, Treasury or other central departments. Second, up to 1981/2 the *price basis* of the RSG *Report* or *Order* and the *White Paper* differed: the *White Paper* referred to prices of November of the previous year whilst the RSG *Reports* and *Orders* referred to prices in November of the present year. This was severely criticised by the Layfield report (G.B. Committee of Enquiry, 1976) and has been overcome by current year pricing since 1981/2. Third, the *coverage* of the *Report* or *Order* and *White Paper* differs: rate subsidies to the Housing Revenue Account were excluded from the *Report* or *Order* but included in the *White Paper*, whilst loans and some charges met from revenues were included in the *Order* but excluded from the *White Paper*. This has also been severely criticised by the Layfield Report and elsewhere, but has been overcome by changing the form of the *White Paper* since 1977/8. Fourth, the RSG *Report* or *Order* settling local relevant expenditure (q.v. below) is too close to the *White Paper* stage: most local expenditure is already committed for the following year by November of the preceding year and so is not amenable to easy modification. This introduces great uncertainty into local expenditure planning and rating decisions. Fifth, parliamentary debate has been clouded by the poor relation of the *Supply Estimates* to the *White Paper* to transfers from and within central government, and some items of revenue and borrowing (see G.B. Government, 1972). A sixth problem has been recognised by Nield and Ward (1976): that central forecasts of local expenditure are inaccurate and have often been affected by central aims of keeping local expenditures under control and rate increases down. Such inaccuracies in central expenditure forecasting have great importance for local government. For example, Harrison and Smith (1978) suggest that cash limits have had the effect of shifting 'the uncertainty and therefore the onus of correcting the errors in economic forecasting from the Treasury to the local authorities'.

The stages of public expenditure planning discussed in the preceding paragraphs serve to define the total level of rate support and other grants. The next stages of decision which are required are the form of the Rate Support Grant itself and how it is to be distributed between local authorities. These are the two most complicated and difficult decisions of all.

4.3 The distribution of the RSG 1967/8–1980/1

Although transferred to local authorities as a single sum of money, the RSG, up to 1980/1, was composed of three separate elements: *needs*, *resources* and *domestic*, each of which had different aims and characteristics. The distribution of the RSG between its components is shown in table 4.5. From this table it is readily apparent that the grant has been subject to a major policy change occurring during the 1974/5 fiscal year. This change in 1974/5 has had the result of diminishing the level of the needs component from about 80 to approximately 60% of the total with the contrasting effect of increasing the resources component from approximately 15 to 30% of the total, and increasing the domestic rate relief grant from about 5 to 10% of the total. Small scale changes between the three elements up to 1980/1 also produce year-to-year variations. The detailed pattern of these changes is discussed for each element in turn below.

(1) *Domestic element (and Domestic Rate Relief Grant)*

This is the smallest component of RSG, but amounting to £600 m. in 1980/1 (at 1978 prices) it is still significant and represents 9% of the RSG. Its effect is simple; to give a differential rate of local property tax between domestic residential property and non-domestic (commercial) property. The domestic element is hypothecated, i.e. the local authorities are obliged

Table 4.5. *Distribution of Rate Support Grant in England and Wales between its elements*

Fiscal year	Total	£m. 1978 prices Block Grant		Domestic	% Block Grant		Domestic
		Needs	Resources	Domestic	Needs	Resources	Domestic
Rate deficiency and General Grant 1966/7	2628	—	—	—	—	—	—
1967/8	4162	3409	678	75	81.9	16.3	1.8
1968/9	4525	3653	720	152	80.7	15.9	3.4
1969/70	4956	4007	725	223	80.8	14.6	4.6
1970/1	5755	4631	817	306	80.5	14.2	5.3
1971/2	5975	4875	819	322	80.9	13.7	5.4
1972/3	6952	5654	925	363	81.3	13.5	5.2
1973/4	6989	5622	932	435	80.4	13.4	6.2
1974/5	9198	5990	2272	936	65.0	24.7	10.3
1975/6	10384	6259	3014	1110	60.3	29.0	10.7
1976/7	9203	5580	2687	936	60.6	29.2	10.2
1977/8	8010	4857	2340	813	60.6	29.2	10.2
1978/9	7810	4767	2297	746	61.0	29.4	9.6
1979/80	7966	4912	2367	687	61.7	29.7	8.6
1980/1[a]	8180	5104	2458	617	62.4	30.0	7.5
1981/2[b]	6840	6313		527	92.3		7.7
1982/3[c]	6797	6328		469	93.1		6.9

[a] provisional at first Increase Order.
[b] before addition of cash limits.
[c] before addition of final cash limits.
Source: G.B. Government, Statutory Instruments, *Rate Support Grant Orders* and *Reports*.

to use this element to reduce the tax rate to domestic ratepayers. Hence it differs from the rest of the Rate Support Grant which is a block grant with a degree of local discretion. As such it represents an anomaly in the RSG. Hence, it has often been suggested that it should be removed to become part of the specific grants category. The motive for the domestic element is usually thought to be largely political. The domestic ratepayer has been a more influential and volatile influence on both council and parliamentary elections than the commerical lobby. However, since 1982/3 special relief has also been suggested for non-domestic ratepayers.

The domestic element was first instituted in 1967/68, when it represented a supplement (or 'abatement') to domestic ratepayers of 2.1p in the £ representing 3.6% of the average rate bill in England (G.B. Committee of Enquiry, 1976, p. 161). It has grown in magnitude since that time to become about 30% of the average English rate bill. Table 4.6 shows the

Table 4.6. *Rate of domestic rate subsidy and subsidy to mixed domestic plus commercial hereditaments*

Subsidy (p in £)	Fiscal year															
	67/8	68/9	69/70	70/1	71/2	72/3	73/4	74/5	75/6	76/7	77/8	78/9	79/80	80/81	81/2	82/3
Domestic rate subsidy																
England	2.1	4.2	6.3	8.3	9.5	10.5	6.0	13.0	18.5	18.5	18.5	18.5	18.5	18.5	18.5	18.5
Wales	2.1	4.2	6.3	8.3	9.5	10.5	6.0	33.5	36.0	36.0	36.0	36.0	36.0	36.0	36.0	18.5
London Boroughs (if differing)																
City	—	—	—	—	—	—	—	23.9	23.9	23.9	22.9	23.9	26.4	25.4	34.3	35.3
Camden	—	—	—	—	—	—	—	19.9	19.9	19.9	19.7	19.9	20.2	20.1	—	—
Kensington & Chelsea	—	—	—	—	—	—	—	18.7	18.7	18.7	18.5	18.7	18.6	18.8	—	—
Westminster	—	—	—	—	—	—	—	22.4	22.4	22.4	21.7	22.4	24.1	23.5	25.6	25.9
Mixed hereditaments																
England	0.8	2.1	2.9	4.2	4.5	5.0	3.0	6.5	9.0	9.0	9.0	9.0	9.0	9.0	9.0	9.0
Wales	0.8	2.1	2.9	4.2	4.5	5.0	3.0	16.5	18.0	18.0	18.0	18.0	18.0	18.0	18.0	9.0

Source: G.B. Government, Statutory Instruments, *Rate Support Grant Orders* and unpublished data from Department of Environment.

differing levels of rate relief offered to domestic ratepayers and residents of mixed hereditaments in different areas. Although uniform for England and Wales up to 1973/4, after that date a major differentiation was introduced to give Wales almost double the subsidy available in England. This was justified on the grounds that with the reorganisation of local government, Welsh ratepayers had to bear a greater burden of the costs of water supply. After 1982/3 the Welsh domestic relief was brought in line with that in England but the overall level of block grant to Wales was increased as a compensation. The net effect of this change was, therefore, to shift the burden of rates away from commercial and towards domestic ratepayers. The uniform domestic element subsidy within England was maintained by the Labour Government in 1974/5 to replace the variable domestic element planned by the outgoing Conservative Government following local government reorganisation. This had the effect of benefiting the metropolitan areas in comparison to the non-metropolitan Counties. In addition, 1974/5 also saw the introduction of variations in the levels of rate subsidy within London, the highest levels of subsidy going to the four Boroughs with the largest rateable value tax bases. This London differential rate subsidy has been justified in terms of the need to provide additional support to domestic ratepayers in those Boroughs with unusually high commercial rateable values who would otherwise be 'penalised' by the low level of support deriving from the resources and needs elements.

Three basic difficulties arise with the form of domestic element. First, the present method of subsidy arbitrarily distorts the tax burden as between domestic and non-domestic ratepayers. Moreover, the effect has been one which, up to 1975/6, has been frequently adjusted from year to year (as shown in table 4.6) and this has induced uncertainty and instability into local finances. As a consequence, the Layfield Report advocated that there should be no change in the domestic/non-domestic ratio of rate poundages between rate revaluations. This recommendation has been largely followed thus maintaining a stability which introduces less uncertainty and distortion into the assessment of commercial costs and competition (see G.B. Committee of Enquiry, 1976, pp. 177–8).

A second difficulty in the form of domestic rate relief arises from the fact that the present form of spatial differentiation of relief bears only a small degree of rationale. For example, many areas in England possess similar problems of low densities and it is not altogether clear why the specific differences in Welsh water supply should be compensated. Moreover, this problem is confused by the equalisation payments due under the *Water Charges Equalisation Act* (1977) and by the fact that many Water Authority boundaries do not coincide with local authority boundaries. For example, many ratepayers in the Welsh border counties receive an additional subsidy but they pay water rate to the English Severn and Trent Water Authority.

The third difficulty with the domestic element, like the problems in the property tax itself, arises from its lack of clear relationship to the ability of local ratepayers to pay taxation. At present it offers the same supplement to rich and poor local authorities, and to rich and poor households, irrespective of differences in household incomes, and irrespective of the local rate poundage levied. Thus although the cash value of the subsidy is directly proportional to rateable value, it bears little relation to the rate bill. For example, ratepayers in local authorities with high rateable values and low rate poundage, will have a much higher proportion of subsidy in their total rate bill than authorities with low rateable values and high rate poundages. As a result it has been suggested frequently that the domestic element should be differentiated spatially to take account of variable household income and variable rate poundage but these proposals have never yet been implemented.

(2) *The resources element*

This derived originally from the Rate Deficiency Grant used from 1959/60 to 1966/7. It formed about 30% of the RSG up to 1980/1, but up to 1973/4 was responsible for about 15% of the RSG. Like the domestic element, it was paid direct to all rating authorities, i.e. all Districts and London Boroughs; but unlike the domestic element, it was differentiated between local authorities, with some authorities receiving no grant at all. The aim of the resources element was to compensate local authorities for their deficiency in tax base (rateable value per head) below a national standard. This element, therefore, sought to overcome differences in local authority ability to raise revenue and to ensure that local authorities were not penalised by their small tax base. A grant was paid to each local authority which fell below the 'standard' rateable value per head set by central government for England and Wales as a whole.

The form of distribution equation employed was given by an adaptation of equation (2.3)

$$G_i = \frac{E_i}{\bar{B}P_i}(\bar{B} - B_i) \qquad (4.1)$$

subject to $(\bar{B} - B_i) > 0$. In this equation E_i is the relevant expenditure, P_i is the local population, and the term $(\bar{B} - B_i)$ is termed the *credited rateable value*. It is easy to see, that for relevant expenditure alone, this equation reduces to that given in equation (2.3).

The standard rateable value chosen in each year is shown in table 4.7. Up to the 1974/5 year the England and Wales average rateable value per head was used as the standard. The raising of the level of the standard above the average from 1974/5 resulted in a much larger number of local authorities receiving resources grants. This was particularly important for

Table 4.7. *Standard rateable value per head used in allocation of the RSG resources element*

		Number of 'need' local authorities above standard				
Fiscal year	Standard rateable value per head	Post 1974 'need' authorities (1)	Counties (2)	Counties and County Boroughs (3)	Counties and County Boroughs which would be Metropolitan Districts (4)	Column (4) plus London Boroughs (5)
Rate Deficiency Grant 66/7	44	—	6	28	8	41
1967/8	·45	—	6	28	10	43
68/9	46	—	6	30	11	44
69/70	48	—	6	30	12	45
70/1	49	—	6	30	11	44
71/2	50	—	7	32	12	45
72/3	51	—	6	31	14	46
Revaluation						
73/4	133	—	6	27	14	46
Reorganisation						
74/5	154	21	—	—	—	—
75/6	170	18	—	—	—	—
76/7	176	17	—	—	—	—
77/8	173	17	—	—	—	—
78/9	177	17	—	—	—	—
79/80	175	17	—	—	—	—
80/81	178	17	—	—	—	—

N.B. The number of local authorities above the standard is not directly comparable before and after the reorganisation of local government in 1974.
Source: G.B. Government Statutory Instruments, *Rate Support Grant Orders*; G.B. Department of the Environment: *Rates and Rate Poundages.*

many London Boroughs and some non-metropolitan Counties in the south east. The resulting grant distribution became much more fully equalising between local authorities with different tax bases since resource differences were compensated almost up to the maximum difference between the richest and poorest local authorites. However, this was at the expense of giving resource grants to relatively rich areas which prior to 1974/5 received *no* such grant. Since 1974/5 ad hoc choices of standard rateable value resulted in two London Boroughs coming in or out of the resources distribution in successive years. In detail, the setting of the standard rateable value derived from finding the right level which, when subtracted from local rateable value and multiplied by local tax rates, gave the total resources element planned in the Aggregate Exchequer Grant. However, because of the effect of variable local tax rates, it was not possible *ex ante* for the central government to know the total level of resources grant.

Hence in practice the central government in Britain used estimates of what the rate poundage was likely to be. These estimates of the local rate poundages were themselves derived from the DoE and CIPFA estimates of total local expenditure, as discussed earlier in this chapter. This total expenditure was then reduced by removing the components funded by specific and supplementary grants, the RSG needs element (but *not* domestic element), and local user charges. This left the estimate of 'relevant local expenditure' used in the resource calculation (so-called 'resources expenditure'). This initial calculation was in the form of an open-ended grant. But the individual allocations were then summed for all local authorities, and the total scaled by a proportional multiplier to equal the aggregate of resources element set aside by central government. This multiplier was equal to 0.939 in 1978/9. Thus, in practice, the resources element was made close-ended.

Because the resource grant calculation was based on local authority estimates of their expenditure and rate poundage, major errors in initial grant allocations occurred, even though there was scrutiny of the estimates by central government auditors. Hence, adjustments to resource allocations were required when the final rate poundages and expenditure levels were known. There were usually two or three such adjustments, with the final resources entitlements usually agreed only in February or March of the year *following* the fiscal year in question, i.e. February 1981 for the 1979/80 fiscal year. These adjustments required taking resources back from those local authorities which levied a lower poundage than that estimated, and giving extra resources to those which levied a higher poundage than that estimated. This difficult political exercise was eased up to the 1979/80 fiscal year by the impact of inflation and steady growth in public expenditure: each year the *Increase Orders* subsequent to the main RSG *Order* allowed corrections to the resource grant without the politically difficult exercise of reclaiming grants from authorities. Sometimes, however, these corrections were sufficiently large that part of the *Increase Order* on other grants (the needs element) had to be used to make the correction. This meant that, for some authorities, the final grant distributions under each element did not necessarily reflect the resource/needs ratio that would obtain under the objective criteria of grant distribution. However the total RSG allocated was correct: i.e. in part, the RSG acted as a unitary grant (see chapter 2). However, since the 1979/80 fiscal year, this adjustment exercise was initially made more difficult due to the imposition, by the Conservative central government, of absolute and severe cash limits. Then in 1981/2 the adjustment process was eased when separate resource and need equalisation were combined into a unitary grant.

There are a number of criticisms of the manner in which the resources element was allocated. First, the undertaking of resource equalisation independent of need equalisation resulted in differences in tax base between local authorities being compensated irrespective of the levels of expenditure they needed to undertake: no advantage was taken of the coincidence of high needs with high resources or low needs with low resources. Hence the *separate* treatment of needs and resources used up to 1980/1 prevented full equalisation. A second difficulty was that the grant was not fully equalising (even after 'clawback') since authorities above the standard rateable value per head could levy expenditure at lower marginal costs than those below the standard. Third, the adjustment process to make the grant closed-ended disadvantaged the less-wealthy authorities and advantaged those authorities levying rates at higher levels than estimated (sometimes called the 'overspenders'). Fourth, since the grant used a per capita allocation, it penalised areas with declining populations.

A fifth, and major criticism related to the method of precepting employed. The non-metropolitan and metropolitan Counties and the GLC each make important precepts on the District rates (which are the tax-gathering authorities) in order to support the services provided by the larger government units. In addition police authorities and other bodies also make small precepts. The pattern of precepting is largely a matter for agreement between the local governments concerned. However, precepting did affect the resources element because it provided resource equalisation independent of expenditure need. At the small scale of Districts a considerable degree of variation led to differential payments of resources elements. However, these payments were given irrespective of whether expenditure need at the local level warranted such a level of support. This feature led to a substantial degree of support going to Districts which did not require it. Hence resource equalisation at District level and need equalisation at County level induced inequity and blurred the lines of accountability for local taxes.

A sixth criticism was that of 'feedback': the size of the resources grant was not independent of local authority behaviour. To some extent, at least, local authorities could perform as rational self-interested optimisers to increase their level of grant by increasing estimated expenditure levels and their rate poundage. In practice the extent of such behaviour was limited by local political constraints, but since RSG gave rise to underpricing of local services, the form of resources element undoubtedly encouraged increases in local tax rates and expenditures.

The phenomenon of London 'clawback' represented a seventh difficulty in the resources element. Clawback was a special resources adjustment, but applied to the needs entitlement of London Boroughs. Since the rateable values in London are on average two-thirds higher than the rest of England and Wales a much lower tax rate can produce the same revenue yield as in authorities outside London. To compensate for this feature, the 'resources adjustment' of clawback was subtracted from the needs grant entitlements of those London Boroughs which had a rateable value per head above the national average. The size of reduction for each of these Boroughs was calculated by determining the difference in rate poundage required to finance local relevant expenditure from the local tax base, and the rate poundage required to raise the same expenditure on the national standard rateable value per head. The total size of the 'clawback' was £200–300 m up to 1980/1 and this contributed to the needs grants of the rest of the country.

An eighth difficulty was that, in addition to clawback, London has also been subject to completely separate treatment of education and non-education services, and an internal equalisation scheme. This latter provides horizontal transfers between London Boroughs and is organised largely independent of Westminster. It has taken a number of different forms (see Jenkins and Rose, 1976*a*, *b*). However, since the London equalisation scheme utilises the considerable tax bases of, in particular, the City of London together with Westminster, Camden, Kensington and Chelsea, it is not independent of the fact that the exceedingly high rateable values of these areas provide insufficient assistance to the rest of the England and Wales authorities outside London (i.e. clawback should absorb these internal London payments for full equalisation to occur). Thus, because of the deficiency payments principle of the resources element, which ignored the resource-rich areas (since 1974/5 all in London) which were above the standard rateable value per head, London as a whole was permitted to partake in higher service levels or reduced tax rates in comparison to the rest of the country (Jackman, 1979).

The resources element was also affected, in a more minor way, by special relief given by the national government to various classes of property. In Scotland, industrial and commercial properties have been subject to special forms of relief dating from 1928. In England and Wales, industrial property was rerated in 1962, but special rating reliefs have been given to charities and other special organisations. From 1979/80 the rateable values of local authorities have been adjusted to allow for 50% of the rate relief to charities. This, and similar *national* policies on rate relief have small but important effects on local finances which often introduced rather perverse effects into the allocation of the resources element.

(3) *The needs element*

This derived originally from the General Grant used from 1959/60 to 1966/7. It formed about 60% of the Rate Support Grant up to 1980/1. As such it was the largest single element of the Rate Support Grant and responsible for allocating nearly 10% of public expenditure in the national economy as a whole. It is not surprising therefore that most attention and comment was directed at this component alone. In addition, the needs element was the most difficult to estimate and was allocated on what was probably the least rational basis of the three RSG elements with the result that most of the criticism of RSG centred on this component alone. Unlike the domestic grant and resources elements, needs element was not paid direct to all rating authorities until 1979/80. Instead it was paid to non-metropolitan Counties, metropolitan Districts and London Boroughs and not to non-metropolitan Counties or the GLC, but their expenditure needs were taken into account in determining the grant allocation. This procedure differed from that in Scotland, where the needs component is paid to all local authorities. The payment procedure was modified in 1979/80, however, in that a proportion of the non-metropolitan County allocation was reallocated to the Districts on the basis of a formula.

The aim of the needs element was to achieve an equalisation between authorities in terms of their ability to meet expenditure needs at the same tax rate. Since the resources element aimed to give to all authorities the same tax base, it was assumed that differences in tax rates resulted only from differences in need to spend. Thus if each authority spent an amount equal to its assessed need, then rate poundages should be equalised between all areas. Hence the overall aim of the RSG to equalise tax rates would be achieved.

The assessed need of each area was derived by a method unique in such grant programmes. A number of needs indicators were sought to take into account variations in expenditure levels arising from the variable distribution of population and economic activity, the variable concentration of different people or different types of economic activity (client groups), differences in the physical and social environment, and differences in scope and quality of service. Table 4.8 summarises the main groups of indicators of need used from 1967 to 1981. In interpreting the table, it should be borne in mind that the definitions of the need indicators were often changed considerably over this period. For example, data were updated and different methods of measuring a given indicator were employed. From this table four periods of evolution can be recognized. For the first period, up to 1973/4, needs indicators were defined *a priori* although their definitions were adjusted slightly. A second period covering 1974/5 and 1975/6 is one of ad hoc changes and experimentation in which new methods of defining 'units' of education and social service need were tried. A third period covers 1976/7 and 1977/8 where a shift was made to using Census variables for measuring social service needs and other new variables were introduced. The fourth period covers 1978/9 up to 1980/1 in which the 1978/9 factors were rolled forward with no change in variables. As such it represented a relatively stable pattern as far as need variables are concerned, but with ad hoc decisions introduced into special area weights which sought to achieve a balancing of need allocations unachievable by use of the needs indicators alone.

The needs indicators were defined *a priori* in the RSG up to 1973/4 and

Table 4.8. *Needs indicators used in the undamped formula for Rate Support Grant in England and Wales (excluding London).*

	General Grant up to 1966/7	RSG 1967/8 –1973/4	Fiscal year 1974/5	1975/6	1976/7	1977/8	1978/9	1979/80	1980/1
1. Population									
Population total	X	X	X	X	X	X	X	X	X
Population decline	X	X	X	X	X	X	X	X	X
2. Area and density									
High-density population[a]	X	X	X	X		X	X	X	X
Low-density population[a]	X	X		X	X	X	X	X	X
Road mileage	X	X							
3. Education needs									
Children under 5	X	X							
No. of pupils, total[b]	X	X							
Education units[c]			X	X					
School meals served					X	X	X	X	X
Nursery school pupils					X	X	X	X	X
Primary school pupils					X	X	X	X	X
Secondary school pupils					X	X	X	X	X
Special school pupils					X	X	X	X	X
Direct grant pupils paid by Council					X	X	X	X	
Further education students							X	X	X

4. Social Services needs

- Persons over 65
- Children in care
- Pensionable age living alone
- Personal Social Services Units[d]
- High-density dwellings
- Lack of exclusive amenities
- Lone-parent families
- Shared households

5. Area Weights

- High-cost Metropolitan[e]
- Non-metropolitan sparsity
- Special area weights[c]

6. Miscellaneous

- Previous year's need entitlement
- Numbers employed
- Dwelling starts
- Labour cost differential[e]

Some simplifications have been made and the dates to which variables refer have been omitted.

[a] Density includes various definitions of acreages and thresholds.

[b] Total number of pupils includes most of other education need variables.

[c] Education units include the bracketed variables for 1974/5 and 1975/6.

[d] Personal social service units include the bracketed variables for 1974/5.

[e] Special area weights and high cost areas are absorbed into the labour cost differential from 1977/8 onwards.

Source: G.B. Government, Statutory Instruments, *Rate Support Grant Orders: Rate Support Grant Regulations.*

were composed of two components: a 'basic payment' per head of population, and a set of 'supplementary payments' which derive from the General Grant employed up to 1966. Each of the indicators (for young and old people, education units, high and low density, declining population and high cost metropolitan Districts) was used in a similar form as in the General Grant; only 'miles of road' was added as a new variable for the period 1967/8 to 1973/4. However, even before 1974/5 changes in definition of variables occurred, for example in the calculation of education units (see G.B. MHLG Circulars 11/69, 56/70). From 1974/5 onwards, however, considerable changes each year became possible under the *Local Government Act* (1974). A new approach was adopted which, instead of employing needs indicators defined *a priori*, used past levels and patterns of expenditure to indicate need. It was assumed therefore 'that the best available indicator of spending need is the pattern of recent past expenditure' (G.B. Government DoE *Circular* 19/74, 1974, p.4). The needs indicators were derived as a weighted average of past expenditures by categories of local authorities with the same characteristics using the method of multiple regression analysis. The *a priori* definition of needs, therefore, was replaced by one which was essentially *a posteriori*. Over two transition years 1974/5 and 1975/6 the variables in the regression analysis were determined *a priori* but the weights were calculated *a posteriori* by regression analysis. The direct result of this change in methodology was the search each year, since 1976/7, for the best regression equation based upon that set of explanatory variables which provided the closest relationship with past levels and patterns of expenditure. Each year a regression equation was estimated which related the actual expenditure per head in each local authority to a specific set of need indicators. Such an equation has the following structure for any specific year

$$N_i = \alpha_0 + \quad \alpha_1 X_i^1 + \ldots + \quad \alpha_j X_i^j + \ldots + \quad \alpha_n X_i^n \quad (4.2)$$

Local authority expenditure/head	first need coeff.	first need indicator	n^{th} need coeff.	n^{th} need indicator

The N_i term defined the regression-dependent variable in local authority i; the X_i^j terms the j-independent variables in local authority i; and the α_j terms are the regression coefficients. The regression formula was used for allocation by entering the current values of the independent variables as need indicators and then scaling the resulting allocation by the total sum of needs grants set aside by central government for all authorities. Before 1975/76 this scaling was achieved in two steps: first, a 'population-based' component (PBC) of the needs element was responsible for between 40 and 55% of the total grant payment; second a needs equalisation component (NEC) derived from formula (4.2) was responsible for the rest. The

Table 4.9. *Relative share of needs element component for population and needs equalisation*

Fiscal year	Population-based component[a]		Needs-equalisation component[b] (%)	Grant threshold (£/head)
	£/head	%		
General Grant 1966/7	33.00	63	37	—
1967/8	49.37	56	44	—
1968/9	50.01	54	46	—
1969/70	54.63	52	48	—
1970/1	59.26	51	49	—
1971/2	63.07	50	50	—
1972/3	64.68	49	51	—
1973/4	66.51	45	55	—
1974/5	58,24	37	63	—
1975/6	—	—	—	179.58
1976/7	—	—	—	170.10
1977/8	—	—	—	142.87
1978/9	—	—	—	154.99
1979/80	—	—	—	198.85
1980/81	—	—	—	169.28

[a] Constituted by basic payment up to 1974/5.
[b] Needs-equalisation component.
Figures relate to final orders and are at 1979 constant prices.
Source: G.B. Government, *House of Commons Papers* and ACC, 1980, table 5.

size of the two components is shown in table 4.9 from which it can be seen that the population component steadily declined from the high level inherited from the 1966 General Grant. The scaling was achieved by multiplying these pre-assigned percentages by the respective population P_i and needs from formula (4.2), i.e.

$$G_i = \frac{PBC}{100} \cdot P_i + \frac{NEC}{100} \cdot N_i \qquad (4.3)$$

After 1974/5 the needs element was scaled by use of a grant threshold expressing standard need $\bar{N}$, as shown in the following equation

$$G_i = N_i - (\bar{N}P_i) \qquad (4.4)$$

The principle underlying the distribution of the needs element was that areas should be fully compensated for differences in their need for rate-borne expenditure, i.e. that all authorities could finance from the rates the same per capita amount of their assessed need. Thus the distribution formula was identical with the needs equalisation formula equation (2.5) except that the constant was chosen equal to the local population (note N_i from equation (4.4) is a population-based assessment (P_iN_i) and hence (2.5) and (4.4) are formally identical).

Since the resources element aimed to give all areas the same rateable value tax base per head of population the aim of the needs element was that each area should have the same rate poundage, provided that each spent according to its assessed need. Thus the declared aim of the RSG to equalise rate poundages could be achieved. The value of the threshold $\bar{N}$ (between 1975–80) is shown in the right-hand column of table 4.9. A high threshold reduced needs grant to low-need authorities and this device was employed especially in 1974, 1975 and 1978 by the Labour central government to shift grant towards the cities.

The aim of the regression equation was to express local authority expenditure as a whole, i.e. to capture as far as possible all of the individual and geographical variables that lead to expenditure differences. This was very much the position advocated by the Association of Metropolitan Authorities and sought to treat expenditure itself as a public good.[3] Hence, it was always stressed by central government that the regression equation had to be viewed as a whole. Despite this aim, however, a large number of criticisms of the needs element relate specifically to the regression procedure for determining the weighting of different need indicators. First, no attempt was made to assess need levels in the fiscal year for which grant allocations applied: the need indicators and expenditure levels used in the regression equation all referred to data for previous years. In some cases, where the decennial Census was used, the data was up to eight or nine years out of date.

A second criticism was the variable and special treatment of London.[4] Prior to 1974/5 London received grant by a separate formula altogether. In 1974/5 and 1975/6 London was not included in the needs formula calculations, but it did receive needs grant using the non-London formula, although there were arbitrary supplements of 3% in 1974/5 and 8% in 1975/6. Since 1976/7 London was included in the general needs and block grant formulae calculation (except for the City of London) but with the special adjustment of 'resources and adjustment' (or 'clawback') made to the needs entitlement up to 1980/1. As a result of its separate treatment, it is usually considered that London gained considerable financial advantage, even after 'clawback'.

A third problem was that the method of statistical estimation employed was open to doubt. The final need equation was usually highly collinear and no attempt was made to allow for this. Moreover, the final regression equation was estimated by a stepwise procedure which was terminated when the explanatory variables (need indicators) did not achieve more than a specified level of significance. The regression estimates of the need weights was not itself thrown into jeopardy by such a procedure (the equation was still an adequate *forecasting* relationship), but significance tests possessed no level of reliability because of the large degree of multicollinearity. A preferable procedure might have been to decide which need indicators were significant in measuring local authority revenue requirements and then, using regression analysis, merely to define the weights for each indicator: thus abandoning the use of both the stepwise procedure and significance levels.

A fourth and related criticism arises from the treatment of variables which entered the regression equation with a negative coefficient. Such variables indicated that expenditure need fell as the level of such variables rose. This could be expected to be the case with, for example, the income level of a local authority, or its per capita employment level. Such negative coefficients would have the effect of subtracting RSG needs element from some local authorities. However, in the regression procedure such variables were omitted and the equation was recalculated. One exception was made for the case of the indicator measuring the number of low-income households, which was considered too significant to omit. However, the general omission of variables with negative coefficients undermined much of the rationale of the needs equation as an explanatory model. To omit selected variables arbitrarily penalised those local authorities with high needs measured on omitted variables as against those local authorities with high needs measured on other variables which were included. This rationale was perverse because, if the need indicators were significant explanations of expenditure requirements, then they should have been included irrespective of whether they entered with negative coefficients.

A fifth criticism of the regression procedure was the manner in which the regression equation was chosen. Apart from the statistical criterion of significance level, the equation also had to satisfy a criterion of political acceptability. Each year a wide range of different regression equations were calculated using different need indicators, or different definitions of the same indicators. Table 4.10 shows the indicators tested and selected up to 1980/1. Each equation was almost equal on statistical criteria, hence the choice of equation was largely political: as that which was most acceptable to the various local authority Associations and to the central government. As noted by Binder (1978) 'The right conditions do not exist for a statistically unchallengeable application of the mathematical technique. So several different formulae can be obtained by varying the mathematical approach. This enables the government to select a formula with results to accord with, or do least damage to, its political objective. Thus it is a subjective system as much as any other'. Hence, the regression procedure possessed all of the characteristics of a pseudo-scientific objectivity used to rationalise and legitimise *ex post* decisions which are, in essence, political in nature: what Caulcott and Hale (1978) termed a political judgment.

A sixth criticism of the regression procedure has been that it reinforced past expenditure patterns, since the estimates were based on data for

Table 4.10. *Independent variables analysed in step-wise regression procedure 1975/6 to 1980/1*

Variable	Year included in undamped needs assessment					
	1975/76	1976/77	1977/78	1978/79	1979/80	1980/1
Over 50 persons per hectare	S	—	—	NT	NT	NT
Persons per acre	NT	NT	NT	S	—	—
Acres over 1.5 per head	S	S	S	NT	NT	NT
Acres over 3.0 per head	S	—	—	NT	NT	NT
Acres per person	NT	NT	NT	S	S	S
Housing starts	—	S	—	—	S	S
Population decline over 10 years	S	—	—	NT	NT	NT
Population decline over 5 years	—	S	—	NT	NT	NT
Elderly living alone	S	S	S	—	—	—
Persons lacking basic amenities	NT	S	S	S	S	S
Overcrowding	NT	S	S	NT	NT	NT
Shared households	NT	NT	S	NT	NT	NT
Lone parent families	NT	S	S	S	S	S
Unemployment	NT	NT	S	—	S	S
Labour cost	NT	NT	S	S	S	S
Primary school pupils	S	S	S ⎫	S ⎫	S ⎫	S ⎫
Secondary school pupils under 16	S	S	S ⎬	S ⎬	S ⎬	S ⎬
Direct grant pupils under 16	S	S	S ⎭	S ⎭	S ⎭	S ⎭
Secondary school pupils over 16	S	S	— ⎫	— ⎫	— ⎫	— ⎫
Direct grant pupils over 16	S	S	— ⎬	— ⎬	— ⎬	— ⎬
Full-time further education students	S	S	—	—	—	—

S, selected variable; NT, variable not tested in that year; —, variable tested in that year but not selected. Brackets indicate a combined factor. 1980/1 is identical to 1979/80 except for a rolling forward of damping.
Source: G.B. Government, *House of Commons Papers* HC-63 1978.

these. As such it tended to favour large spenders and penalise small spenders and this had important implications for cumulative growth and decline (Godley and Rhodes, 1972; SPNW, 1973). Moreover, use of expenditure per head alone as the dependent variable ignores important differences in the scope and quality of services. As a result of these effects it became clear that a group of authorities making similar expenditure decisions, e.g. to overspend, could affect the needs formula in their favour. This was claimed to work in favour of the London Boroughs and some metropolitan Districts.

The fact that much needs data were dependent upon local discretion to provide a service introduced a further major difficulty into the regression procedure. Since total expenditure data of a local authority depend on the size and range of local services offered, choices by local authorities to provide services at different levels reduce or increase their total expenditure per head. Hence total local expenditure imperfectly reflects local need. Similarly, for any given service category, the identification of the size of client groups is often affected by local discretion. This affects most particular special schools, provision of private and direct grant school places at local authority cost, the size of most personal social services, and the level of local transport spending. Hence, the present size of client groups is not an objective measure of need. The 1974/5 formula was particularly criticised for this problem. Since 1975/6, Census and other 'objective' indicators of social living conditions largely replaced discretionary data in the regression formula. As a result of the problems with the regression procedure the formula for needs element allocation and the weights placed on variables changed markedly from year to year. Table 4.11 summarises the weights applied to some of the major variables entering the regression equation from 1974/5 to 1980/1. Although there are some general patterns such as the steady increase in weight for lone-parent families, most of the weights vary a great deal from year to year. This made the grant far from intelligible, induced instability, diffused lines of accountability, and modified the criteria used for equalisation from one year to the next. For example no regression weight could be associated with any form of unit cost, no account could be taken of variable standards or efficiency, and major changes in grant could result from small changes in the data.

Because of the instability resulting from the inherent defects of the procedure and the use of different formulae, it was necessary to employ 'damping' as a partial hold-harmless provision to limit the extent of variation in one year's entitlement from the previous, and hence to stabilise each local authority's grant. No damping was necessary up to 1974/5 when the same needs indicators were employed in each year, but after that date the annual search for a new regression equation discussed above led to a need to prevent allocations varying too much from one year to the next. The levels of damping employed are shown in table 4.12. The effect of using damping was, on the one hand, to limit the destabilising effect of annual changes in definition of needs indicators, but on the other hand, it introduced a new dimension of instability which permitted year on year *ad hoc* decisions of the past to be summed and weighted together. The result of damping was therefore a more certain and stable distribution of RSG, but at the expense of using an almost indecipherably complex weighted sum of *ad hoc* decisions. This had the effect of making the grant very unresponsive to known changes which did occur, e.g. in population concentration and salary levels. The contribution of the formula for each financial year to the damped formula for 1980/1 is shown in the right-hand column of table 4.11. Damping was applied by summing and then averaging the regression coefficients calculated for each formula for the previous year involved in the damping procedure; i.e., it was applied to the regression coefficients and not to the needs indicators. This created a number of difficulties in comprehensibility. Regression coefficients calculated from needs indicators averaged over a series of years would clearly

Table 4.11. *Variation in non-London weights (in £ at 1979 prices) placed on major needs equation variables 1974/5—1980/81 (The table also shows the contribution of the 1975/6 to 1979/80 formulae to the damped formula for 1980/81)*

Need variable	Fiscal year (£)						1980/81 (damped formula)
	1974/5[a]	1975/6	1976/7	1977/8	1978/9	1979/80	
Acreage	—	—	—	—	1.9805	0.9175	2.898
Acreage over 1.5 per head	—	0.5540	1.8386	2.0734	—	—	4.466
Acreage over 3.0 per head	—	1.584	—	—	—	—	1.584
Density times population	—	—	—	—	0.4997	—	0.4997
Persons over 50 per ha.	—	3.900	—	—	—	—	3.900
Housing starts	—	—	102.07	—	—	217.43	319.5
10-year population decline	—	19.32	—	—	—	—	19.32
5-year population decline times population (000's)	—	—	0.03108	—	—	—	0.03108
Retired living alone	—	173.72	149.48	73.60	—	—	396.80

	(%)						
Persons lacking basic amenities	—	—	12.374	32.033	27.388	25.305	97.10
Persons over 1.5 per room	—	—	57.19	74.61	—	—	131.8
Shared households	—	—	—	41.87	—	—	41.87
One-parent families	—	—	446.8	1052.9	1655.4	1600.9	4756.0
Primary school pupils	19.1	146.83	111.31	94.83	100.35	118.68	572.0
Secondary pupils under 16	⎫ 92.1	246.67	187.00	94.82	100.34	118.67	747.5
Secondary pupils over 16	⎭	383.25	290.55	—	—	—	673.8
Direct grant under 16		172.67	130.90	94.82	100.34	118.67	617.4
Direct grant over 16		268.25	203.35	—	—	—	471.6
Further education students	71.7	425.81	322.79	—	—	—	748.6
Registered unemployed		—	—	151.97	—	190.23	342.2
Labour cost differential		—	—	—	—	—	
times population		—	—	1.1384	1.3412	2.5864	5.066

[a] For 1974/5 Education weights are obtained by allocating the percentage of the education unit to the £ weight given.

Table 4.12. *Values of damping employed in RSG needs determination outside London*

Year	Damping (% of needs allocated on present year's formula)	Value placed on previous year's formulae (% of)				
		1 year previous	2 years previous	3 years previous	4 years previous	5 years previous
1975/6	50[a]	50	—	—	—	—
1976/7	33.33[b]	66.66	—	—	—	—
1977/8[c]	33.33	33.33	33.33	—	—	—
1978/9[c]	25	25	25	25	—	—
1979/80[d]	20	20	20	20	20	—
1980/80[d]	—	20	20	20	20	20

[a] Includes a needs indicator of 71.3% of 1974/5 allocation.
[b] Includes a needs indicator of 83.86% of 1975/6 allocation.
[c] Requires averaging where old formulae do not contain new indicators.
[d] No new formula employed, data updated for previous formulae.
Source: G.B. Government, *House of Commons Papers.*

relate needs grants to average objective needs. However, an average of regression coefficients leads to no such clear relationship. Since the needs indicators used each year varied, the coefficients related to them varied depending upon which other variables were incorporated into the calculation of the conditional expectation of the dependent variable; i.e., the calculation is not commutative since the coefficients for each year varied: for example, with coefficients b_1, b_2, and b_3 and three-year damping of data on indicators X_1, X_2 and X_3 we have the relation

$$(X_1 b_1 + X_2 b_2 + X_3 b_3)/3 \neq \left\{ \frac{X_1 + X_2 + X_3}{3} \right\} b.$$

In addition to damping a safety net operated after 1978/9 which prevented authorities suffering unreasonably high grant losses in a given year (no net operated to reduce large gains, however!). The safety net operated in later years through the use of a national pool into which a small component of each area need allocation was paid on a per capita basis (this was equal to £0.61 m in 1980/1). The pool was then redistributed to those areas which would have suffered unreasonably high losses of grant. In 1980/1 any loss greater than a 1p in the £ rate (in real terms) was compensated in this way. The pool arrangement had much in kin with the pooled grant proposal of Department of Environment to Layfield (see equation 2.17).

Further criticism of the needs element related to the way in which data were constructed. Some local authorities claimed that statistics for their areas were incorrect (especially the population and social service variables). In addition many data were directly or indirectly provided by local authorities: e.g., Registrar General's Population Estimates and School Pupils. As a result, a considerable number of protracted negotiations often characterised grant distribution with special adjustments being made to the indicators for individual authorities in a rather *ad hoc* fashion. Because of this feature, since 1978/9 the RSG included a provision to 'freeze' data after a particular date, usually one year from publication of the *Order*, to prevent negotiations and adjustments being carried on indefinitely.

Other criticisms relate to the introduction of arbitrary forms of area bias. Apart from the separate treatment of London, a further component of area bias was first introduced explicitly into the needs formula in 1974/5 when a special and largely arbitrary weighting for both the south east and West Midlands was employed. Since then various special weights were employed and the equivalent grant per head that these generated is displayed in table 4.13. After 1977/8 these weights resulted from the joint operation of the labour cost differential and the 'safety net' discussed above. The labour cost differential was constructed from data in the *New Earnings Survey* (NES). An average index for all local authorities was set at 100 and percentage differences of wage rates from the average index for given authorities were calculated. The resulting labour cost percentage differences were then applied to the actual average wages for that area (e.g., an average wage of £70 per week in a given local authority which is 2% above the group average produced a labour cost differential of 2% × £70, equal to 1.4). This procedure incorporated wage inflation. These differentials were then entered into the regression calculation as independent variables for those areas in London and the south east considered to have particularly high labour costs; other areas had values of zero entered. Hence, the labour cost differential acted as a form of dummy variable in the regression calculation. The procedure was complicated by three features. First, additional data on wages to that contained in the *NES* were used in 1980/81. This related to data for teachers' salaries for which information is more accurately known. A second complication was the instability and variable accuracy of the *NES*. The *NES* is a sample survey which for small areas had large sampling errors. As a result considerable instability affected the labour cost differential. A third complication arose from the use of damping in the regression formula (q.v.). Since the final regression weights were averages of sums of present and previous year's weights, the relation to the calculated labour cost differential was very weak.

It is clear from table 4.13 that the two types of area favoured by these weights were, first, the high-cost authorities of the south east, and second, some of the sparsely populated rural areas. In later years, in 1978/9 and 1979/80, an explicit bias to favour inner cities was introduced by the Westminster Labour Government. This was entered into the RSG needs element not as an explicit variable, but instead by choice of the population density thresholds and weights on other need factors which favoured city

Table 4.13. *Grants per head in £ for special areas outside London resulting from the operation of the labour cost differential after 1977/8, safety net from 1974/5 and arbitrary weights in 1974/5*

Area	Weight applied to population of area in fiscal year (£ per head)						
	74/5	75/6	76/7	77/8	78/9	79/80	80/81
Berks		—	—	**0.534**	**7.834**	**5.344**	**6.434**
Bucks		—	—	**0.534**	**2.542**	—	—
Essex	5.99	—	—	**0.512**	{**2.805** / **0.627**}	**7.931**	**9.119**
Herts		—	—	**0.979**	{**3.242** / **9.297**}	**16.594**	{**2.677** / **14.387**}
Kent		—	—	**0.334**	{**1.575** / **0.418**}	**3.375**	{**0.358** / **0.963**}
Surrey		—	—	**0.689**	**5.293**	**8.887**	**12.665**
Beds		—	—	**0.156**	**2.333**	**2.981**	**7.954**
Hants		—	—	**0.578**	**3.586**	—	**2.736**
Oxon	1.04	—	—	**0.133**	**2.333**	**8.100**	**8.156**
E. Sussex		—	—	—	—	—	—
W. Sussex		—	—	—	**0.498**	—	—
I. of Wight		—	—	—	—	—	—
Heref. & Worc.		—	—	—	—	—	—
Salop	1.44	—	—	—	—	—	—
Staffs		—	—	—	—	—	—
Warwicks		—	—	—	—	—	—
Gwynedd	—	—	—	—	**0.167**	—	—
Powys	—	—	—	—	**0.298**	**9.121**	**7.575**
Districts							
Kirklees	—	—	—	—	**0.164**	—	—
Knowsley	—	—	—	—	**0.387**	—	—
Sheffield	—	—	—	—	**1.762**	—	—
Sunderland	—	—	—	—	**0.238**	—	—
Wigan	—	—	—	—	**1.980**	—	—
Manchester	—	—	—	—	—	—	**0.247**

The labour cost differential is in bold type and is in per capita terms for each area.
Source: G.B. Government, Statutory Instruments, *Rate Support Grant Orders.*

areas. In practice, however, this method of introducing area bias clouded the extent of political choice and was much less effective than specific weights would have been since the aid was often diffused to other areas in addition to those intended to receive it.[5]

A further issue of controversy in the form of the needs element was the treatment of areas with differing needs position. Full need equalisation required full compensation of differences in expenditure need of local authorities above the authority with the lowest need, except insofar as different levels of need could be supported by different levels of local tax base. However, for that part of the needs element termed population-based component no such explicit equalisation occurred and this formed a considerable proportion of the total up to 1976/7. The consequence of basing so much of the need distribution on per capita terms was that low need authorities were compensated through the needs element as well as high need authorities; thus undermining full equalisation and setting off 'vicious circles' of enriched or impoverished service levels. After 1975/6 the use of the full equalising formula (4.3) reduced this effect.

The distribution of the needs element between the different precepting authorities in non-metropolitan areas also introduced difficulties since needs were assessed at the scale of the larger unit of the non-metropolitan County, whilst it is at the District level that taxes are raised and many expenditures are made. Between 1966 and 1978/9 the needs grant was allocated directly to non-metropolitan Counties and metropolitan Districts, with no reallocation to the other level. After 1979/80, however, the needs grant was divided between the Districts of the metropolitan Counties by a complex sub-formula. Seventy five per cent of the needs element was allocated to Districts using their share of the lowest mid-year population estimates. The remaining 25% was allocated to the Districts using the three-step procedure to assess District need within the County. The first of the steps measured each District's share of the previous three years' County expenditures revalued at constant prices. The second step scaled the grant to each District in accordance with the difference of each District above that with the smallest expenditures per head in the County, thus seeking to achieve full needs equalisation. The third step involved paying the District needs grant in full if it amounted to less than the remaining 25% of total County needs grant. However, if the sum of the District needs grant was greater than 25% of total County needs, the 25% of needs grant was distributed only in *proportion* to differences in the expenditure needs of the Districts. In addition in 1979/80 there was a safety net or damping to limit the poundage value by which the needs element falls short from the extra precept Counties will make. This safety net stood at 2p. It was implemented only in 1979/80 as it was deemed necessary to limit the effects on ratepayers of the once-and-for-all shift of grant from Counties to Districts. Clearly the variable method of precepting, although improved since 1979/80, considerably blurred lines of accountability and added additional complexity to grant distribution. For example, a metropolitan County received a grant reflecting not only its own need, but also that of the Districts contained within it. As a result, the precepts the County levied did not reflect just its own expenditure, and nor did the rates levied by the Districts reflect their expenditure. The change in needs allocation in 1979/80 also had important distributional effects between areas. In particular, most of the old County Boroughs were favoured, especially Merthyr Tydfil, Blackburn and Nottingham (see Bucknall, 1978).

4.4 Conclusion

As a result of many of these criticisms of the needs element, and of the criticisms of the resources element, the procedure for distributing the RSG between areas was modified in 1981/2. As stated by M. Heseltine, Secretary of State for the Environment (1979) the RSG distribution up to 1979/80 was

'based on the assumption that need is demonstrated by authorities' expenditure. Resources element provides the same marginal rate of grant support to a local authority's expenditure regardless of how extravagant that expenditure might be. Furthermore, high-spending authorities can actually attract to themselves a larger share of the resources grant at the expense of other more prudent authorities. Needs element is distributed on the basis of an analysis of past expenditure patterns. The consequence of this is that if authorities with high levels of expenditure all decided to maintain or increase their levels they could create a feed-back that enhanced their measured needs.

Within such a system it is very difficult to convince authorities that it is in their interests to economise, for to do so might over a period reduce their eligibility for central government support. At its simplest, this phenomenon is known as the 'problem of the over-spenders'. This is a notoriously difficult area. But so great is the volume of public expenditure, and so urgent the need in the interests of the ratepayer and the taxpayer to exercise proper disciplines, that the Government have decided that action is required'.

As a result of these problems, after 1981/2 a change in philosophy occurred to combine resources and need equalisation using a unitary grant formula similar to equation (2.16), although the form of determination of the Aggregate Exchequer Grant remained largely unchanged. This development is discussed in the next chapter.

CHAPTER 5

THE DISTRIBUTION OF THE RSG SINCE 1981

5.1 Introduction

The method of allocating the Rate Support Grant employed up to 1980 aroused an increasing body of criticism both against the form of the resources element and, most importantly, against the needs element. Moreover, it became seen increasingly by central government as a method which undermined its ability to manage the economy as a whole, particularly to control public expenditure. As a result, the Conservative Government which attained office in 1979 sought first to modify the three-element system by imposing new cash limits, adjusting relevant expenditure, and a number of other devices. As a second method of control, however, they introduced, from the 1981/2 financial year onwards, a new form of RSG. This was composed of two elements: a *domestic rate relief grant* and a *block grant*. Under this grant structure domestic rate relief is identical to the previous domestic element, and the block grant is a combination of the old needs and resources elements, but with important changes in definitions and formulae. Despite these changes the method of determining total grant magnitude, the Aggregate Exchequer Contribution and relevant expenditure remained the same after 1981, as summarised in the previous chapter. The important change after 1981 was, therefore, in the manner of grant distribution resulting from the use of the block grant.

The *block grant* is the largest component of RSG, amounting to £7823 m in 1981/2 (at 1980 prices). This is equivalent to 92% of the RSG, 49% of total local spending, and 8% of public expenditure in the nation as a whole. As a result it is an enormously significant element of public policy. Although a unitary grant, it is concerned with the same two essential issues to which the separate resources and needs elements of the old RSG were directed: resource and need equalisation. As a unitary grant, however, it has sought to achieve simultaneous need and resource equalisation using a formula similar to that given in equation (2.16) discussed in chapter 2.

The block grant since 1981/2, in contrast to the preceding system, is distributed direct to all authorities, i.e. to both District and County level and the GLC, ILEA and Metropolitan Police Authority, thus extending the procedure initiated in 1979/80 of allocating needs as well as resource grants to Non-Metropolitan Districts. In addition it introduces, for the first time, separate methods of assessment for England and Wales. This RSG block grant takes a form which is closely related to the theory of unitary grants developed by Musgrave (1961) and Boyle (1966) and discussed in chapter 2. The general form of the distribution equation used to determine each local authority's entitlement is given by equation (2.16), i.e.

$$G_i = E_i - B_i \bar{t}. \tag{5.1}$$

G_i is the level of block grant per head, E_i is the relevant expenditure per head, B_i is the rateable value per head, and $\bar{t}$ is the standard rate poundage termed 'grant-related poundage' (GRP) defined relative to a standard expenditure, termed 'grant-related expenditure' (GRE). All terms are for local authority i; K, the tapering multiplier in equation (2.16) which determines the rate at which 'excess' local spending is progressively penalised by reduced grant is absorbed in $\bar{t}$ in the RSG (see below). This formula combines need equalisation with resource equalisation, but at a standard tax rate (or expenditure).

The block grant contains two essential features. First, there is the GRE for the authority. This is equivalent to the previous needs element in attempting to represent the expenditure requirements (or needs) of an area with specified characteristics which are weighted by the estimated £ per unit cost for servicing those needs. The second feature of the block grant is a measure of the extent to which an area can find its assessed spending needs (as measured by GRE) from its own sources. This is a form of resource equalisation component which measures the local tax base and multiplies it by a standard rate poundage of grant-related poundage. (GRP).

The block grant thus requires three preliminary decisions prior to grant calculation: first, specification of the standard expenditure (GRE) per head; second, the standard rate poundage (GRP) must be determined; finally, the tapering multiplier and any thresholds attached to it must also be defined. Each of these terms is discussed in detail in the rest of this chapter.

5.2 Grant-related expenditure (GRE)

The use of GRE seeks to achieve for the Block Grant what had been aimed for with the previous RSG, an equalisation of rate poundages. As stated to the Grants Working Group (quoted by Welsh Office, 1980, p. 3) 'The aim of the standard expenditure concept when allied to a resource equalisation mechanism, is to enable each authority, if it so wishes, to provide all services appropriate to its area to some common standard for the same rate poundage'.

The form of GRE was inspired directly by many of the criticisms of the old needs element. Since the needs element was deemed to reward the high spenders by including a large amount of their discretionary spending in need assessment, a means had to be found for eliminating this effect. The needs element did not allocate directly to non-metropolitan Districts, ILEA or GLC and treated London separately, and this created anomalies. There was also no direct relation of need to costs of particular services, and the grant was highly unstable and hence required damping. GRE attempted to overcome these criticisms by separating discretionary from non-discretionary spending, allocating grants to *all* spending authorities in proportion to their service responsibilities, and using a direct service cost analysis. As stated by M. Heseltine, Secretary of State for the Environment, the aim of the new grant system was

'to be more comprehensible, stable and equitable as between authorities. It should therefore encompass plausible and intelligible factors which do not change unduly from year to year; it should exclude as far as possible differences in expenditure which are the result of local preferences; and it should subordinate the use of expenditure data, where possible, in any analysis to evidence of the factors which affect local authority costs, and should allow scope for informed judgement' (Ministerial guidelines to Grants Working Group, 1980).

The calculation of GRE begins by taking the national figure for relevant expenditure (after adjustments for specific grants, debt charges, interest receipts, fee income, etc.). This total is then distributed between authorities by, first, identifying measurable factors which influence the costs to authorities of providing services (e.g., the number of schoolchildren needing education, the number of premises requiring refuse collection, etc.). These factors approximate to client groups and define the number of units of service need. A second step is the assessment of the relative importance of different factors for each service which becomes expressed in a set of weights. These weights approximate to the unit costs of providing each service need, i.e. a £ per unit costs for each service need factor.

Like the previous needs element five main factors in GRE include

(1)   the population of an area;
(2)   the physical features of an area;
(3)   social and environmental characteristics of an area which may constitute
        particular problems;
(4)   differences in service costs between areas;
(5)   and special requirements for service provision in different areas.

Within each of these categories a £ per unit of service need is specified which is then multiplied by the number of units of that service need in each relevant local authority. Thus, for example, in 1981/2 a figure of £46.054 per unit of children under five is applied to the number of such children in each local authority responsible for education and care of the under fives (namely, the non-metropolitan Counties, metropolitan Districts, ILEA,

Outer London Boroughs and Isles of Scilly). This operation is carried out for all indicators relevant to the service functions provided by a given local authority and the result summed for all services; i.e.

$$\bar{E}_i = \sum_{k}^{n} b_k X_{ik} \qquad (5.2)$$

where $\bar{E}_i$ is the standard expenditure (or GRE), X_{ik} are the various need factors ($k = 1, 2, \ldots, n$), and b_k are the weights applied to each factor. As we shall see in chapter 6, this approach is similar to the 'average' or 'representative' approach to the need determination.

The indicators, and the weights applied to them in 1981/2, are listed in the RSG *Reports* (G.B. House of Commons Papers, 1980, 1981). The number of factors taken into account in GRE is very large indeed. Compared with the previous system it is fair to say that the formula for the block grant has added very significantly to, rather than diminished, the complexity of need assessment. For the English Counties, which are a level of authority permitting direct comparison, the number of need indicators has increased from 22 in 1980/1 to 52 in 1981/2. For Outer London the equivalent increase is from 30 to 49; and for Inner London (excluding ILEA) the increase is from 26 to 39. Contrary to its aims, therefore, GRE has tended to increase the complexity of needs assessment.

The method of assessing client group size and the unit costs of each group differ considerably between services. Six main approaches were employed in 1981/2 (see Bramley and Evans, 1981) for the services shown below.

	Assessment method	*Services 1981/2*
(1)	Simple per capita	Transport, minor social services, and most minor miscellaneous services.
(2)	Simple client size or service unit	Libraries, museums, nursery and further education, school meals, careers, concessionary fares.
(3)	Clients weighted by expert judgement	Schools, police.
(4)	Clients weighted by research study results	Social services for the elderly and children, and social work.
(5)	Past expenditure levels based on regression analysis	Fire, courts, planning, refuse, environmental health, parks, sports, other housing.
(6)	Actual expenditure (after adjustments)	Council housing, future debt charges, advanced further education, rent and rate rebates, transport administration, land drainage.

Client group-size and population form the bulk of the six factors employed, but regression-based assessment similar to the old needs element is especially important for the non-metropolitan Districts. Since the initial year, various changes have been made to the detail of each service assessment method. These have had important consequences for individual authorities. For example changes to sparsity weights have increased allocations to the rural areas. For 1982/3 inner cities were aided by increasing the weight applied to socio-economic factors, non-white school-children, the mentally handicapped and mentally ill. In addition there has been a small shift towards using actuals or near-actuals rather than population cohorts, e.g. the 1982/3 GRE used bad housing as an index of the mentally ill, rather than a fixed percentage of the population; nursery education GRE was based in part on socio-economic factors and not merely on the number of under fives, as in 1981/2. There was also specific aid for the young unemployed in 1982/3.

GRE has had the advantages of directing grant to the relevant spending authority, has attempted to overcome the rewarding of high spenders, and has achieved a more direct relation of grants to clients and service costs. This has, however, occasioned a complete shift in the philosophy of the RSG. Whereas the RSG was seen as a means of general income support, the use of GRE directs support only to national services; it seeks to provide only limited support to discretionary services. In addition, RSG is a block grant which is an aggregate of needs over a wide range of services leaving local discretion as to exact funding levels in any particular service. GRE, by giving clear cost assessments in each service category contains more of the spirit of an aggregate of specific grants rather than a general block grant (chapter 2, q.v.).

Given these changes in underlying philosophy and the considerable difficulties in assessment of GRE, it is not surprising that this has been, like the old needs element, the centre of most controversy and criticism. The approach used to assessment draws on previous research by the Department of the Environment and the local authority associations on client group and unit cost assessment (see, e.g. ACC *et al.* 1979), but also includes some need indicators to capture geographical variations in service costs (e.g., sparsity and social problems).

The major set of criticisms with respect to the metropolitan and Labour-controlled areas relates to the treatment of discretionary services and the distinction between actual and potential client numbers (see Bennett, 1982*b*). The GRE discounts discretion in four important ways. First, in England GRE applies the England average of costs to many services. This means that authorities which have historically high levels of discretionary spending are pooled with the low spenders (even those which provide no service at all). Second, the choice of need factors in GRE tends to emphasise population totals or age cohorts rather than actuals of a particular client group. For example, with children in care the number of actuals is given light weight in comparison with the child cohort of 5–17 year olds. Most other personal social services employ population totals rather than actuals, e.g. for the number of social workers and old people's homes. This approach also affects nursery education, further education and concessionary transport fares. It penalises areas which provide the services at high levels. A third discounting effect is introduced by special area weights in the GRE, e.g. the use of economies of scale to reduce the need for old people's service expenditures in large authorities; and the use of density and sparsity terms to increase the entitlements in rural areas. Such effects may indeed be present but the choices used in GRE are essentially arbitrary. The fourth main cause of discounting is introduced by the treatment of the Housing Revenue Account (HRA). In GRE, deficit and surplus from the HRA enter need assessment in 1981/2, and deficit alone in 1982/3. This entry is modified by multiplying it by the England average council house rent level. This has the effect of decreasing the assessed need of areas having large services, i.e., those charging low rents (mainly in London and the cities). Each of these four sources of discretionary spending penalises the high spenders relative to low spenders. In practice, this penalises the cities, not because they are cities, but because these are the areas which have usually chosen to provide many services at relatively large levels. Moreover, it is also these areas which are predominantly Labour Party-controlled local councils. It seems, therefore, that the central government's aim of rigorous control of 'overspenders' accords well with a policy of diminishing assessed spending need, and hence Rate Support Grant allocation, to many Labour-controlled areas.

A further area of criticism which derives from these treatments of discretionary difference between actual and potential client numbers is the different incentives GRE gives to spending in different areas. Those authorities with small or zero expenditure in certain services nevertheless receive grant for those services and have immense incentives to increase their spending on these services up to the England average cost figure, whilst those areas with large expenditures have incentives to cut services or to raise from the rates the expenditure required above the England average. From one point of view, then, GRE has been a major stimulus to equalisation of service levels by stimulating the expansion of services in the authorities which have been low providers (mainly Conservative-controlled rural Counties). From another point of view, however, the form of GRE has stimulated the expansion of local spending which the central government sought to limit or reduce. For many non-metropolitan Counties this effect was used to inflate their poundages for 1981/2 so as to increase flows to balances which could then be used when expected changes and reductions in grant occurred in the following year.

Another area of criticism surrounds the data used in GRE. As already noted, its data demands are very heavy, especially for areas in which data are often sparse. Because the 1981 Census was not available for the first

GRE for 1981/2, and because the Census, when it was available for 1982/3 onwards, does not include sufficient social and economic data, extensive use had to be made of Office of Population, Census and Surveys (OPCS) estimates, local estimates (e.g. of school rolls), and the *National Dwellings and Housing Survey*. Based on small samples, the reliability of the NDHS is very questionable for small geographical areas. The other data sources derived from special surveys and professional judgements both suffer from the drawback that they are difficult to replicate. These sources were aimed at overcoming the relations to past expenditure levels but, as a result, the reliability of GRE, and the ability to extend it for every year in the future without radical changes, has been undermined.

Moreover GRE has not been able to make a departure from what is essentially the political nature of need assessment. The treatment of services, but most especially the unit cost assessments applied to each service, were still the subject of negotiation between the local authority associations and the central government, with many alternative options being calculated and their distributional effects assessed. In these negotiations all associations reported central government being less willing to compromise than in the past. It is clear that the final choice of the form of GRE was in large measure a political decision influenced by central government's desire: first, to control and to reduce total local spending; second, to place specially strong barriers in the way of certain councils which they perceived to be 'overspending'; and third, at all costs to avoid regression analysis and the use of past expenditure as the basis of grant.

A further source of criticism of GRE has been the danger that the service-by-service estimates of spending need can be used (either by central government or local political groups) to make authorities conform with national norms. On the one hand, central government has denied that such pressure will ever be used; however, all the incentives (as noted above) are to conformity. On the other hand, if local political groups do use GRE in this way, this will be to the benefit of local participation and information; and is in the direction of improved monitoring and accountability as 'value for money' by local government. However, local authority associations all resisted the publication of GRE for individual services since they claimed this undermined the aim of RSG to provide general unhypothecated revenue support.

5.3 Grant-related poundage (GRP)

The standard rate poundage (termed grant-related poundage or GRP) is determined by central government. In setting this poundage, as with previous national standard rateable value per head for the resources element, central government chooses a level at which it is expected most authorities will receive benefits and as means of closed-ending. However,

Table 5.1. *Grant-related poundages (GRP) for England and Wales and the percentage assigned to each level of government*

| | Grant-related poundage GRP and % assigned to each level | | | | Threshold (£ per head above GRE) | |
| | 1981/2 | | 1982/3 | | 1981/2 | 1982/3 |
	(%)	(GRP)	(%)	(GRP)		
England average	100.0	134.42	100.0	151.34	36.60	39.95
Wales average	100.0	144.59	100.0	144.99	24.32	22.19
(A) England						
Non-metropolitan Districts	13.0	17.5	13.4	20.35	4.77	5.36
Non-metropolitan Counties	87.0	116.92	86.6	130.99	31.83	34.49
Metropolitan Districts	81.6	109.7	81.3	122.98	29.87	32.38
Metropolitan Counties	18.4	21.72	18.7	28.36	6.73	7.37
City of London	44.7	60.01	34.4	68.65	16.34	18.17
Inner London Boroughs	36.2	48.6	35.8	54.22	13.23	14.28
ILEA	42.5	57.14	41.2	62.38	15.56	16.43
GLC	12.8	17.27	13.4	20.31	4.70	5.35
Metropolitan Police	8.5	11.41	9.5	14.42	3.11	3.80
Outer London Boroughs	78.7	105.74	77.1	116.61	28.79	30.70
Isles of Scilly	100.0	134.42	100.0	151.34	36.60	39.85
(B) Wales						
Non-metropolitan Districts responsible for libraries	18.5	26.79	19.0	27.52	4.10	4.21
Metropolitan Districts not responsible for libraries	17.4	25.11	17.8	25.87	3.84	3.96
Non-metropolitan Counties	82.6	119.48	82.2	119.12	18.28	18.23
Dyfed County Council	82.3	119.05	81.9	118.73	18.22	19.17
Mid-Glamorgan County Council	82.1	118.75	81.7	118.46	18.17	18.13

The bracketed terms sum to 100%.
The poundages shown are for local expenditure equal to GRE. These poundages are modified by the thresholds when expenditure exceeds GRE. All terms at Initial *Report* stage.
Source: House of Commons Papers, *Reports*.

unlike the previous resources element, the level chosen is not in terms of the maximum tax-base equalisation possible for dividing a given grant sum, but instead is in terms of the maximum expenditure differences from standard that can be supported on differing tax bases; i.e., need and resource grants are combined together by use of the GRP. The level of GRP set in England for 1981/2 as a result of this decision was 134.42p in the £, and for 1982/3 151.34p. In 1982/3 GRP shifted to favour non-metropolitan Districts, Metropolitan Counties, the City of London and the Metropolitan Police. In Wales GRP also shifted to favour the Districts at the expense of the Counties (see table 5.1).

The first stage in the definition of GRP is its national level. This is then divided between different levels and types of authorities in terms of their

differing responsibilities for services. Two approaches to the disaggrega-
tion of GRP are permitted in the legislation: either (a) the level of services
in different classes of authorities can be employed; or (b) the average GRE
of authorities in different classes may be used. In practice the calculation
has been made using the second definition, i.e. the level of relevant
expenditure (GRE) accepted for each level of government in the relevant
year has been used, with adjustments for anticipated cost and price
changes. The relative percentages of the national GRP assigned to each
level are shown in table 5.1. As can be seen, the level of GRP set in
England and Wales differs, as does the assignment between metropolitan,
non-metropolitan areas and London Boroughs. Distinction is also made in
Wales for Districts which provide library services and those which do not
such that special treatment is accorded to Dyfed and Mid-Glamorgan
because they do not provide library services. Also, the Isles of Scilly,
because of its remoteness and small size, is treated separately.

5.4 Thresholds, tapering and safety nets

The third definition required in the new block grant is of the thresholds and
tapering multipliers to be employed. These tapering terms are important
since they are used to penalise those local authorities which are 'overspending' in the sense that their expenditure exceeds the GRE assessed by
central government as typical of authorities with similar objective characteristics of needs. For those local authorities for which this 'excess'
expenditure is beyond a specified threshold the level of grant is progressively reduced so that the 'marginal excess expenditure' is borne increasingly by local ratepayers. However, it is implicit in this procedure that the
threshold is set fairly high so that authorities will not be penalised unless
their planned expenditure is well in excess of their assessed need to spend.

   The legislation of the *Local Government, Planning and Land Act* (1980)
permits four variations in the thresholds and multipliers
(1)   Different thresholds can be set for different areas.
(2)   A simple schedule or a graduated schedule can be applied to grant
       reductions over the threshold.
(3)   Expenditure over the threshold can be measured as either (a) £ per head
       above GRE, or (b) percentage of GRE.
(4)   Different thresholds and schedules can be applied in England and Wales.
   For 1981/2 two different approaches to this decision were applied, one in
England and one in Wales.

England

In England in 1981/82 and 1982/3 the threshold was set at 10% above the
average GRE per head, with the tapering multiplier set to achieve a 25%
increase in the poundage cost of expenditure above that threshold. In the

legislation the multiplier is used to adjust the GRP schedule. As a result, the relationship between GRP and expenditure is different below and above the threshold, and hence the grant level also differs. Thus for expenditure in England up to the threshold (i.e. up to 110% of GRE) GRP will be at a constant level. However, for expenditure above 110% of GRE, the GRP is adjusted to make the marginal poundage rise by 25%. This calculation is further complicated by an attempt to keep marginal poundages in line with previous marginal poundages. Thus in 1981/2 an additional multiplier was used to adjust GRP and to keep marginal poundages in line with their level in 1980/1. This was set at 0.5618 pence, the calculated marginal poundage cost for 1980/1. As a result of the operation of these two multipliers, one to induce tapering of excessive expenditure and another to keep marginal poundages relatively constant, the final GRP schedules for England were determined as follows

$$t = t^* + 0.5618\left[\frac{(E_i - \bar{E}_i)}{P_i}\right] + E_i^t \qquad (5.3)$$

and

$$E_i^t = \begin{cases} 0 \text{ when } \dfrac{E_i - \bar{E}_i}{P_i} \leqslant \text{threshold } \gamma_i & (5.4) \\[4mm] k \text{ when } \dfrac{E_i - \bar{E}_i}{P_i} > \text{threshold } \gamma_i & (5.5) \end{cases}$$

In this equation t is the standard poundage (GRP), the term t^* is the GRP schedule for the relevant class of authority (the national uniform tax rate) shown in table 5.1, (i.e., a percentage of 134.42p in 1981/2), E_i and $\bar{E}_i$ are the respective total expenditure and standard expenditure (GRE) as before; and P_i is the local population. The term E_i^t is the threshold adjustment, k is the tapering constant applied where expenditure is above the threshold, and after 1981/2 the threshold γ_i was defined in terms of expenditure but for each class of authority rather than for individual authorities. As shown in table 5.1, the threshold was set to £36.60 per head of GRE in 1981/2 for England as a whole and thresholds for individual authorities were derived for their class of authority in the same proportions as the poundages. Thus, as shown in table 5.1 the threshold above GRE for the non-metropolitan Districts was 13.02% of £36.60, equal to £4.77 per head. The tapering constant for 1981/2 was set so as to increase the marginal poundage costs of expenditure above the threshold γ_i by 25%. Hence the tapering constant was defined by

$$k = 0.7023\left(\frac{E_i - \bar{E}_i}{P_i} - \gamma_i\right) \qquad (5.6)$$

The multiplier 0.7023 scales the excess to 125% of the marginal poundage for 1980/1 (i.e. 125% of 0.5618). Thus for expenditure above the threshold the GRP equation (5.3) is modified to become

$$\bar{t} = \bar{t}^* + 0.5618\left(\frac{E_i - \bar{E}_i}{P_i}\right) + 0.7023\left(\frac{E_i - \bar{E}_i}{P_i} - \gamma_i\right) \qquad (5.7)$$

where the first bracketed term applies only for expenditure up to the threshold.

The role of tapering can be understood from figure 5.1. For expenditure

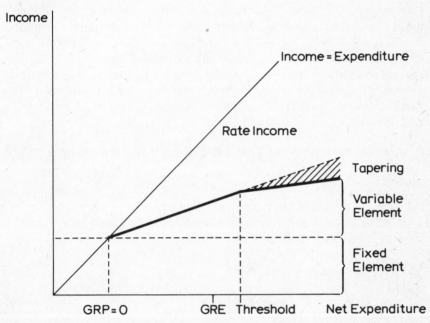

Fig. 5.1 Tapering of Rate Support Grant above the GRE threshold in England and the division of grant between a fixed (needs) element and variable (resources) element.

up to GRE, and up to the threshold above GRE, grant income is in a constant ratio to expenditure: the so-called 'variable element' (SCT, 1981). For expenditure over the threshold, grant is reduced as a proportion of expenditure: although it still rises with expenditure, the rate of change is much lower than below the threshold. The tapering above threshold is at a constant rate, however, which is in contrast to Wales (see below figure 5.2). For expenditure up to the level of GRP = 0 a second component of grant is attracted, the so-called 'fixed element' which represents the support of central government towards the maintenance of minimum standards on a standard tax base. The fixed element is akin to the old needs element, whilst the variable element is akin to the old resources element; this is shown in more detail by comparison with equations (2.14) and (2.15) in chapter 2.

Wales

For Wales the procedure was the same as in England in that the threshold was still set at 10% and the marginal poundage cost was also kept the same as 1980/1, although for Wales this required a multiplier of 1.53 instead of 0.5618. There were two important differences from England, however. First the tapering multiplier differed. A *non-linear* structure was adopted: the multiplier increased as a power function for expenditure over the threshold, thus successively higher expenditures were successively more strongly penalised. The second difference from England was that the percentage and not the per capita adjustment (3(b) on p. 111) was used, i.e. the total expenditure was divided by the standard expenditure, rather than subtracted from it. These two differences gave a final set of GRP schedules for Wales as follows

$$ t = t^* + 1.53t^* \left[1.10\gamma_i^k - 1 \right] \tag{5.8} $$

where

$$ k = \begin{cases} 0 \text{ when } E_i/\bar{E}_i < \text{threshold } \gamma_i \\ 1.35 \text{ (in 1981/2) when } E_i/\bar{E}_i > \text{threshold } \gamma_i \\ 1.45 \text{ (in 1982/3)} \end{cases} \tag{5.9} $$

and the threshold is defined by

$$ \gamma_i = E_i \Big/ \left\{ \bar{E}_i + \frac{(10\bar{E}_i)}{100} \right\}. \tag{5.10} $$

The terms are defined as before except that the tapering multiplier k is now a power function which differed in 1981/2 and 1982/3. Also the comparison of local expenditure with the standard is now a division rather than a subtraction as it is for England. The poundage impact of increased local tax burden of this Welsh non-linear tapering is shown in figure 5.2. The difference in choice of tapering between England and Wales has differential effect on the marginal grant received with different levels of overspend above GRE. With per capita measurement of the GRE above the threshold (as in England) negative marginal grants are experienced by those overspending authorities with rateable value per head higher than the average. With expenditure above GRE measured as a percentage of GRE (as in Wales) negative marginal grants are experienced by those overspending authorities with low GRE relative to their rateable value.

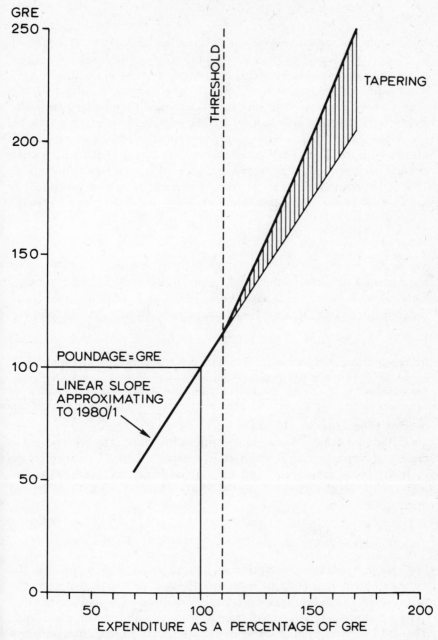

Fig. 5.2 Tapering of Rate Support Grant in Wales: non-linear tapering as used in 1981/2.

116   *Central grants: theory and practice*

London

In addition to these components, special treatment was accorded to London to take account of its high rateable value tax base and its high rate poundage. This was incorporated by using a combination of special area weights and a development of the old system of London 'clawback'. These changes were introduced in one step via a special multiplier which also included the London equalisation scheme of horizontal transfer between Inner and Outer London. However, a special adjustment was made in the case of both the City and Westminster so that they contributed the same amount in current poundage terms in equalisation payments as they did in 1980/1. Domestic ratepayers in these two areas were also protected by special treatment in the domestic rate relief grant, as discussed earlier.

Safety nets

Safety nets have been included in the grant allocation since 1977/8, and were carried forward in the block grant. These have differed between England and Wales and have been introduced into the final grant entitlements via multipliers. Thus final grant levels are given as follows

$$G^*_i = \bar{E}_i - pkB_iE_i \qquad (5.11)$$

from extension of equation (5.1), where p is the safety net multipier. This is greater than unity when a local authority would make excessive grant gains relative to the previous year, and is less than unity when it would bear excessive losses of grant. A safety net was also employed to limit overspending penalties above the threshold.

For the first year of block grant entitlement, 1981/2, the procedure was made more complicated by the need to reassess the 1981/2 entitlement on both the block grant basis and the old resource and needs basis and required the recalculation of the 1981/2 allocation on the basis of the old grant (so-called base year).

Transition from 1980/1 grants

For 1980/81 a special one-off, transition arrangement was employed at the *Increase Order* stage (i.e. in November 1980). This provided assessments of standard expenditure based on the difference of a local authority's 'actual uniform rate poundage' from a 'national uniform rate poundage'. The latter derived from an expenditure assessment of need to spend, and the former was the actual rate poundage levied in relation to its spending needs. The national uniform rate poundage was set at a rate of 119p. If governments exceeded this rate by a significant threshold, their resources element of the previous RSG was to be reduced; or, in the case of those

London Boroughs which received no resource element, their needs element was to be reduced. In practice, although a number of local authorities exceeded a 119p tax rate, very few were penalised for overspending. However this feature did create a great deal of controversy since it was the first demonstration of the power under the 1980 *Local Government, Planning and Land Act* to penalise individual authorities.

5.5 The experience of 'overspending' since 1981

The threshold and tapering mechanisms available to the central government in the 1980 *Act* proved to be insufficient on their own to limit increases in local spending. The predicted overspend for Councils for 1981/2 was £900m or 5.6% (CIPFA, 1981*b*). This arose for two main reasons: first, councils are traditionally cautious and overbudget in estimates, and the uncertainties that resulted from the transition to the new RSG especially over the form of 'clawback' for closed-ending the grant, heightened this; second, because of the way in which GRE was assessed, especially the treatment of discretionary spending, for many councils their GRE was far above their previous spending levels, actually encouraging increased spending.[1] This second feature characterised most of the rural Counties, particularly where they had been under Conservative control for some years. In addition, local authorities believed that the central government's assumptions on inflation and interest rates were too optimistic. As a result of both of these features, but especially the second, the central government made three changes for the 1981/2 year

(1)   In January 1981 warning was given of a new threshold penalty criterion in terms of increases in expenditure volume. This was formalised into the Increase Order stage in early 1982, and has become a permanent feature of block grant.

(2)   After June 1981 central government threatened reductions in grant ('holdback') for authorities which continued to overspend.

(3)   In the *Local Government Act* (1982), introduced in December 1981, the central government introduced legislation to abolish the right of local authorities in England and Wales to levy supplementary rates.

The volume threshold target

This threshold introduced an additional criterion to tapering of expenditures above the GRE threshold: tapering was employed for any council with 1981/2 expenditure more than 5.6% above their 1978/9 outturns (repriced to the relevant November 1980 prices). 1978/9 was chosen as the base year for this volume target because it offered the most recently available data on *actual* expenditures. Hence this choice once again reflected the central government's desire to exclude local authority actions through estimation of expenditures which might feed back to increase grant levels.

The 5.6% volume target was seen as necessary to prevent some councils being able either to make reductions in poundage or to increase spending or balances. The councils in this fortunate position were mainly the non-metropolitan Counties. This threshold was combined with a number of possibilities which modified the original tapering proposed in G.B. Government, House of Commons *Papers* (1981) as shown below

% of expenditure above target of 105.6% in 1978/9	Penalty term	
< 0.1	0	'Full protection'
0.1–2.0	25%	'Partial protection'
2.1–4.1%	60%	
> 4.1%	100%	'No protection'

Within these levels of protection the penalties imposed were as follows

(1)   Increased cost of spending in rate poundage terms by increasing GRE by 9 p from 134.42 to 143.42 for England as a whole, divided between authorities according to table 5.1.

(2)   Increased cost of spending in rate poundage terms by increasing the 1980/1 marginal poundage ratio from 0.568 to 0.600 per £ (see equation 5.3).

(3)   Increased rate poundage cost of spending above threshold from 0.7023 to 0.7500 per £ (see equation 5.7).

(4)   Resetting the other multiplier terms for the authorities wholly or partly protected.

It should be noted that these penalties were imposed by resetting the GRP schedules, whilst the level of penalty to be imposed was derived from a volume target. The effect of the legislation was to create three classes of authorities. The first of these was for authorities spending within their 5.6% volume increase expenditure targets. For these the original thresholds and tapering (equations (5.6) and (5.7)) applied. Thus although they spent within the volume target they could still be penalised. The second group of authorities was for those spending at more than 4.1 above the 5.6% increase in expenditure volume (i.e. more than 9.7% of the required 1978/9 level). For these authorities a completely new set of thresholds replaced equation (5.7) as follows

$$\bar{t} = \bar{t}^* + 0.600\frac{(E_i - \bar{E}_i)}{P_i} + 0.7500\left\{\frac{E_i - \gamma_i}{P_i} - \gamma_i\right\} \qquad (5.12)$$

where the first bracketed term applied only to expenditure up to the threshold. Additionally $\bar{t}^*$ was increased to the appropriate percentage of 143.42. The third group of authorities lag between equations (5.7) and

(5.12) with either 25 or 60% weight applied to equation (5.12) to give the appropriate multipliers.

The distribution of areas suffering different levels of penalty is shown in figure 5.3. Since the level of penalty does not necessarily correspond with level of grant cuts this map can be taken as only an approximate guide to the level of penalties actually imposed. Bearing this problem in mind, the map shows that the areas bearing the highest penalties are mainly in Inner London and the metropolitan Districts. However, many non-metropolitan Counties also suffer high levels of penalty. In contrast with most of the metropolitan areas, however, the Counties were mostly able to reduce expenditures in line with volume targets. Given that many of their expenditure estimates and hence poundages were set in line with GRE (which awarded grant to the Counties even though they did not undertake some expenditure) in many cases the conformity to the target could be achieved with little effort. In the metropolitan areas, however, the targets could be achieved only with greater difficulty. In Wales, no penalty threshold in terms of a volume target was set in 1981/2 since all authorities came below the level of a 2% increase on 1978/9 outturns and this was deemed acceptable by the central government.

For 1982/3 the interaction between the volume and GRE targets was formalised into a more rigorous and clearly defined procedure (G.B. DOE, 1982). Authorities spending both above GRE and above volume targets up to a maximum of 5% of the target were subject to a complex set of penalties to be described below. It is noticeable that for the post-1981/2 period there is no longer any pretence by central government about the use of grant to modify local expenditure decisions. The terms target, control and overspending are liberally used in the legislation and the very real concerns of local authorities regarding the effect of grants on local democratic decisions seem to have been strongly justified. These effects were heightened by court decisions at the close of 1981 declaring that large levels of subsidy to metropolitan area transport authorities were illegal: that the *Local Government Acts* obliged local authorities to run their services in an 'economic and efficient' manner. For 1981/2 these subsidies were derived in the main from supplementary rates as a result of metropolitan Counties controlled by Labour councils making precepts upon metropolitan Districts or London Boroughs, some of which were controlled by Conservative councils. Having been declared illegal, many of these supplementary rates had to be repaid or credited against future rate levies. The most notable examples were the GLC and West Midlands. Further restrictions on local discretion were introduced by the termination of the right of local authorities to levy supplementary rates in England and Wales in the *Local Government Finance Act* (1982).

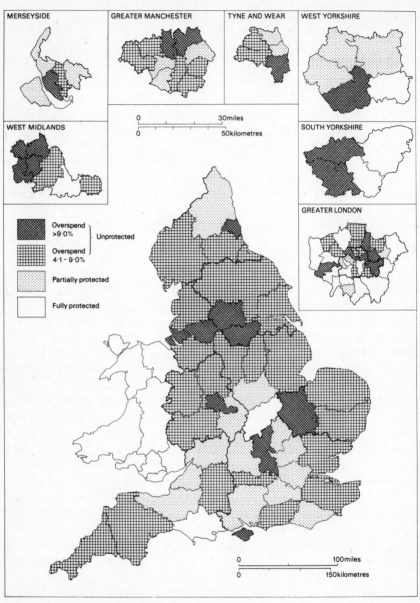

Fig. 5.3 Levels of penalty in the 1981/2 Rate Support Grant as at first announcement in mid-1981. Penalties were divided between areas in terms of overspend
relative to expenditure volume targets and scaled at various levels of 'protection'.
Subsequently penalties were defined also with respect to the GRE target so that
spending at or below GRE was not penalised. This removed a number of
non-metropolitan Counties from the penalty category. ILEA, which has 6%
overspend, is not shown (SOURCE: *The Times*, 3 June 1981).

Reductions in grant (Abatement)

The need to impose reductions in grant was pressed on the central government by two factors. First, in England, many councils refused to conform to spending guidelines even with penalties of grant reductions deriving from the imposition of thresholds and tapering. Instead they sought to make good the grant reductions by raising yet more income from the rates (mainly through supplementary rates raised mid-year) and hence could maintain, or even expand, their expenditures. A second factor, in Scotland, was that because the ability to levy supplementary rates was unavailable, one council (Lothian) proposed to maintain expenditures by going first into debt and then into bankruptcy. The objective then was to force central government either to give in on grant penalties, or to take over local administration completely.

In England the threatened reduction in grant (additional to threshold and tapering penalties) was £450 m and in Scotland £30 m. The reduction in almost all cases fell on to Labour-controlled councils which continued to seek rate rises or increased expenditures despite central government pressures. In England the main councils concerned were metropolitan councils in Greater London, Merseyside and the West Midlands (these were specifically named by Heseltine: see *Local Government Chronicle*, 11 September 1981). In Scotland the threatened £30 m grant cut affected one council only, that of Labour-controlled Lothian containing the Edinburgh region. In practice, under this pressure many councils gave in and did eventually agree to some reductions in expenditure such that the final cut in grants in England was rather less than the original £450 m proposed.

The whole procedure for grant reductions (termed abatement) was developed in an extremely *ad hoc* fashion during the 1981/2 financial year with a consequently high degree of uncertainty for the local authorities. However, for 1982/3 the method of grant reduction was codified into a structure the aim of which was to allow its use for successive years. In this scheme local authority expenditure, which was above both GRE and volume targets, occasioned a progressive reduction in grant. This reduction was restricted to a maximum of 15p in the £ divided between tiers of authorities in the same ratio as GRP (see table 5.1). The level of reduced grant was determined from the following equation which would be applied at the RSG Supplementary Report stage by amendment to GRP and/or multipliers:

$$\text{Grant penalty (for expenditure in excess of target)} = \left[\frac{\text{Actual expenditure} - \text{Target (GRE or volume)}}{\text{Target (GRE or volume)}} \right] . 100\%. \, A \quad (5.13)$$

where *A* is an abatement factor applying up to 105% of target. The definition of the targets was one of the most crucial and controversial stages in this determination of grant penalties. This required seven steps. First, a scaled minimum volume budget was defined by imposing adjustments upon the actual expenditure outturn estimates for the local authorities in 1981/2. Second, this budget was scaled by special 'cash' and inflation factors. Third, the scaled budget was then compared with both GRE and volume (5.6% increase on 1978/9 plus allowance for inflation). Fourth, deviations resulting from this comparison (positive or negative) were further scaled by weights of 0.2 for each 1% deviation of the scaled budget from both GRE and volume. Fifth, the resulting grant sum for all local authorities was compared with the total level of grant available (defined in terms of GRE); sixth, a series of constraints was then imposed to limit the level of grant penalty and changes in grant: no grant reduction would exceed 70%; no target would imply an increase in expenditure; and expenditures for local authorities at or below GRE *and* volume targets would have to be reduced by no more than 1% in real terms. Seventh, after imposing the constraints the whole process was repeated from the fourth step to allow deviations from grant totals to be eliminated (i.e. the grant was closed-ended). In addition, authorities spending within GRE were exempted from penalty.

This procedure allowed central government very considerable control and has become termed 'targetry'. Particularly controversial was the allowance by inflation factors to the minimum volume budget of only 4% increase in local authority pay levels, and a 9% increase in general price levels, both considered by the local authorities to be underestimates of the likely levels which would obtain. The further adjustment of the scaled minimum volume budget by 'cash' scaling was intended to reflect 'the level of cash provision in 1982/3 for the mix of services which each class of authority provides' (G.B. DoE, 1982, Annex B). This was thus a form of cash limit for each service category reflecting further controls on local discretion. As a result of these controls, plus the method of 'closed-ending', the determination of grant penalties represents a considerable change from previous practice in grant allocation. Explicit targets are defined for local expenditures and deviations are penalised by successively reducing the grant percentage. Aggregate deviations from target by all authorities are used to adjust targets for individual authorities such that individuals are penalised by a scaled overall overspend. The use of cash limits within service categories is a considerable change towards a form of aggregate specific grant for each service; and the inflation factor expresses some optimism in wage and price movements such that further penalties in grant levels are exacted in an attempt to make local authorities toe the line in wage settlements and absorb price changes. Hence, the whole process is riddled with uncertainty for the local authorities since precise detail of changes in grant cannot be calculated until well after year-end. The effect of grant penalties on the post-1981 grant has, therefore, been greatly to increase uncertainty in the grant systems, and massively to increase central control of spending, even down to the level of service-by-service GREs and cash limits.

Reductions in grant were aimed at adjusting the marginal incentive to spend by severely limiting grant levels for expenditure above targets, thus placing the increasing burden of high levels of spending on to local tax payers. However, whilst it might be wished that there would be a smooth transition in grant adjustments for spending up to, and beyond, targets, the result of the block grant legislation has been to induce a range of different marginal incentives to spend. Figure 5.4 shows four local authorities typical of the main administrative groups and tiers. For the three authorities shown lowest in the figure, the West Midlands representative of the metropolitan Counties, Manchester representative of the metropolitan Districts, and Lambeth representative of the London Boroughs, grant abatement commences at the volume target and ceases at the maximum of 5% above target. Spending up to the threshold (10% above GRE) in these three cases has no effect on reducing grant levels. However, for spending between the threshold and the target, grant is reduced according to equation (5.12), and for spending above target, grant is reduced (abated) according to equation (5.13). The abatement ceases at the point at which expenditure is equal to 105% of target expenditure subject to grant loss not exceeding the equivalent of a 15p rate (divided between tiers of authority according to table 5.1). After that point, the marginal rate of grant equals the rate which obtains between the threshold and target levels. For many of the non-metropolitan Counties, of which Avon is typical, the pattern is completely different. As shown in figure 5.4, the target is lower than GRE and the threshold above GRE exceeds the point at which grant abatement ceases. For these areas three different marginal rates of grant operate above GRE, and for very high levels of expenditure the level of grant which would have obtained at the level of GRE can be recovered. Indeed, for some non-metropolitan Districts and Counties there is no disincentive to spend through grant reductions and the block grant then behaves like the old needs and resources elements.

The composite effect of the differing marginal rates of grant is to induce widely varying incentives to spend in differing local authorities, and this detracts considerably from the aims of equalisation and certainty in grant allocation. Moreover, the marginal incentives vary in an arbitrary way depending upon the actual level of spending chosen. For some authorities, mainly those in inner London, the level of grant abatement is so high that virtually no grant is obtained; hence expenditure is set mainly in accordance with what ratepayers will bear. A good example of this is Camden. Whilst this feature of the block grant has the meritorious feature that it

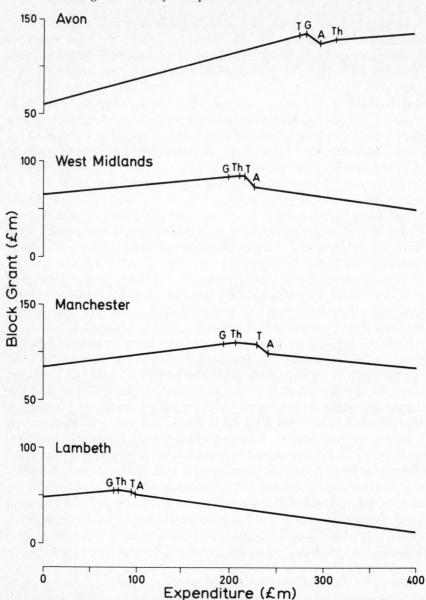

Fig. 5.4 Levels of block grant to four typical local authorities shown with respect to levels of expenditure in relation to GRE (G), threshold above GRE (Th), target (T), and the point at which abatement ceases (A).

encourages accountability, this is achieved at the expense of considerable central control, major uncertainty in grant levels, and a high degree of arbitrariness between areas in the relative schedules of the marginal cost of expenditure above targets or GRE.

5.6 Assessment of the block grant

The block grant instituted since 1981/2 emerged directly from the previous two elements of the Rate Support Grant: needs and resources. The simple device of combining needs and resources into one single element allowed significant improvements in the degree of equalisation in grant allocation to be achieved. The unitary grant defined by the *Local Government, Planning and Land Act* (1980) emerged most directly from the Layfield Report (see especially chapter 2 and Annex 28), the evidence to Layfield by the Department of Environment, and the Labour Government Green Paper (G.B. Government, 1977), but in essence derives from the theoretical work by Musgrave (1961), developed by Boyle (1966) as discussed in chapter 2.

The major advantage of the block grant is that it overcomes the major impediment to achieving full equalisation within the previous RSG system: the wide disparity between tax bases. Full equalisation to the rateable value per head of the richest local authority (the City of London) under the previous RSG would have required an enormous increase in grants overall, could have led to serious underpricing of local services in most local authorities, and would undermine local accountability through the revenue system. The use of separate, fully equalising needs and resource elements required each local authority to be paid a grant which put them in the same financial position simultaneously with that authority with the highest rateable value per head *and* with that authority with the lowest expenditure needs per head. By combining the two elements as a unitary block grant three major improvements in equalisation were achieved. First, advantage was taken of coincidences of high needs and high resources on the one hand, and low needs and low resources on the other hand. Second, arising from such coincidences, tax-rich low-need authorities were penalised by exacting negative payments from the resources element by the device of reducing their needs element. Hence, the block grant has permitted resource transfers from resource-rich to resource-poor areas by a sleight of hand which has been rendered relatively politically feasible in that virtually all local authorities were given some grants. Third, the unitary block grant approach has the advantage that it allows full equalisation, but at a lower total grant level than the previous RSG system by taking advantage of high-resource low-need coincidences.

In addition to these advantages, the unitary block grant offered particular attractions to a Conservative government. First, it reinforced what

Boyle (1966) has noted, that Conservative governments have traditionally favoured equalisation of fiscal capacity at standard levels of need since this permits greater central control. In contrast, Labour governments have traditionally favoured equalisation of fiscal capacity alone since this allows freedom for local authorities in setting their rates and choosing expenditure levels: Labour governments have favoured especially the freedom for local authorities to choose to spend at high levels by expanding the supply of public goods (particularly education and housing) thus reducing the supply of private goods. A second advantage has been that the unitary grant suggests the possibility of gaining greater control of local expenditure, and hence of total public expenditure by methods of progressively reducing grants for local authorities with high spending levels in relation to assessed needs. Third, it suggests a more realistic 'market' pricing of locally-provided public goods by shifting a greater burden of their support towards the local taxpayers and away from the national taxpayer. Thus, the unitary grant provides the potential for greatly increased central control of local government in terms of the level of grants set. In addition to the choice of standard tax base, total level of grants and choice of need indicators available up to 1980/81, the Rate Support Grant since 1981/2 has introduced three new instruments of control: first, the fixing of the standard expenditure; second, the choice of penalty multipliers used to taper grant to 'overspending' authorities; and third, the setting of the 'overspending' threshold.

The block grant also has other advantages. It has produced a more direct grant system since grant levels are directly related to the service responsibilities of the level of government involved. Hence there is a more direct relation of grant levels to spending need. Moreover, this has also permitted much more complete integration of London into the total system; although some special treatment is still retained. As stated by G.B. Government (House of Commons *Papers*, 1980, HC–52, p. 8): comprehensible indicators of spending requirements are used 'that directly affect local government expenditure – rather than being simply correlated with it – . . . which take account of their different service responsibilities'. Moreover, this approach probably goes as far as is presently possible in implementing unit cost assessment for differing client groups.

A further advantage is that overall economic efficiency should be improved. The setting of standard tax rates together with tapering and thresholds does not encourage limitless underpricing of local services by grants, and thus stimulates more efficient setting of local tax rates and hence should encourage greater rationality and quality of local budgetary decisions and greater local accountability.

Despite the advantages of the block grant as a means to achieve better equalisation of local authority needs and resources, it has been widely criticised, especially by all local authority associations. These criticisms centre on eight technical issues. First, the grant was seen as greatly undermining local discretion and independence in setting expenditure levels and tax rates. In particular, it was seen as a partisan control by a Conservative central government to reduce spending in Labour-controlled councils, on social services, education and housing. Indeed, this was, after all, an explicitly stated aim.[2] Second, the block grant was seen as an overreaction to the problems of 'feedback' and 'open-endedness' in the previous system. Third, the publication of the standard expenditure (GRE) estimates and service cash limits for each of the 456 district authorities has been seen as an unwarranted intrusion into local affairs by setting a specific service standard from external, partisan and national priorities rather than local ones. Fourth, and as a result of the desire to define standard expenditures, central government has demanded much more detailed information on expenditure and manning levels (see G.B. DoE, 1979*a,b*, 1980*a,b*) and this has been seen as both placing unnecessary burdens on local authorities, and inducing further limitations of local independence. Fifth, the setting of standard expenditures totals and grant penalties can only be finally achieved after the end of the financial year. Hence, an adjustment process akin to the old resources clawback is still required and this introduces uncertainty into local planning. A sixth problem has been the continuing difficulty of removing the effects of bias from previous expenditure levels (the poverty trap). The method of assessing some of the needs in the block grant suffers from this problem almost as much as the previous system (see Jacobs, 1980; and chapter 8 below). Seventh, the attempt to assess grants at District level has been widely criticised as unviable: data are too coarse, information on costs and wages too sparse and inadequate, and the wide range of expenditure levels means that any central tendency measures (such as the mean used in regression analysis) are very gross. Eighth, some data are still derived from out of date sources such as the 1971 Census and the National Dwellings and Housing Survey, although a very much smaller part than under the previous system.

In addition to these technical issues, the block grant can also be criticised on the grounds that it is no easier to understand than the previous system and also involves a large number of the decisions which are still largely political.[3] The assessments of GRE, GRP, the threshold term and the tapering multiplier are each largely political choices by the Secretary of State for which few objective criteria are available. In addition the block grant, like the separate resource and needs elements before it, still requires multipliers to adjust for London equalisation, safety nets, close-ending the total grant, as well as tapering, and grant penalties. The political biases built into the block grant have been widely interpreted as aimed at penalising particular Labour-controlled areas. For example, Taylor (1980) suggests that

'One has to start by looking at history. During 1974–75 the shadow environment team had endless consultations with Tory leaders in local government at every level. And the great cry that emerged, particularly from the shires, was that socialist authorities mainly in the metropolitan areas, were spending far too much money. Nobody ever quantified that, but it became an article of faith. This block grant . . . is the Secretary of State's way of saying they will control (these) overspending authorities.'

A further major criticism centres on the method of assessing GRP. This single term contains an immense complexity of decision. There is the division of relevant expenditures between levels of government, the effects of tapering and thresholds, adjustments to GRP to induce closed-ending of the grant after the year-end, and the multipliers used to keep marginal poundages in line with previous years. As a result of the complexity of these terms, the final assessed GRP is far from easy to interpret. Hence, it could be greatly improved if it could be kept separate from the issues of tapering penalties, thresholds and marginal poundages.

In addition the specific methods of penalties used in 1981/2, and introduced rather hastily, have considerably confused not only the aims of the grant itself, but also have introduced constitutional issues; issues which have been further confused by the abolition of supplementary rates from 1982/3. In both England and Scotland the £450 m cuts in grant in 1981/2 allowed economies in central grants to be made and helped to keep overall local spending under better control. However, the cuts considerably undermined local autonomy and led to a flurry of concern for constitutional issues related to appropriateness of the government level and to the power of the voters who should decide on local spending levels.

The use of the 5.6% threshold was heavily criticised, particularly by the non-metropolitan Counties who were the ones most adversely affected. They were introduced as a panic measure in the middle of the planning period (January 1981), when authorities were making their expenditure estimates, without detailed thought and final details were not available until January 1982. As a result the effects were largely arbitrary, they contradicted the rationale of the GRE spending threshold, they penalised (the mainly Conservative) authorities which had been frugal in 1978/9, they confused volume spending targets with poundage targets, and in the choice of which authorities were fully or partially protected they created a framework of extremely rough justice. Perhaps the major criticism arose, however, because of the contradiction between the rationale of GRE and GRP used in grant calculation with the volume targets used in determining penalties. As a result small volume overspends could lead to massive grant penalties, and *vice versa*. For example, Wealden District council with a volume overspend of £2000 (0.1%) received a grant penalty of £42 000, and Islington Borough with overspend of £2.31 m (6.0%) received a grant penalty of £3.38 m. Moreover, these massive grant reductions occurred even with partial protection in the case of Wealden. In contrast, some unprotected councils had small grant penalties for massive overspends: e.g., Harrogate District with volume overspend of £2.21 m (45.2%) received a grant penalty of only £0.3 m and South Yorkshire County with volume overspend of £22.3 m (17.8%) received a penalty of only £5.1 m (examples from Hale, 1981, p. 694). Although volume and GRE targets were subsequently aligned, the overall result of the 1981/2 penalties was to introduce a great confusion and to penalise areas with high rateable values per head, i.e. mainly London and the metropolitan Districts. These unsatisfactory results arose because the penalties were calculated by modifying the GRP schedules, but the threshold derived from a volume spending figure in the past. A poundage modification bears differently on areas with high rate yields, i.e., those with large rateable value tax base. Moreover, because of the arbitrariness of the penalties imposed, there was frequently no incentive for authorities to modify expenditure to meet targets. In addition the penalties on County expenditure became translated for many Districts into large increases in their poundages. This was particularly controversial for 'frugal' Districts within high spending Counties. For example, Conservative-controlled Fylde District had the choice of either levying an 18p supplementary rate and also suffering a cut in RSG, or cutting its own expenditure in order to pay for increases in Lancashire County expenditure consequent upon the shift from Conservative to Labour Party control on 1981 (quoted in *The Times*, 10 August 1981). Similar effects also affected Districts in metropolitan areas. Apart from such political controversies the method of imposing penalties also confused lines of accountability as to who was responsible for increases in rates or cuts in expenditure (the District, the County or the central government).

As a result of these effects, and despite its declared aims, the block grant has introduced a large element of uncertainty into local expenditure planning. Under the old RSG, the authorities could be fairly certain what their grant entitlement would be once the settlement was announced at the *Order* stage. Under the block grant the central government has much greater discretion to make changes: it can adjust tapering and thresholds, it can introduce alternative or additional allocation criteria (as it did in 1981/2 with additional volume limitations on any local expenditure increases), and it can change the whole basis of GRE and GRP using multipliers. This is a result partly of the increased powers of the Secretary of State under the 1980 *Act*, but it also arises from the technical reason that, at year-end, adjustments to GRP have to be made to close-end the grant total. These have differential distribution effects depending upon the actual poundage and expenditure of each authority. In 1981/2, for example, most authorities over-budgeted at the estimate stage to cover inflation and other categories; GRPs, as a result, had to be raised at the *Supplementary Report* (*Increase Order*) stage (January 1982) and are likely to be reduced at the final

Supplementary Report stage (December 1982). These changes and other causes of uncertainty in the block grant increase the uncertainty under the new RSG and hence led to severe doubt as to whether it succeeds in its aim of improving planning and control of local expenditure.

Because of these problems the local authority associations as a group have seen the block grant as 'a complete irrelevance', a *'deus ex machina'* the complexity of which is 'mind-bending' (see Martin, 1980). They have further viewed the grant as providing the central government with far wider powers than it required to achieve its objectives of improving the previous system, and introduced a massive degree of arbitrariness into final allocation decisions; in fact, it allowed 'any government, through secondary legislation, to structure any year's Rate Support Grant distribution in almost any manner they wished. There are no effective safeguards against the use of these practices to discriminate in favour of, or against, particular types of authorities, or even some individual authorities' (ACC, 1980*a*, p. 3).

5.7 Conclusion

This chapter has concluded the discussion of the long and complex history of the Rate Support Grant in England and Wales; from 1967 until 1974 it had major emphasis on needs equalisation; from 1974 to 1980 resource equalisation became increasingly important; and from 1981 onwards greater rigour of central control has been introduced. Particularly important also has been the ever-increasing role of specific grants and rate relief for domestic tax payers. The analysis of the aggregate decisions regarding the RSG, however, disguise important distributional consequences which have resulted from employing differing distribution formulae. It is these social, economic, geographical and political distribution effects which are the subject of the following chapters of parts two and three of this book.

PART II

DISTRIBUTIONAL EFFECTS OF BRITISH LOCAL FINANCE

CHAPTER 6

LOCAL EXPENDITURE NEED

6.1 Introduction

The purpose of the chapters in part two of this book is to describe the variation in the aspects of need, tax base, expenditure, tax rates and levels of grants in individual and groups of local authorities. This gives a full background discussion to the problems of local finance in Britain and the distributive effects between areas which have resulted from the way in which the Rate Support Grant has been allocated. Most of the analysis of these chapters concerns the period since reorganisation of local government in 1974, but where possible, the analysis is taken back to the beginning of the RSG in 1967, and in some cases the distribution effects deriving from the financial arrangements existing before 1967 are also discussed. In each chapter also a brief analysis is given of the relation to local party control and to inner city status (as defined in chapter 1). This allows preliminary assessment of political and urban factors in local finances.

The present chapter is concerned with differences in the needs of local authorities to spend on different services. As a concept need for public goods has an approximate equivalence to the demand for private goods, but is also mediated by other factors. Needs vary between local authorities depending upon the number of services they provide, the number of clients in each of their service groups, the costs of providing their services, the level of their services, and many other factors. Hence, need is a very controversial and confused topic, and in general terms can be defined only against the background of political or ideological judgements as to what constitutes need. In Britain the statutory constraints and obligations on local authorities limit the range of variation both in the services that they provide and the quality at which they are available. As a result, it is fairly easy to define the services for which need is assessed, even though it may be difficult to assess what constitutes need within any particular service category. In this chapter are discussed first, in section 6.2, a standardised index of expenditure need; second, in section 6.3, the varying need positions of local authorities are assessed using this index as a yardstick; and third, in section 6.4, the relation of need to local tax base and expenditure is briefly described. Because of the difficulties of comparison before and after 1974 analysis is restricted here, unlike the following chapters, to developments since 1974.

6.2 Need assessment and a standardised index

Assessment of differences in the expenditure need positions of local authorities is perhaps the major research problem in analysing local government finance and central grants. The difficulties which impede the solution of this problem are that there are as many ways of assessing need as there are values as to what public goods should be provided in any area and in what quantity. From the point of view of analysing need as a basis for allocating central grants, however, the position is simplified a little. What is then required is a means of measuring need which permits assessments of differences in standardised revenue requirements.

If one aim of central grants is to place local authorities in an equal position to provide given services, then what is required is a means of separating that local need which is general to all authorities from that local need which is a result of local authority choice or discretion to provide services. To do this the main source of data which is available is present and past data on expenditure and output levels. The problem with such data, however, is that observed expenditure and output data contain the influence of local discretion. Although subject to central controls, local authorities have a level of discretion in connection with services as to whether these services are provided or not, or if provided, at what level of quality or quantity they are available. Expenditure outturn differences between local authorities can differ, therefore, not just because of differences in service needs, but also because of the extent to which local governments have chosen to provide a service. In Britain police and fire expenditure are least affected by discretion, and education, transport, personal social services, and housing are most affected; as for example in the provision of children's and old people's homes, home helps, or the level of local council house building. To overcome this problem requires a measure of service need which is independent of measured levels of local supply.

Assessment of the effects of the level of this local discretionary spending can be undertaken by three routes. First, unit costing can be employed for the services actually supplied. Second, some indication of 'latent need' can be estimated by the use of 'need indicators', e.g., the sparsity of population can be used to indicate need for roads, or the number of old people used to estimate need for old-age personal social services. This has been the approach of the Rate Support Grant needs formula from 1974 to 1980 and has also been used by Jackman and Sellars (1977c); both employ regression analysis. However, each latent need method is ultimately arbitrary. As a result, a third approach, which is that employed in this book, is to determine average service costs over groups of local authorities. This should allow identification of a mean over different levels of discretionary spending.

The approach to need assessment employed below utilises the *representative need index* approach. The representative needs index is a special case of a composite index but where the need levels of each local government are weighted by the extent to which governments differ in the level or quantity of the services they provide. The need for each local service is determined by multiplying a standard weighting function for that service by the size of the client group (or workload) requiring access to that service, and then summing the results for all services, i.e.

$$N_i = \Sigma_j X_{ij} b_{ij} \qquad (6.1)$$

where

X_{ij} are workloads or size of *client groups* for service j in area i,

b_{ij} are *unit cost* weights expressing the costs per unit of workload.

Defined in this way the representative index is a special form of unit cost index. The weighting term can be chosen to equal the load per capita service cost, the national average per capita expenditure on each service, a standard per capita cost for each service, the highest per capita expenditure of any local government, or (that used below) the average per capita service costs in relevant groups of authorities.

There is, of course, a wide range of alternative methods for assessing need. These have been summarised by Davies (1968), Owens (1980) and Bennett (1980, 1982*b*) and fall into five main groups as alternatives to the representative index: (1) individual indicators;[1] (2) composite indicators;[2] (3) standard service input indices;[3] (4) standard service output indices;[4] and (5) unit cost measures.[5] Each of these approaches, together with variants and extensions of them, has advantages and disadvantages. For the purpose of defining a new index which can be assessed from existing British data, however, most of the approaches must be ruled out. The individual need index is not appropriate when used for block grants (where general spending across a range of service categories is being considered): the composite indicator approach is that used in the RSG needs formula up to 1980/1 and adoption of a particular form and set of weights would be as arbitrary as any of the previous needs formulae; the standard service input approach ignores the effects of both local discretionary spending, and the consumption elasticity effect of differing tax bases representing different demand constraints; the standard service output approach, on the other hand, ignores the effect of differential supply conditions; finally, the unit cost approach, although the most direct and theoretically most satisfying method of assessment, to be more than an arbitrary tool, requires detailed appraisal of differences in local discretionary and non-discretionary spending and the tracing of each service input category to each output level. Within the constraints of the present book and existing British data, such a unit cost approach must be ruled out.

Any analysis of local need is likely to be partial in some respects and the present index has inadequacies which result from not controlling for local variations in quality of service, and in largely ignoring the distinction between demand as a true need concept and actual demand as revealed by local provision. Attempts to control for service quality are very difficult but have been attempted.[6] Attempts to control for potential real need in comparison to revealed need are notoriously difficult and controversial; the usual method is to employ a principal components analysis of latent need indicators and then combine these as a set of orthogonal underlying need determinants.[7] Each of these complications is ignored here for the present in favour of a fairly direct and simple index using the representative needs index approach. This relatively simple method allows many of the benefits of unit costing to be obtained, but possesses a level of simplicity and generality which can be implemented with existing British data bases. It also approximates fairly closely to both GRE and to the G.B. Treasury (1979) methods of need assessment and hence is useful for comparative purposes.

The representative index requires, for each local government, measurement of the cost per unit of providing each service; then, using this cost measure as a weight, needs are defined as the sum of the costs over all service groups. The representative index,[8] then, requires determination of two main features, relevant client size, and unit costs, which are then combined to give the final index.

Client groups

The choice of data to measure client group varies according to the service category being studied. Within the relatively uniform system of local government in Britain, the degree of variation of the number of services provided by each government is fairly small. Table 6.1 summarises the level of expenditure in each major category of locally-provided services. As can be seen, the major item of local spending is that on education which consistently accounts for over one half of the total. Moreover, education plus the four other service categories of personal social services, law and order, transport, and environmental services, consistently account for over 90% of the total local expenditure. In constructing a representative needs index, therefore, the majority of local client variations can be captured by five sets of measurements of needs for each of these spending categories. In practice of course, each category is itself a composite of a large number of subcategories, each of which requires separate measurement. For the purposes of this book 30 service sub-categories are employed.

Table 6.1. *Annual expenditure by sector in Local Authorities in England and Wales (1978 PES survey prices)*

Year	Trade and employment	Roads and transport	Housing	Other environmental services	Law and order	Education etc.	Health and PSS	Other	Total
				£m at 1978 prices					
1964/5[a]	—	1003	249	1282	1009	5319	1003	1160	11025
1973/4	50	796	255	1120	1160	5632	748	68	9829
1974/5	56	979	408	1278	1234	5926	858	93	10828
1975/6	63	1042	416	1317	1312	6092	936	100	11278
1976/7	65	961	398	1289	1378	6152	959	94	11296
1977/8	66	894	387	1298	1340	6162	976	97	11220
1978/9	70	906	421	1340	1394	6318	1011	107	11561
1979/80	77	911	452	1370	1427	6406	1041	108	11792
1980/81[b]	76	880	462	1283	1646	5954	1015	145	11462
1981/2[c]	71	822	331	1231	1659	5749	1009	125	11013
				% of each year's expenditure					
1964/5[a]	—	9.1	2.3	11.3	9.2	48.2	9.5	10.5	100
1973/4	0.5	8.1	2.6	11.4	11.8	57.3	7.6	0.7	100
1974/5	0.5	9.0	3.7	11.8	11.4	54.7	7.9	0.9	100
1975/6	0.6	9.2	3.7	11.7	11.6	54.0	8.3	0.9	100
1976/7	0.6	8.5	3.5	11.4	12.2	54.5	8.5	0.9	100
1977/8	0.6	8.0	3.4	11.6	11.9	54.9	8.7	0.9	100
1978/9	0.6	7.8	3.6	11.6	12.1	54.6	8.7	0.9	100
1979/80	0.7	7.7	3.8	11.6	12.1	54.3	8.8	1.0	100
1980/81[b]	0.7	7.7	4.0	11.3	14.4	51.9	8.9	1.3	100
1981/82[c]	0.6	7.5	3.0	11.2	15.1	52.2	9.2	1.2	100

[a] some headings not comparable in detail.
[b] estimated outturn.
[c] planned.
Source: Public Expenditure White Papers.

The choice of appropriate indicator for each client group is a complex issue. For some services, such as those to school children and old people, direct client group information is available which directly relates to the service provided. For other services the total population (as in the case of refuse collection and disposal and cemeteries) or the number of households (as for recreation and planning) provides a good measurement of client group size. Similarly miles of road and vehicles licensed are good measures of need for transport spending. Again port tonnage is a reasonable surrogate for port health needs. For some services, however, satisfactory statistics on client need are more difficult to obtain because of the effects of local discretionary variation. This affects many personal social services, e.g. children in care, other residential accommodation; and also nursery education and concessionary fares. For these it is necessary to distinguish the object of the analysis at the outset. If it is desired to assess need on the basis of services as actually provided, then service statistics for client groups receiving the service are the relevant ones to use. However, if it is

Table 6.2. *Client group indicators employed in constructing needs index together with their sources*

1	Population (OPCS estimates relevant year)
2	Nursery schoolchildren (DES statistics relevant year)
3	Primary schoolchildren (DES statistics relevant year)
4	Secondary schoolchildren (DES statistics relevant year)
5	Further education (DES statistics relevant year)
6	All schoolchildren (DES statistics relevant year)
7	Old people (Pensioners OPCS estimate relevant year)
8	Port tonnage (National Ports Council *Annual digest of port statistics*)
9	Miles of road (Ministry of Transport, *Transport Statistics*)
10	Vehicles licensed (Ministry of Transport, *Transport Statistics*)
11	Average no. of children in care (*Personal Social Statistics*, actuals; CIPFA)
12	Serious offences (*Statistics of the Criminal Justice System, England and Wales*; London: HMSO)
13	Average no. of in-care actuals: mentally handicapped adults and children, young physically handicapped, mentally ill, and elderly (CIPFA)
14	Number of households (NDHS, 1978–9)
15	Old people living alone (CIPFA and OPCS)

sought to assess need independent of local discretion, as sought for GRE in the Rate Support Grant philosophy since 1981/2, then other data independent of local provision have to be employed. This then drives the analyst towards 'latent need' indicators. In the present analysis, however, the sizes of client groups are assessed from those actually supplied with services, but a distinction made later in the analysis is between discretionary and non-discretionary services. This differs from the approach used in GRE. The categories of client group employed, and the data source from which they are derived, are shown in table 6.2.

Unit costs

The measurements of unit costs to provide the weights in equation (6.1) can be undertaken by various methods. That employed here uses the average costs per client group derived from three years of unpublished Department of Environment data on the *actual* expenditure outturn statistics for local authorities. The three years employed were 1976/7–1978/9 converted to 1980 outturn prices. Such actuals include the effects of discretionary spending. Since the level of analysis employed is that of the 'needs' authorities, expenditure data for the non-metropolitan Districts had to be aggregated to County level and data for the metropolitan Counties, GLC and ILEA had to be disaggregated respectively to metropolitan District and London Boroughs. The scaling multipliers used by the Department of Environment for RSG up to 1980 for this purpose were employed. These derive from a complex negotiation between DoE and the local authorities and are based upon population, rateable value and service-level differences between authorities.

The resulting unit costs were derived separately for each major class of administrative area. This differs from GRE which only produces differences for England and Wales. The present approach rests on the philosophy that the division of services between subunits differs (e.g., the role of the ILEA and GLC in London is different from any other area), and that there are historical differences in treatment of services and hence of expenditures between administrative groups. However, the main reason for separate treatment is that different classes of authority differ in the nature of their service organisation. The main contrast is of course between the urban and rural areas. But in addition the very low population densities of Wales demark it from other non-metropolitan areas. Similarly the metropolitan Districts, Inner London and Outer London also differ greatly as to the nature of their service delivery problems. Clearly any classification of authorities could be used to control for these differences, and one based on population density, or socio-economic status might be appropriate. However, the choice of administrative class conforms with the actual differences in statutory status of areas.

This method seeks to control for a level of discretionary spending by attributing to each local authority the average costs of the group to which it belongs; e.g., the standardised costs of Bedfordshire providing homes for the aged is assumed to be the same as the average cost of providing this service which has been attained in the group of all English non-metropolitan Counties. Differences in the costs of providing a service in one area from the average of that area's group are discounted (i.e. differences are attributed to discretionary choices to provide the service at a higher or lower service quality). However, these average service costs are affected by the discretion of a whole group of authorities choosing to provide services at a high or low quality and this may affect, for example, Inner London.

The resulting average service costs per client group are shown in table 6.3. From this table it is clear that locally-provided further education, residential children in care, other residential accommodation, and community care account for the largest costs per client. For these services a very small difference in the size of client group will result in a large difference in spending requirement. Estimates of the variance of these cost estimates (not reported) show that most variation, and hence uncertainty, is concentrated in a very small number of service categories, the chief of which are (4) further education, (8) children in care, (9) other residential accommodation, and (10) community care. Variation is also much higher in Wales than the other administrative areas. Each of these services is significant in total cost terms and large variation is to be expected since these are also the services for which there is a high level of local authority discretion in service levels which at this stage has not been taken into account. Despite the relatively large variance for these four services,

Table 6.3. *Service costs (£ per client at 1980 prices) in each main administrative area derived from the three years of expenditure outturn statistics for 1976/7 to 1978/9*

Service need category	Need indicator employed	Non-met. Counties, England	Non-met. Counties, Wales	Metropolitan Districts	Inner London	Outer London
1 Nursery schooling	2	0.0355	0.0364	0.0200	0.0261	0.0159
2 Primary schooling	3	0.5311	0.6436	0.5470	0.6665	0.6124
3 Secondary schooling	4	0.8344	0.8425	0.8106	0.8843	0.9067
4 Further education	5	5.1415	5.9228	5.4404	7.0009	6.4733
5 Special schools	6	0.0476	0.0430	0.0521	0.0862	0.0684
6 Other schools	6	0.0531	0.0751	0.0546	0.1307	0.0741
7 School meals and milk	6	0.1001	0.1012	0.0962	0.0938	0.1014
8 Residential children in care	11	13.8303	21.1068	3.2193	4.7046	3.6045
9 Other residential accommodation	13	12.1736	13.7309	5.7197	5.5999	4.8596
10 Community care	15	91.3863	11.3565	139.5741	242.9886	147.9552
11 Day care training	1	0.0049	0.0061	0.0083	0.0219	0.0089
12 Fieldwork R. & D.	1	0.0047	0.0049	0.0062	0.0146	0.0071
13 Port Health	8	0.0001	0.0001	0.0001	0.0001	0.0001
14 Magistrates Courts	12	0.0418	0.0479	0.0294	0.0154	0.0143
15 Other Courts	12	0.0044	0.0196	0.0034	0.0029	0.0027
16 Civil Defence	1	0.0001	0.0002	0.0001	0.0001	0.0001
17 Probation	1	0.0016	0.0016	0.0023	0.0012	0.0011
18 Refuse collection and disposal	14	0.0066	0.0099	0.0081	0.0150	0.0088
19 Environmental health	14	0.0032	0.0047	0.0037	0.0076	0.0040
20 Cemeteries	1	0.0007	0.0013	0.0011	0.0018	0.0009
21 Crematoria	1	0.0002	0.0004	0.0004	0.0003	0.0002
22 Recreation	14	0.0109	0.0226	0.0162	0.0251	0.0169
23 Town & Country Planning	14	0.0075	0.0112	0.0080	0.0117	0.0092
24 Highways	9	0.0175	0.0191	0.0486	0.1703	0.0537
25 Public Lighting	9	0.0018	0.0025	0.0102	0.0229	0.0089
26 Parking	10	0.0018	0.0015	0.0018	0.0167	0.0019
27 Public safety	9	0.0001	0.0002	0.0006	0.0031	0.0011
28 Transport admin.	10	0.0042	0.0062	0.0098	0.0296	0.0117
29 Public passenger transport	1	0.0018	0.0034	0.0134	0.0133	0.0123
30 Concessionary fares	7	0.0060	0.0110	0.0251	0.0117	0.0107

The needs indicators employed and their sources are listed in table 6.2.

however, the variance for most other expenditure categories is very small, and this should give some confidence in the final index derived from them.

Need index

Once the client groups and unit costs have been obtained for each service category the final need index is constructed by multiplying the cost per unit of each subcategory of client group in the relevant group of local authorities by the size of the client group in the relevant local authority, and then summing the result. The resulting index is derived here after excluding two areas, the City of London and the Isles of Scilly from the analysis since their per capita resource and expenditure positions differ so greatly from all other areas.

The final need index can be rendered in two forms: first as a £ expenditure requirement which depends on the size of client groups in each area; and second, as an index which has been scaled to per capita or similar terms. For grant purposes it is the first, expenditure requirement, which is the most important and is that given by GRE in the RSG. For comparative purposes a descaled need index is more useful, however. But because expenditure is a composite of all client needs and not just those related to population, per capita scaling is not necessarily the most useful and expenditure level or some other criterion may be preferable. It is noteworthy that in the calculation of penalties in the post 1980 RSG, provision is made for calculating comparisons between GRE as both GRE per head and GRE per £ of actual expenditure (see page 111). Different choices markedly affect distribution as between areas.

6.3 The need position of local authorities

The need position of local authorities in England and Wales can be assessed on a relatively uniform basis at the level of 'needs authorities' since each one involved is providing almost entirely the same range of services (although service output qualities will differ). In the discussion below the need position is detailed in two forms: first, variation in major client groups; and second, variation in total expenditure need weighted by average costs, using the representative index outlined above.

Client groups

The variation of numbers of major client groups per head of population in different areas is shown in tables 6.4 to 6.6 for the main service categories. Education is the major client-group category since this alone is responsible for measuring client-need differences which account for about one half of total expenditure. Education need can be broken down into various subcategories and table 6.4 shows change in need in six categories. In line with the England and Wales decline in birthrate and ageing of the population, Primary School pupils show a general decline after 1976/7 with the English Counties and Outer London registering this decline one year earlier. Secondary pupils less than 16 years old show increase up to 1979/80 and decline thereafter. Other pupil numbers display the interaction of demographic change whith political decisions. Direct grant students (publicly-funded places in semi-private schools) have steadily declined whilst further education places have stayed approximately steady. Differences

Table 6.4. *Means and standard deviations (in parentheses) of education need variables by administrative area (per 1000 local population)*

Variable	Non-met. Cos., Eng.	Non-met. Cos., Wales	Met. Districts	Inner London	Outer London	Total
Primary school pupils						
1974/5	105.1(9.7)	113.7(7.0)	117.7(7.3)	120.8(25.8)	99.2(9.6)	108.9(14.1)
1975/6	105.5(9.8)	115.3(6.3)	117.1(8.6)	115.2(24.4)	97.7(9.2)	108.3(15.5)
1976/7	104.9(9.9)	115.8(6.4)	118.1(7.8)	112.1(23.8)	97.1(9.2)	109.9(13.5)
1977/8	95.8(8.5)	97.8(4.9)	103.5(7.3)	94.8(19.4)	85.9(8.6)	96.5(11.2)
1978/9	95.4(8.5)	97.7(4.9)	103.9(7.2)	87.1(19.7)	86.6(8.8)	96.9(11.3)
1979/80	93.7(8.3)	96.3(4.8)	101.2(7.4)	89.6(18.2)	83.4(8.4)	94.0(11.1)
1980/1	91.6(8.1)	94.2(4.4)	97.3(7.5)	83.7(17.0)	79.7(8.2)	90.7(10.9)
Secondary school pupils <16						
1974/5	61.7(5.8)	65.2(2.8)	66.9(7.5)	75.0(16.0)	58.5(6.7)	64.4(9.1)
1975/6	68.7(7.2)	72.7(3.1)	75.6(7.4)	80.2(16.9)	64.3(7.4)	72.2(9.9)
1976/7	71.2(6.3)	76.6(3.2)	77.7(7.4)	82.0(17.4)	66.3(7.5)	73.7(9.6)
1977/8	77.5(6.6)	80.7(3.5)	84.3(7.7)	87.1(17.8)	71.9(7.7)	79.8(9.9)
1978/9	77.1(6.8)	80.3(3.9)	84.7(6.7)	87.6(17.8)	73.0(7.0)	80.1(9.7)
1979/80	78.6(6.6)	81.1(3.7)	85.6(5.8)	84.5(17.2)	73.6(7.4)	80.7(9.1)
1980/1	79.4(6.4)	81.2(3.4)	85.2(5.1)	82.1(16.7)	73.3(7.7)	80.5(8.7)
Secondary school pupils >16						
1974/5	7.0(1.3)	9.2(1.1)	6.6(1.4)	11.5(2.4)	8.8(1.4)	7.8(2.0)
1975/6	9.5(1.3)	11.6(1.0)	9.5(1.3)	13.3(2.8)	10.9(1.4)	10.3(1.9)
1976/7	9.8(1.4)	11.8(1.0)	9.8(1.4)	13.9(2.9)	11.3(1.4)	10.6(2.1)
1977/8	5.9(1.3)	7.4(1.1)	5.3(1.2)	9.1(1.8)	8.0(1.5)	6.5(1.9)
1978/9	5.9(1.3)	7.4(1.1)	5.4(1.1)	8.7(1.7)	8.3(1.6)	6.6(1.9)
1979/80	5.9(1.3)	7.3(1.0)	5.5(1.1)	8.1(1.6)	7.9(1.9)	6.4(1.7)
1980/1	6.1(1.3)	7.5(0.9)	5.7(1.1)	8.1(1.6)	8.0(1.8)	6.6(1.7)
Direct grant pupils <16						
1974/5	0.6(0.5)	0.1(0.2)	1.7(1.4)	0.72(0.15)	0.7(0.6)	0.9(1.0)
1975/6	0.6(0.6)	0.1(0.2)	1.7(1.4)	0.73(0.15)	0.7(0.6)	0.9(1.0)
1976/7	0.5(0.5)	0.1(0.2)	1.6(1.4)	0.06(0.14)	0.7(0.6)	0.9(1.0)
1977/8	0.5(0.5)	0.1(0.2)	1.3(1.4)	0.04(0.008)	0.5(0.6)	0.7(0.9)
1978/9	0.5(0.5)	0.1(0.2)	1.3(1.4)	0.05(0.009)	0.5(0.6)	0.7(0.9)
1979/80	0.4(0.4)	0.1(0.2)	0.7(1.2)	0.02(0.004)	0.5(0.5)	0.5(0.8)
1980/1	0.3(0.3)	0.1(0.2)	0.5(0.9)	—	0.3(0.4)	0.3(0.6)
Direct grant pupils >16						
1974/5	0.3(0.3)	0.05(0.09)	0.8(0.7)	1.23(0.27)	0.3(0.3)	—
1975/6	0.3(0.3)	0.05(0.09)	0.8(0.7)	1.21(0.26)	0.3(0.3)	0.5(0.6)
1976/7	0.3(0.3)	0.05(0.09)	0.8(0.8)	1.19(0.25)	0.3(0.2)	0.5(0.6)
1977/8	0.2(0.3)	0.04(0.09)	0.8(0.8)	0.54(0.11)	0.3(0.3)	0.4(0.5)
1978/9	0.2(0.3)	0.05(0.09)	0.8(0.8)	0.59(0.12)	0.3(0.3)	0.4(0.5)
1979/80	0.2(0.2)	0.03(0.05)	0.4(0.5)	0.05(0.01)	0.3(0.2)	0.3(0.3)
1980/1	0.2(0.2)	0.02(0.04)	0.3(0.4)	—	0.2(0.2)	0.2(0.3)
Further education students						
1974/5	6.9(1.1)	6.2(1.4)	7.6(2.6)	13.2(2.1)	7.0(2.7)	—
1975/6	7.5(1.3)	6.6(1.0)	8.2(3.0)	13.3(2.8)	8.3(2.8)	—
1976/7	8.3(1.3)	7.9(1.0)	9.0(3.1)	11.2(2.3)	8.8(3.1)	—
1977/8	8.6(1.4)	8.0(1.1)	9.3(2.9)	14.3(2.9)	9.1(2.8)	—
1978/9	8.8(1.2)	8.4(1.2)	9.0(2.1)	11.7(12.4)	8.6(2.0)	—
1979/80	8.9(1.2)	8.6(1.1)	9.0(2.1)	11.3(2.2)	8.9(1.9)	—
1980/1	8.9(1.2)	8.6(1.1)	9.0(2.1)	11.3(2.2)	8.9(1.9)	—

Table 6.5. *Means and standard deviations of selected personal social service need variables by administrative area (per 1000 local population)*

Variable	Non-met. Cos., Eng.	Non-met. Cos., Wales	Met. Districts	Inner London	Outer London	Total
Lone-parent families with dependent children (1971 Census)	10.4(1.1)	19.7(1.4)	12.6(2.3)	16.7(2.9)	11.1(1.9)	11.9(2.7)
OAPs (1971 Census)	139.1(31.8)	144.7(19.8)	121.2(15.8)	132.8(7.2)	132.0(14.5)	132.0(23.3)
OAPs (OPCS 1978 estimate)	148.6(28.9)	154.8(15.0)	138.9(14.8)	160.6(8.1)	156.3(15.5)	148.6(21.6)
OAPs living alone (1971 Census)	38.8(8.8)	39.3(6.3)	40.0(7.9)	51.6(6.1)	37.3(5.9)	40.3(8.3)
Old and handicapped in NHS homes (1971 Census)	2.6(0.4)	3.6(0.5)	2.6(0.4)	4.1(0.3)	2.9(0.4)	3.4(0.4)
Households lacking basic amenities (1971 Census)	113.2(8.4)	27.1(12.3)	7.8(3.2)	7.4(2.4)	4.9(1.9)	10.5(6.5)
Children in care (1971 Census)	0.8(0.4)	0.6(0.4)	0.7(0.3)	0.8(0.4)	1.0(0.4)	0.8(0.4)
Personal Social Unit						
1974/5	14.1(2.2)	14.7(2.3)	18.3(4.7)	37.3(7.2)	16.3(4.2)	18.3(7.9)
1975/6	14.1(1.2)	14.7(2.3)	18.4(4.8)	37.9(7.3)	16.4(4.2)	18.4(8.1)
1976/7	15.2(2.5)	21.2(2.4)	20.9(5.5)	38.7(7.6)	17.9(4.9)	20.3(8.2)
1977/8	15.5(2.7)	18.5(2.5)	22.0(5.9)	39.7(7.5)	18.6(4.3)	20.4(10.5)
1978/9	15.7(2.8)	12.9(1.7)	22.7(6.2)	40.6(8.0)	19.1(5.0)	20.8(9.0)
1979/80	15.8(2.7)	12.9(1.4)	22.8(6.2)	40.5(8.0)	19.3(5.1)	20.9(8.9)
1980/1	15.8(2.7)	12.9(1.4)	22.8(6.2)	40.5(8.0)	19.3(5.1)	20.9(8.9)

between areas show, for services exhibiting discretionary effects, slightly higher client numbers in the urban areas for direct grant and further education places. However, when aggregated for all pupils (figure 6.1), the main spatial differences are of the relatively high education needs in all the metropolitan Districts and some of the Inner London and the rural Counties, compared with Wales, the south west, Kent-Sussex-Surrey, East Anglia and the rural north.

Personal social services (pss) needs are composed of a wide variety of client groups, e.g., children in care, the handicapped, old people in homes, home helps, etc. To capture this wide spread of needs, a variety of indicators has been used in RSG. In the needs formula from 1975/6 to 1980/1 various Census data were heavily employed for lone parent families, people in shared households, and households with density of more than 1.5 persons per room. The variation of some of these indicators is shown in table 6.5 together with the personal social services unit used for RSG allocation in 1974/5 which has been reconstructed for later years. The variation of individual pss needs in almost all cases gives higher incidence in the urban areas, especially Inner London. The distribution of one of these, lone parent families with dependent children, is shown in figure 6.2.

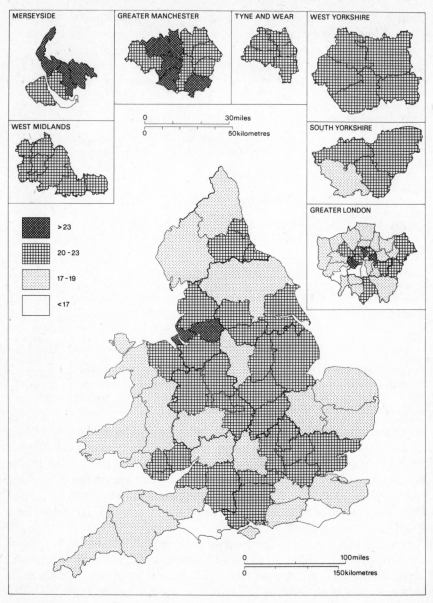

Fig. 6.1 Numbers of pupils in all levels of maintained education per 1000 of local population in 1978 (median 19.8) (SOURCE: unpublished data from Department of Education and Science).

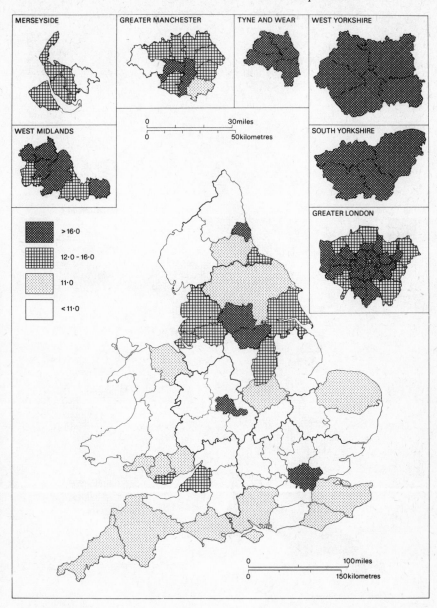

Fig. 6.2 Numbers of lone-parent families per 1000 of local population in 1971 (median 11.3) (SOURCE: G.B. *Census*).

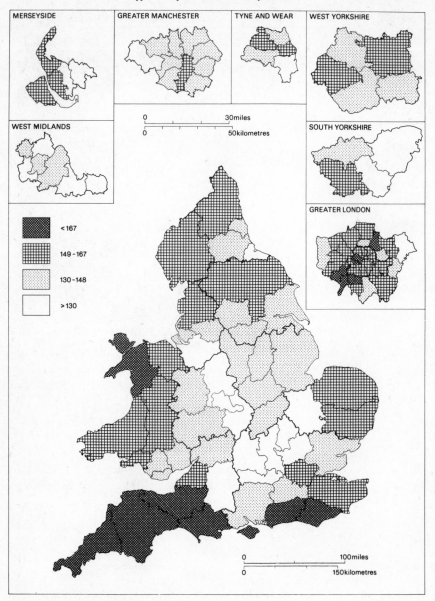

Fig. 6.3 Numbers of old age pensioners per 1000 of local population (median 148) (SOURCE: OPCS mid-year population estimates).

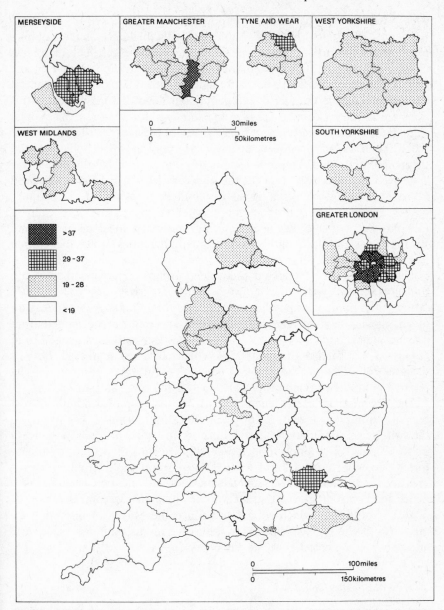

Fig. 6.4 Personal social service unit used for Rate Support Grant purposes in 1974 updated to 1978 per 1000 of local population (median 19) (SOURCE: DHSS).

This variable was particularly important in distribution of the RSG between 1975/6 and 1980/1 (see table 4.11). Its incidence almost completely reflects the urban areas, especially the central cities. The major pss client need which differs from this urban concentration is that of old people. As shown in figure 6.3, although the cities, and especially central cities, have a large number of old people per head, the major concentrations are in the retirement areas of the south, south west and Wales. The aggregation of pss needs into a single group of services for the purpose of the RSG in 1974/5 produced an index which is shown in table 6.5 and mapped in figure 6.4. This index was highly controversial at the time it was employed and was only used for the one year 1974/5. It heavily favoured the cities placing strong emphasis on children in care and old people's homes, and placed very little weight on potential client groups especially of the old. As a result this need indicator reflected local discretion and strongly favoured those authorities, mainly in the cities, which had chosen to provide a high level of pss spending.

The relative size of selected client need groups for environmental and transport services is shown in table 6.6. Many of these services reflect, fairly directly, the number of people or households in an area. A good measure of change in need for the environmental services can be gained from housing starts; and need for transport services is usually best gauged from either miles of road or the number of vehicles licensed in an area. Both of these sets of indicators show highest need in the rural areas. In addition the relative numbers of unemployed also favour the rural areas.

The variation in size of client groups between types of administrative areas is an important feature for grant allocation purposes. For example, the switch from population-based indicators and miles of roads in the RSG up to 1973/4 (which often reflected *potential* client group size), to indicators which reflected the level of actual client group size of children in care etc., for the RSG after 1974/5, led to a marked shift and increase in assessed need, and hence of allocated grants, to the city areas. In contrast, the shift of GRE in 1981/2, which sought to reduce the effects of discretion on measures of client group size led to a massive reduction in assessed need, and hence in allocated grant, to the cities.

Representative needs index

The combined effect of each of the client groups discussed above can be judged from the relative size of each authority on the representative need index (equation 6.1). The average values of this need index, together with GRE are shown in table 6.7. A full list of the need index values is given in Appendix 2. In this index total need per head is highest in the metropolitan Districts, Wales and Inner London, and increases slowly for most areas in the early 1970s but remains static after about 1976. Aggregate need in

Table 6.6. *Mean and standard deviation (in parentheses) of selected environmental, transport and miscellaneous service need variables by administrative area (per 1000 head population)*

Variable	Non-met. Cos., England	Non-met. Cos., Wales	Met. Districts	Inner London	Outer London	Total
Environmental Services Housing starts 3-year period						
1974/5	23.6(6.3)	19.4(6.3)	14.1(4.8)	13.7(6.7)	10.1(4.5)	16.9(7.6)
1975/6	23.5(6.2)	19.3(6.3)	14.1(4.9)	13.9(6.9)	10.2(4.6)	16.9(7.6)
1976/7	23.4(6.1)	19.3(6.1)	14.2(4.9)	14.2(7.0)	10.3(4.6)	16.9(7.5)
1977/8	19.6(4.2)	17.5(5,8)	12.9(3.7)	13.8(5.4)	10.3(4.8)	15.1(5.7)
1978/9	19.5(4.1)	17.5(5.8)	12.9(3.7)	13.9(5.5)	10.3(4.8)	15.1(5.6)
1979/80	20.2(4,0)	16.7(5.4)	13.4(3.9)	14.2(5.6)	9.6(3.7)	15.3(5.6)
1980/1	19.1(3.4)	14.9(3.4)	12.9(3.7)	12.6(4.1)	8.8(2.8)	14.4(5.1)
Transport Services Miles of road (000s) (Trans.	6120.2	4448.6	1057.2	201.9	518.3	2826.7
Statistics)	(800.1)	(481.6)	(252.5)	(80.6)	(89.7)	(421.2)
Vehicles licensed (000s) (Transport						
Statistics)	295.8(40.2)	123.1(17.7)	106.7(26.8)	34.2(7.4)	87.1(28.4)	160.9(26.3)
Miscellaneous Unemployed						
1974/5	7.8(2.1)	6.4(2.1)	4.7(1.6)	4.6(2.3)	3.4(1.5)	5.6(2.5)
1976/7	7.1(1.4)	5.7(1.0)	4.8(1.9)	3.2(1.5)	3.3(1.9)	5.2(2.2)
1978/9	5.7(1.3)	4.5(1.4)	4.0(1.4)	3.2(2.2)	2.5(1.2)	4.2(1.9)

Inner London has, however, begun to decline since 1977. The spatial pattern of need as measured by the index per head in 1978/9 is shown in figure 6.5. This shows the heavy concentration of spending need in the inner cities and Wales, with generally low spending need in the English Counties, Outer London and many metropolitan Districts outside the central cities. The index shown includes all the major services subject to a high degree of discretion. However, recalculation of the needs index after removing these discretionary services produces only minor changes (see Bennett, 1982*b*, fig. 2).

The change in spending need recognisable from the aggregate needs index is shown in figure 6.6. Because of the number of client groups being averaged together the magnitude of change is overall very small. The main increases in aggregate need are concentrated in the English and Welsh Counties, and the outer metropolitan areas of the metropolitan Counties and London. The areas experiencing declines in aggregate need are the central cities in Merseyside, Manchester, Birmingham and Newcastle, plus

Table 6.7. *Mean and standard deviation of need index and GRE at 1980 prices*

Variable	Non-met. Cos., England	Non-met. Cos., Wales	Met. Districts	Inner London	Outer London	Total
Need index						
1974	229.9 (99.2)	144.7(49.1)	117.2(65.5)	116.6(13.5)	81.6(18.9)	150.8(97.0)
1975	240.7(104.6)	149.6(49.6)	120.8(65.6)	114.6(10.7)	84.0(19.9)	156.2(97.2)
1976	246.9(106.6)	153.2(50.6)	122.1(65.8)	108.8(10.2)	84.6(20.2)	158.7(97.4)
1977	245.5(106.3)	149.0(48.3)	119.4(63.7)	111.0(10.7)	83.2(19.9)	157.0(97.5)
1978	246.1(106.2)	149.0(47.8)	118.3(63.2)	103.7(10.7)	83.1(19.8)	156.4(97.6)
1979	245.6(106.2)	149.0(47.8)	116.6(60.1)	103.2(10.7)	82.9(21.3)	155.6(97.8)
1980	245.7(106.1)	148.4(47.2)	115.6(59.4)	102.0(10.7)	82.0(21.0)	154.7(97.9)
1981	245.8(106.1)	148.4(47.3)	115.3(59.2)	102.1(10.7)	82.0(21.0)	154.4(97.9)
1982	245.8(106.1)	148.4(47.3)	115.2(58.9)	102.1(10.7)	81.9(21.0)	154.2(98.7)
GRE						
1981	245.6(112.1)	135.7(49.2)	118.7(56.6)	104.4(11.6)	92.5(18.8)	156.9(97.9)
1982[a]	236.3(103.0)	105.2(38.1)	112.4(52.1)	97.0(11.3)	95.4(16.9)	149.6(58.1)
Need index per head						
1974	0.23(0.02)	0.43(0.05)	0.36(0.02)	0.55(0.07)	0.36(0.02)	0.37(0.03)
1975	0.34(0.02)	0.45(0.06)	0.38(0.03)	0.55(0.07)	0.37(0.02)	0.37(0.03)
1976	0.35(0.02)	0.46(0.06)	0.38(0.03)	0.54(0.07)	0.37(0.02)	0.38(0.03)
1977	0.35(0.02)	0.45(0.06)	0.37(0.02)	0.55(0.07)	0.37(0.02)	0.38(0.03)
1978	0.35(0.02)	0.45(0.06)	0.37(0.02)	0.55(0.07)	0.37(0.02)	0.38(0.03)
1979	0.35(0.02)	0.45(0.06)	0.37(0.02)	0.51(0.06)	0.36(0.02)	0.38(0.03)
1980	0.35(0.02)	0.45(0.06)	0.36(0.02)	0.50(0.06)	0.36(0.02)	0.38(0.03)
1981	0.35(0.02)	0.45(0.06)	0.36(0.02)	0.50(0.06)	0.36(0.02)	0.38(0.03)
1982	0.35(0.02)	0.45(0.06)	0.36(0.02)	0.50(0.06)	0.36(0.02)	0.38(0.03)
GRE per head						
1981	0.34(0.02)	0.40(0.03)	0.38(0.03)	0.52(0.08)	0.41(0.03)	0.39(0.03)
1982[a]	0.32(0.02)	0.34(0.03)	0.36(0.03)	0.53(0.08)	0.43(0.04)	0.27(0.03)

Gross figures are £m, and per capita figures are £000s per head.
N.B. For comparison purposes with the 'needs' authorities used in this book GRE for the non-metropolitan Districts are aggregated to County level and the metropolitan Counties, GLC and ILEA are disaggregated to respective metropolitan District or London Borough.
[a] 1981 prices.

the majority of Inner London boroughs. Hence there has been over the period 1974–79 a significant relative shift of need away from the central urban areas. Other analyses show this to be in main part a consequence of the suburbanisation of population and industry (see Drewett *et al.*, 1976; Hall *et al.*, 1975).

The representative need index developed here is a fairly simple and straightforward method of assessing both need and standard expenditure requirements. The basic philosophy of the approach using service-by-service analysis of expenditure requirements and developing unit costs for major client groups is a very generally applicable one and has been used for

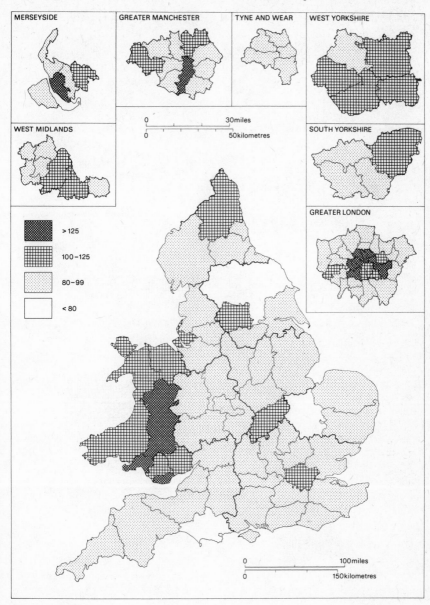

Fig. 6.5 Representative need index per head of local population in 1978 including discretionary services.

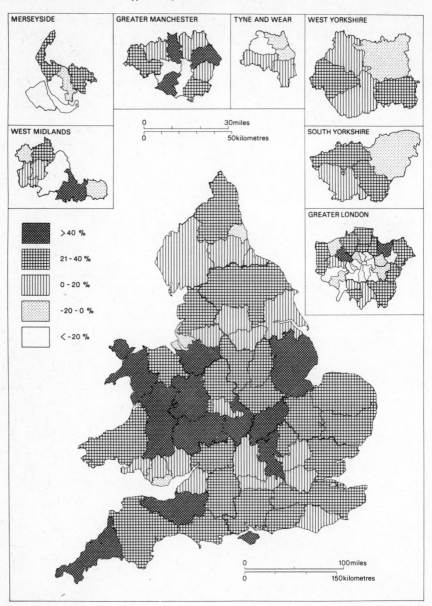

Fig. 6.6 Change in representative need index 1974–79; percentages multiplied by 100.

allocating the RSG in the calculation of GRE since 1981 and also (in G.B. Government, 1977, p. 24) by the Labour Government who proposed the assessment of local authority need on a national analysis of expenditure in separate service categories. As a result, as shown in table 6.7, the scales of GRE and the representative needs index differ only a little and their relative weight in different areas is in close agreement. This is borne out by figure 6.7 which displays the regression relation between GRE and the index. The relationship between the two need measures is strong (R^2 = 0.989 for 1981). However, there is a small number of important authorities which have major discrepancies on the two indexes.

All of the Welsh authorities are above the regression line and hence are over-estimated by the index relative to GRE. Since the GREs for Wales have been defined by central government using a different set of procedures than for England, they cannot be expected to accord closely with the index of this book which treats England and Wales jointly. The effect of the central government decision to give separate treatment has been to reduce Welsh allocation from what would have been given using the index of this book and England and Wales totals. The other major discrepancies are nearly all Metropolitan Districts and especially Inner London Boroughs (with the exception of three English Counties), and many are central cities. In an analysis of the major services in some of these authorities (Bennett, 1982b), it was clear that many of the discrepancies are accounted for by the treatment of discretionary services in the GRE. It will be recalled (see p. 106) that GRE uses English averages for most services whilst the representative index here uses averages of client group size for types of administrative authorities. This difference in method of assessing client groups has the effect of averaging the large client groups (of the authorities providing services at a high level) with the small client groups (of authorities providing modest services or none at all). Hence, the particular central government choice of the treatment of services subject to a high degree of local discretion has had the effect of reducing expenditure need in a large number of Metropolitan Districts and Inner London Boroughs, many of which are central city areas.

Despite discrepancies between assessed need which arise from the method of need assessment employed, whichever method is used shows the highest needs per head to be concentrated in Inner London, a number of other inner city and metropolitan areas, and Wales. Of these needs social services and education are the major components. Lowest need levels are in the non-metropolitan Counties, Outer London and many of the more suburban metropolitan Districts. The needs in these areas, although arising from social services and education, also have more major transport components.

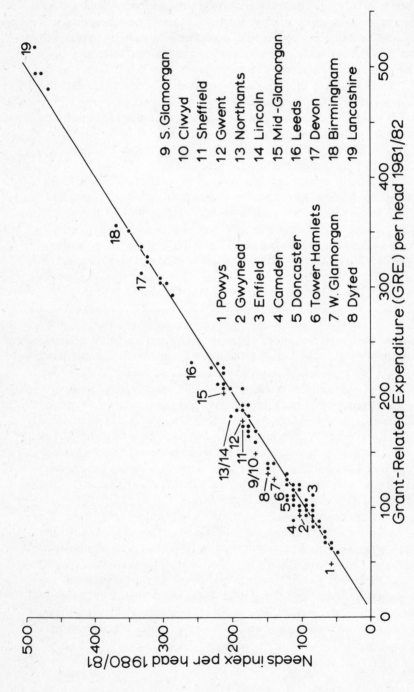

Fig. 6.7 Scattergram of representative need index against GRE for 1981/2 together with authorities distant more than two standard deviations from the least squares regression line, and all Welsh authorities show by a +. The equation of the regression line is $Y = -1.286 + 0.999\,X$; the R^2 value is 0.989. (SOURCE: Bennett, 1982b.)

6.4 The interrelation of needs, expenditure and resources

The measurement of expenditure need is only a starting point in the determination of expenditures and other behaviour of local authorities, and in the distribution of grants. Ideally, as a form of demand surrogate, we may expect need and expenditure to have a close interrelationship. However, other factors intercede to prevent this interrelation being perfect. A major factor, of course, is the extent to which grants are sufficiently good at equalising differences in needs and resources to provide services such that expenditure can rise perfectly in line with need. Another factor is the extent to which expenditure discretion by local authorities, even under conditions of perfect need and resource equalisation, creates variations above and below assessed need levels as a result of expressions of local preference or variations in the party ideology of different councils towards service provision.

In the absence of equalisation grants the comparison of need and tax base allows assessment of the extent to which need can be supported from the local tax base. Figure 6.8 shows that high need is associated with relatively large tax base in many inner city areas but especially London. Low need and relatively small tax base, however, characterise many rural areas and some metropolitan Districts. For each of these types of area we would expect relatively small levels of RSG transfers in a unitary grant system (this is only partly borne out in practice: see chapter 8). In contrast, areas with high need and low rateable value should be the ones with the highest levels of grant receipt. These areas are mainly concentrated in Wales, West Yorkshire, Manchester and Merseyside. The fourth group of areas, those with high tax base and low need, are the ones in the most fortunate positions that, although they will receive low grants, they have relatively high capacity relative to need which can be used to provide higher levels of service or reduced local tax rates. Although very simplified, figure 6.8 shows these areas to be concentrated mainly in the south east, west midlands, Newcastle and Outer London.

Although the spatial distributive effects of the RSG have not yet been discussed, it is instructive to compare assessed need with actual expenditure to see how far they do accord. This is shown in simplified form in figure 6.9. Expenditure and need do accord in many areas. However the two most interesting cases are where there is relatively high need and low expenditure, and where there is relatively low need but high expenditure. These cases conform in simplified terms to ones in which services are, respectively, 'under-provided' or 'over-provided'. The 'under-providers' are concentrated in the metropolitan Districts, Outer London and the Welsh borders. Since these are areas which, in many cases, also have high rateable values, it is here that relatively low rate poundages relative to assessed need for local taxes are to some extent possible, although the

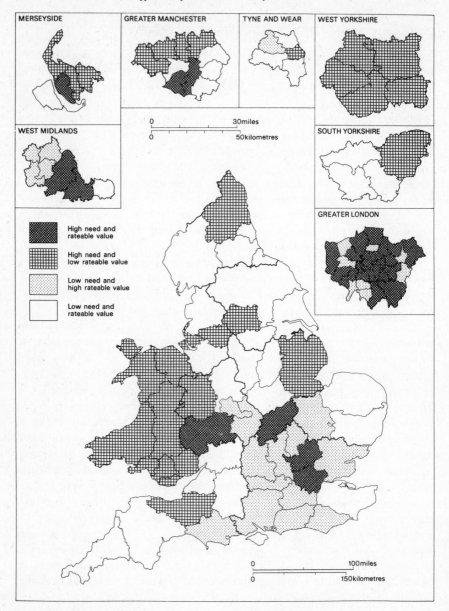

Fig. 6.8 Variation of representative need index in relation to rateable value in 1978 in categories defined relative to the medians of the need index and rateable value.

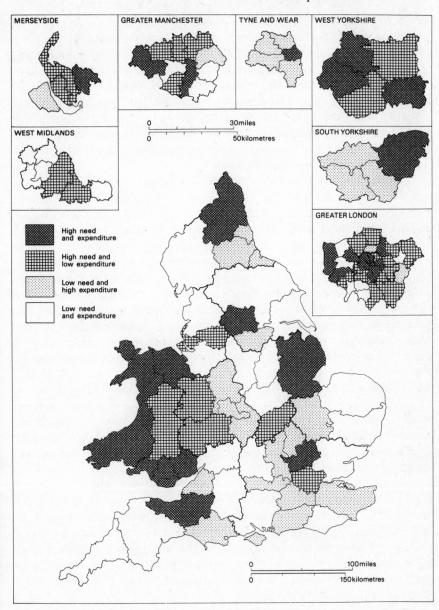

Fig. 6.9 Variation of the representative need index in relation to local relevant expenditure in 1978 defined in categories relative to the median need index and expenditure.

effect of the RSG has yet to be taken into account. In contrast, the 'over-providers' are also in the metropolitan Districts, especially the north east and South Yorkshire, but are also in the south east and the midlands. This unlikely combination of areas falls into two groups. First, it should be expected that (excluding the effects of RSG) poundages would be relatively lower in the south east and midlands areas, since these also have high rateable value tax base. Second, it should be expected the poundages should be relatively higher in the metropolitan Districts in the 'over-providing' group since low need and high expenditure are usually associated with relatively low tax base.

It is tempting to relate the patterns of so-called 'over-provision' and 'under-provision' to differences in the party colour of the local council: the Conservative south east and outer metropolitan suburbs tending to 'under-provide' and the Labour inner cities, north east and South Yorkshire tending to 'over-provide'. Subsequent analysis serves to confirm some effect on expenditure levels from such party positions (see chapter 8) but the effect of the RSG in particular tends to complicate the position. Moreover, needs and political party are themselves fairly strongly inter-related so that different local party control does not necessarily lead to a predictable pattern of over- or under-provision. Table 6.8, for example, shows that higher needs are often associated with Labour control (e.g. in Inner London and the metropolitan Districts) and other party control (in Wales and the metropolitan Districts), whereas low need is usually associated with Conservative control (e.g. most urban areas). Hence, in discussing the political effects on over- or under-provision relative to need, there is the effect of two inter-correlations which must be taken into account: first of need and socio-economic status with party-voting support; and second, of local council party ideology and the level of discretion which has not been excluded from assessed need.

Tabulations of need (table 6.8) suggest that further exploration of the political correlates of the relation of expenditure and need is justified. But the ANOVA results quoted in that table suggest that need differences do not relate significantly to the differences between urban areas identified in the Inner City areas legislation in 1978. This important feature is a result partly of the high tax base of these areas and partly of the operation of the RSG over the period in question; a conclusion which is justified by the results of the following chapters.

6.5 Conclusion

This chapter has described the differing need levels for the major local services in Britain. Most attention has, however, been concentrated on aggregate spending need which has been assessed using a representative needs index. This index is relatively simple to construct and is closely

Table 6.8. *Average and standard deviation (in parentheses) of level of need index and GRE for administrative areas and party colour*

Variable	Non-met. Cos., England	Non-met. Cos., Wales	Met. Districts	Inner London	Outer London
GRE/head 1981/2					
1	0.34(0.01)	—	0.37(0.03)	0.55(0.02)	0.39(0.03)
2	0.34	0.38(0.01)	0.38(0.03)	0.51(0.03)	0.43(0.03)
3	0.35(0.01)	0.43(0.04)	0.38	—	—
4	0.36(0.02)	0.39(0.01)	0.38	—	0.41(0.01)
Index/head 1980/1					
1	0.34(0.02)	—	0.37(0.03)	0.52(0.04)	0.35(0.01)
2	0.33(0.004)	0.40(0.01)	0.36(0.02)	0.50(0.06)	0.37(0.02)
3	0.37(0.03)	0.51(0.08)	0.34	—	—
4	0.37(0.02)	0.43(0.01)	0.34	—	0.35(0.02)

Key: 1, Conservative; 2, Labour; 3, other; 4, party change.
A three-way ANOVA of this table, with inner city status as a third category shows that administrative area is significant at the 99% level and party colour at the 90% significance level. Inner city status is not statistically significant and neither is the interaction term (even at the 70% level).

related to the GRE used for RSG purposes since 1981. The use of this index combined with the individual need analysis allows the conclusion that the areas with the highest need per head are concentrated in Inner London, a number of other inner cities and metropolitan areas, and Wales. Of these needs, those for social services and education are the major components. The areas with the lowest needs per head are in the non-metropolitan Counties, outer London and the more suburban metropolitan Districts. In these areas, although education and social services are important, transport services are often major contributors to need as a result of large road mileage. In addition to the marked geographical differences in need per head the preliminary analysis reported towards the end of this chapter demonstrates that important differences in the pattern of spending as related to need do occur, and to some extent the resulting pattern of 'over-providers' and 'under-providers' of services can be related to the party controlling the local council. This analysis also suggests that although need and local party control may be strongly interrelated, need and the inner city status accorded under the 1978 legislation are not related. These preliminary results are extended in the following chapters.

CHAPTER 7

LOCAL TAXES AND EXPENDITURES

7.1 Introduction

This chapter is concerned with two issues: first, evaluating the tax and expenditure positions of local authorities; and second, assessing how these positions have been related to allocation of the Rate Support Grant. In common with the following chapters, most detailed analysis is given to the years since local government reorganisation in 1974. Less detailed analysis is given of the earlier years of the Rate Support Grant from 1967 to 1973, and for comparative purposes the results of brief analysis of the General Grant in 1962 and 1965 are also included.

In the following analysis two features are given particular attention: first, the annual statement of local expenditure, tax rate and tax base; and second, the direction and rate of change in expenditure and resources. The combination of the two aspects (level and rate of change) accords with the US Advisory Commission on Intergovernmental Relations definition of fiscal 'blood pressure' of local governments (US, ACIR, 1977). This definition is a measure of the ratio of an authority's financial position in the year of interest relative to the change in position from a base year. This ratio can be used as an index of fiscal pressure or fiscal stress. Alternatively the numerator and denominator can be plotted as two axes of a graph giving four categories of local authority: first, those with high and rising fiscal pressure; second, those with low and falling pressure; third, and fourth, two transitional categories of low pressure and rising, and high pressure and falling. This approach is used below to map authorities in differing positions with respect to pressure on tax base, tax rates or expenditures.

The rest of the chapter is divided into five main sections which are concerned, respectively, with the revenue position in section 7.2, the expenditure position in section 7.3, the tax rate position in section 7.4, and in section 7.5 with the debt position of local authorities. Initially in each subsection the changing position of local authorities with respect to revenues, expenditures and rate poundage is described; then the results of analysing differences relating to inner city status and party colour are discussed.

7.2 The revenue position of local authorities

The revenue position of local authorities in England and Wales is relatively easy to assess since only one local tax is available and user charges represent only a small proportion (about 14%) of locally-raised revenue. The single tax source of the rates accounts for about 80% of locally raised revenue for current expenditure purposes. Variations in the level of receipt from the rates result from the interaction of two factors: local tax base and local expenditure levels expressed through the tax rates. Attention is restricted for the moment to the tax base.

It is the rateable value of a local authority which measures the base available for taxation (the tax capacity). Differences in this property tax base between areas arise from the variable concentration of residential and commercial property and it is the commercial tax base which accounts for the major part of the variation in rateable values in Britain. For example, whereas the range of total rateable value per head in 1978/9 varied from £41 000 in the City of London to £68 in mid-Glamorgan, the range of variation in residential rateable value per head was much less, from £232 in Westminster to £33 in mid-Glamorgan. Over the period since reorganisation of local government in 1974 up to 1978/9 total rateable values for England and Wales have increased from £6 659 m. to £7 058 m. or from £136 to £147 per head. However, there have been very important differences in magnitude of changes in the size of the tax base between areas. Southwark, Wakefield, Doncaster, Northamptonshire, Northumberland, North Yorkshire and the Isles of Scilly registered the largest rates of change, at over 12%. In contrast, a small number of jurisdictions have registered absolute declines. Amongst the 'needs' areas analysed only the City of London and the London Boroughs of Westminster and Hammersmith have registered such declines; but in addition the non-metropolitan Districts of Watford, Scunthorpe, Blaenau-Gwent and Neath have suffered absolute decline which is masked within non-metropolitan County totals. In each case the main cause of loss of tax has been closure or movement of industrial and commercial premises.

The problem of these inequalities in the local tax base has been a major underlying thread to the arguments relating to the whole system of local finance in Britain. The only major local tax source is the property tax of 'the rates' which has three main disadvantages. The first of these is that there is a great inequality between the size of tax base between areas. This is clear from figure 7.1 which shows the variation of total tax base and per capita tax base: the absolute size of tax base differs greatly, but the per capita differences are dominated by the small number of very rich London boroughs (particularly the City, Westminster and Camden). A second disadvantage of the rates is that they are relatively regressive, taking a higher proportion of incomes from lower income than from higher income

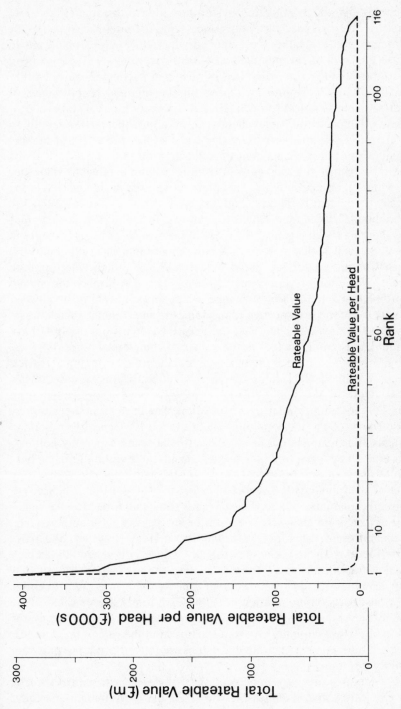

Fig. 7.1 Rateable value per head of local population and rateable value total against rank for local authorities in England and Wales at the level of needs authorities.

households. Although mitigated by the initiation of income-related rate rebates to lower income households since 1967, the rates are still a largely flat-rate tax and this limits the extent to which they can be employed to increase the level of revenue raised by local government. The third main disadvantage of the rates is their lack of buoyancy in comparison with other taxes. The rates in Britain are based on the nominal rental values of property which is subject to revaluation at five-year intervals (at the most frequent). Hence, rating assessments cannot adjust themselves rapidly to rises in values, incomes or prices; this has also had the effect of making local authorities more dependent on central grants.

Table 7.1 shows the changing distribution of total rateable value per head within each major type of administrative area since 1962. What is most clearly evident is the very high tax base of London, followed by the metropolitan Districts and English Counties. In addition it is clear that whilst there has been a general increase in tax base in all areas (except the City) since 1962/3, this increase has been most rapid in London and the English Counties, and least rapid in outer Wales and the metropolitan Districts. For more recent years figures 7.2 to 7.4 display the spatial variation in percentage rates of change in rateable value per head between 1974 and 1978 for the respective total, domestic and commercial tax bases. From these figures it is clear that the greatest increases in tax base have resulted mainly from changes in the values of commercial property in the non-metropolitan Counties and in a small number of metropolitan Districts (especially Solihull, St. Helens, Wigan, Wakefield, Barnsley and Doncaster). In contrast, the major losses in tax base have also resulted from changes in the commercial property position. The declines in the inner city areas of most of the metropolitan Counties, the East and Inner London Boroughs and the industrial areas of South Wales are especially notable.

It is interesting to explore how far the considerable variations in tax base relate to areas of different local party control and inner city status. The results of such analysis are reported in Bennett (1982b) and, hence, are only briefly summarised here. These results show that total rateable values per head vary significantly between different types of local authority and between different categories of urban stress area. However, the party colour of local councils was not statistically significantly related to tax base at the 90% significance level. Further analysis of the tax base data for 1980 shown in tables 7.2 and 7.3, however, shows some important features. First, whilst party has no relation to tax base for total rateable values there is a significant relation for industrial rateable value per head. Second, the relation of tax base to inner city status arises from the role of the domestic tax base. Third, commercial rateable values are related neither to party nor inner city differences.

These features are very important since they suggest, with respect to inner city status, that in contrast to the USA, the British inner cities have

Table 7.1. *Changes in the mean and standard deviation (in parentheses) of rateable value per head for different administrative areas in England and Wales*

Fiscal year	English Counties	Welsh Counties	County Boroughs or metropolitan Districts	Inner London	Outer London	Total
1962/3	13.2	11.1	15.1	32.4	19.4	16.0
	(2.5)	(2.2)	(3.5)	(28.6)	(2.5)	(9.3)
1965/6	36.7	27.9	40.3	68.1	58.4	99.2
	(7.5)	(6.3)	(8.8)	(29.6)	(8.8)	(735.4)
1967/8	38.3	29.8	41.9	100.4	60.8	106.2
	(7.8)	(6.3)	(9.1)	(99.9)	(9.2)	(774.7)
1969/70	39.8	31.0	44.0	107.8	63.1	117.2
	(8.3)	(6.4)	(9.3)	(111.9)	(9.8)	(887.6)
1971/2	41.7	32.6	46.0	113.7	65.7	112.4
	(8.4)	(6.7)	(9.8)	(119.9)	(10.2)	(796.6)
1973/4	110.5	86.4	116.3	308.0	165.6	447.7
	(19.7)	(13.5)	(22.7)	(313.1)	(25.5)	(4092.1)
1974/5	116.5	87.7	106.5	340.1	173.5	114
	(19.7)	(15.3)	(22.6)	(363)	(27.5)	(135)
1975/6	118.3	88.9	108.2	345.4	174.7	146
	(19.9)	(15.5)	(23.1)	(364)	(27.8)	(136)
1976/7	119.8	89.5	109.7	350.1	175.8	148
	(20.3)	(15.5)	(23.4)	(361)	(27.9)	(135)
1977/8	123.5	92.3	112.2	353.3	177.1	150
	(20.5)	(16.4)	(23.2)	(364)	(28.2)	(136)
1978/9	126.7	94.5	113.9	355.9	178.3	153
	(20.8)	(16.2)	(23.3)	(363)	(28.5)	(136)
1979/80	129.4	96.2	115.7	358.7	179.6	155
	(21.8)	(16.3)	(23.5)	(366)	(28.8)	(137)
1980/81	128.9	96.8	118.8	370.6	182.4	158
	(20.4)	(15.6)	(25.0)	(367)	(27.5)	(139)

Note that revaluation of rateable values took place in 1963 and 1972 and reorganisation of local government in 1974.

high tax base in comparison with other areas and this is mainly the result of the even spread of commercial activity between authorities; i.e., there has not been the decentralisation of retailing and commerce from central cities which has occurred in the USA. With respect to local party colour, these results suggest that it is the areas with large industrial tax bases which are largely Labour party controlled; i.e., there is a strong correlation of party and social status resulting from a high local industrial workforce. This has important political implications: the suggestion of November 1981 to place a ceiling on the level of industrial rates would, in effect, restrict the rate yield of Labour as opposed to Conservative-controlled areas.

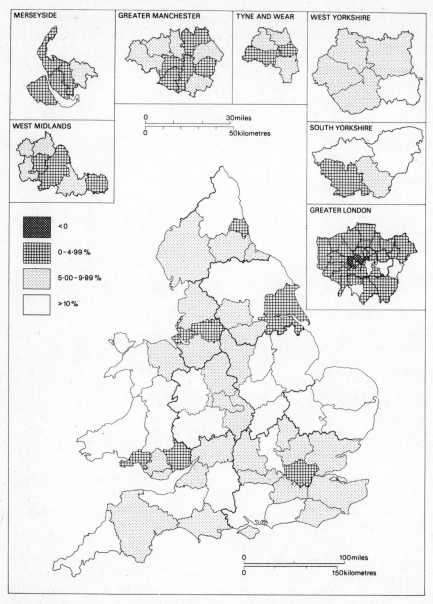

Fig. 7.2 Change in total rateable value per head; percentage change 1974–78.

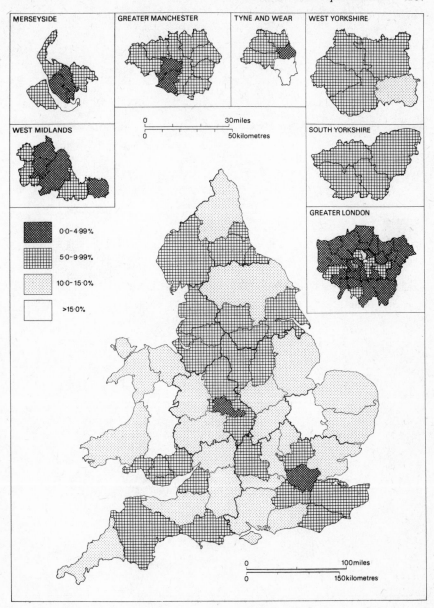

Fig. 7.3 Change in domestic rateable value per head: percentage change 1974–78.

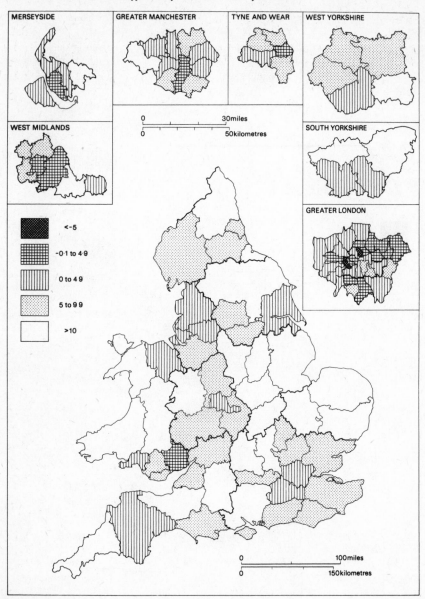

Fig. 7.4 Change in commercial rateable value per head: percentage change 1974–78.

Table 7.2. *Mean and standard deviation (in parentheses) of domestic, commercial and industrial rateable value per head in 1980 for different administrative areas, party control and urban stress areas*

	English Counties	Welsh Counties	Met. Districts	Inner London	Outer London
Local political party					
DRV: Cons.	73.8(12.5)	–	63.9(15.3)	236.5(4.3)	111.8(9.4)
Lab.	48.1 –	41.9(18.2)	56.8(9.3)	100.5(15.6)	91.4(8.8)
Other	64.3(4.4)	43.5(4.5)	59.3 –	–	–
Change	58.2(6.3)	56.4(1.9)	70.2 –	–	97.5(15.7)
CRV: Cons.	60.2(9.8)	–	53.9(14.1)	726.5(705.8)	72.1(17.7)
Lab.	47.9 –	50.1(12.2)	59.4(17.7)	151.5(113.6)	94.0(32.1)
Other	45.7(3.6)	47.8(7.4)	76.7 –	–	–
Change	60.6(9.3)	56.1(12.9)	91.8 –	–	63.9(13.3)
IRV: Cons.	14.2(5.3)	–	16.1(6.0)	1.9(0.6)	11.7(6.8)
Lab.	14.4 –	19.2(7.2)	17.4(8.1)	16.4(10.7)	27.3(11.3)
Other	8.7(3.6	8.7(5.3)	14.9 –	–	–
Change	18.7(6.2)	12.5(5.2)	45.5 –	–	19.3(5.2)
Urban stress					
DRV: High stress	–	–	61.4(6.8)	96.6(8.3)	72.7 –
Stress	–	–	57.1(10.7)	94.6(2.7)	96.2(5.4)
Other urban	–	–	65.5(16.4)	284.5(55.5)	103.9(12.7)
Non-met.	70.1(13.0)	46.2(7.6)	–	–	–
CRV: High stress	–	–	76.1(17.3)	129.8(59.5)	86.3 –
Stress	–	–	53.5(15.0)	84.4(29.6)	87.5(17.8)
Other urban	–	–	56.3(14.6)	630.2(526.2)	79.6(29.4)
Non-met.	58.8(10.1)	50.7(9.7)	–	–	–
IRV: High stress	–	–	16.6(4.1)	18.5(12.4)	22.3 –
Stress	–	–	16.5(8.9)	11.6(3.7)	23.8(8.2)
Other urban	-	–	19.5(9.2)	5.3(5.9)	18.5(12.4)
Non-met.	14.4(5.6)	13.6(7.1)	–	–	–

CRV, commercial rateable value.
DRV, domestic rateable value.
IRV, industrial rateable value.

The change in tax base is translated into a 'pressure' movement by comparing it with the absolute magnitude of the tax base. Figure 7.5 displays the spatial variation in the absolute tax capacity per head, compared with the percentage change in tax base from 1974/5 to 1978/9. These changes are classified into groups according to whether an area is above or below the median rateable value and median rate of change in rateable value for England and Wales as a whole. The spatial distribution of changes in rateable values displays a remarkably structured pattern. The authorities with high tax bases are largely concentrated either in London and the south east, or in a small number of metropolitan Districts. In contrast, the areas with low tax bases are the non-metropolitan Counties

Table 7.3. *F statistics calculated from three-way ANOVA of differences in rateable values resulting from administrative, political party, and urban stress factors*

	DRV	CRV	IRV	TRV
Main effects	17.41*	3.57*	2.28*	5.07*
Admin.	44.49*	10.92*	2.02	14.81*
Party	2.56+	0.21	4.11*	0.38
Inner	4.18*	1.99	0.67	2.49+
Interaction	–	–	–	–

*, significant at 95% level.
+, significant at 90% level.
CRV, commercial rateable value.
DRV, domestic rateable value.
IRV, industrial rateable value.

outside the south east, and the bulk of the metropolitan Districts except London and the west midlands. Turning to the areas with more rapidly increasing tax bases, these are also concentrated in most of the more central non-metropolitan Counties, and about half of the London Boroughs, and also characterise older city areas such as Manchester, Stockport, Newcastle-upon-Tyne, N. Tyneside, St. Helens, Wolverhampton, Walsall, Leeds, Wakefield, Doncaster and Cleveland. Authorities with only slowly increasing tax bases, in contrast, are the more peripheral and rural non-metropolitan Districts, half of the London Boroughs, and the majority of metropolitan Districts. Comparing this general pattern with the distributions for domestic and non-domestic rateable values in Figures 7.2 and 7.3, it can be seen that the high rateable value levels in London and the south east result from a combination of both high domestic and commercial property rate bases, but that many of the highest rates of increase in tax base have resulted from both domestic and commercial rate base increases in the less-populated areas.

The variable distribution of the tax base is a first measure of the inequity in ability of local authorities to fund services, before central grants are taken into account. (The second measure of inequity, that of expenditure need, has been discussed in the previous chapter.) The extent and geographical distribution of inequities in tax base can be seen in figure 7.6. This displays the pattern of those areas with various deficiencies in tax base (credited rateable value) from the standard rateable value used for Rate Support Grant purposes in 1978/9. A very clear pattern is evident in which the high tax base of London, the south east, west midlands and the inner city areas of Liverpool, Manchester, Newcastle, Leeds and Sheffield stand out. In contrast the areas with lowest tax capacity are in Wales and the north. For 1978/9 the areas which were above the standard rateable value

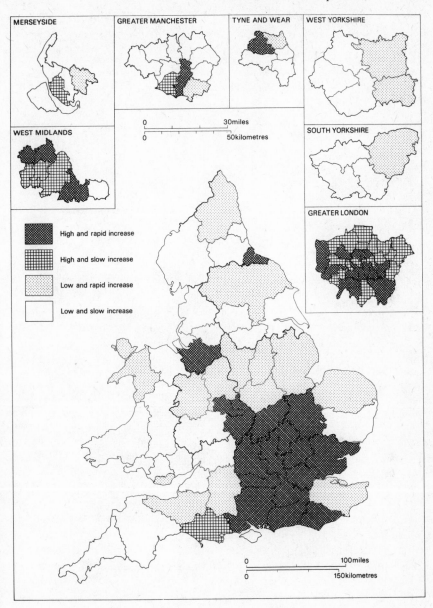

Fig. 7.5 Level of rateable value per head and change in total rateable value per head 1974–78 in categories relative to the median values.

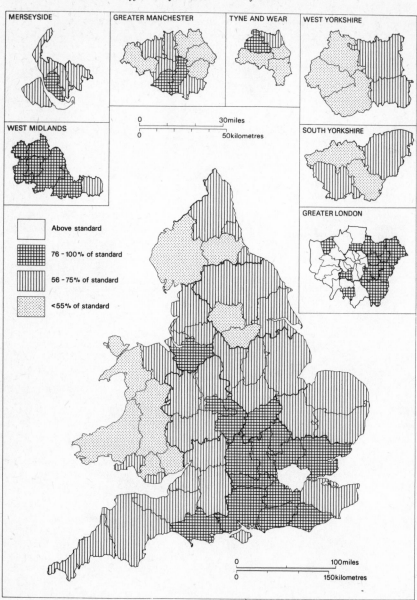

Fig. 7.6 Total rateable value per head in England and Wales relative to the standard rateable value per head used for Rate Support Grant purposes in 1978; areas above the standard received no resources element.

per head, and which hence received no resources grant for that year, are all concentrated in west London.

7.3 The expenditure position of local authorities

The expenditure position of local authorities in Britain is relatively easy to assess since, as a centralised state, the level and range of services is fairly uniform in similar categories of local authorities. For the purposes of allocating central grants (such as the Rate Support Grant), central government chooses to define 'relevant expenditure' as that expenditure which local authorities necessarily undertake, on behalf of central government, and which falls on the property tax. Relevant expenditure excludes expenditure supported by income received from user charges, sales, interest receipts and specific grants. Hence it is the expenditure which falls directly on the rate bill. This normally represents between 70 and 90% of total local expenditure (see table 4.2), and it is this definition of expenditure which is used exclusively below.

The expenditure of local authorities measures their total revenue requirements that must be met either by the local tax base or by grant transfers from central government. Whereas the rateable value of a jurisdiction measures the size of the tax base, the level of expenditure of a jurisdiction measures the size of the demands placed on that tax base, and it will be unusual for the two financial components to be in balance.

Between 1974/5 and 1980/1 the level of local relevant expenditures in England and Wales remained relatively stable, rising from only £12 046 m. to £12 389 m. (at 1978 prices). The level of total local expenditure declined from £16 175 m. to £13 833 m. at 1978 prices, but as shown in table 4.2, the percentage of local expenditure allowed by central government as relevant expenditure for Rate Support Grant purposes rose markedly from 74.5 to 86.8%. As a result of these changes total local expenditure reverted to roughly its 1972 level, but far more is taken into account for grant purposes.

The changing distribution of expenditure per head between different groups of local authorities is shown in table 7.4. This shows a great deal of variation both in terms of absolute values and rates of growth. The highest levels are in Inner London, but the highest rate of increase has been in the non-metropolitan Counties. In recent years this may be partly explained by party bias to be discussed later.

For recent years the spatial distribution of expenditure levels is shown in figure 7.7. The resulting pattern is complex. High expenditures per head characterise the south east, South Wales, much of Inner London, and many of the metropolitan Districts (especially in Tyne and Wear, South Yorkshire and part of Manchester and North Yorkshire). In most cases the areas with high 1978/9 expenditures are also the areas which have a high

Table 7.4. *Changes in the mean and standard deviation (in parentheses) of total expenditure per head for different administrative areas in England and Wales at constant 1980 prices*

Fiscal year	English Counties	Welsh Counties	County Boroughs or metropolitan Districts	Inner London	Outer London	Total
1962/3	101	138	120	114	17	103
	(15)	(28)	(21)	(31)	(4)	(39)
1965/6	138	177	137	224	126	149
	(11)	(36)	(40)	(151)	(13)	(73)
1967/8	140	187	196	251	185	248
	(178)	(40)	(166)	(168)	(14)	(724)
1969/70	150	189	181	104	160	200
	(8)	(33)	(19)	(22)	(12)	(447)
1971/2	160	203	220	305	220	335
	(9)	(35)	(163)	(236)	(18)	(1640)
1973/4	180	230	227	169	203	248
	(10)	(37)	(27)	(41)	(21)	(522)
1974/5	376.8	442.7	412.2	648.8	425.6	429.4
	(48.2)	(42.3)	(47.1)	(360.9)	(43.6)	(143.5)
1975/6	431.2	516.1	479.1	769.9	515.5	502.1
	(44.5)	(43.5)	(52.8)	(496.2)	(52.3)	(187.9)
1976/7	380.7	457.5	422.9	705.2	460.3	446.9
	(49.4)	(28.8)	(43.8)	(447.3)	(47.6)	(172.8)
1977/8	341.6	408.4	390.9	649.9	414.5	406.5
	(44.2)	(21.7)	(44.5)	(397.9)	(41.8)	(156.9)
1978/9	326.1	386.9	373.3	623.1	398.8	388.7
	(22.5)	(18.3)	(45.1)	(351.5)	(41.6)	(142.3)
1979/80	336.7	407.5	387.1	654.5	415.1	404.2
	(24.4)	(22.5)	(41.4)	(357.3)	(47.5)	(147.3)
1980/1	333.3	389.6	393.9	671.4	406.0	404.1
	(21.8)	(22.3)	(45.4)	(396.6)	(56.3)	(160.7)

rate of increase in expenditure. This may well reflect the frequently-observed feature of considerable inertia in spending patterns exhibiting cumulative expenditure growth and cumulative decline.

A comparison between the expenditure per head and rateable value per head, discussed in previous pages, gives an indication of the level of tax rate or central grant transfers required. Figure 7.8 makes such a comparison for the 1978/9 year. The authorities are grouped into four categories along the lines of the suggestion by Hicks and Hicks (1945)

(i)   *Rich spenders* authorities with large tax base and large expenditures, 'spending up to their income' either by providing for local needs, or overproviding for a low local need to spend.

(ii)   *Poor spenders* authorities with low tax base but large expenditure levels, reflecting either that they are 'spending beyond their means', or that there is a service resource/expenditure imbalance.

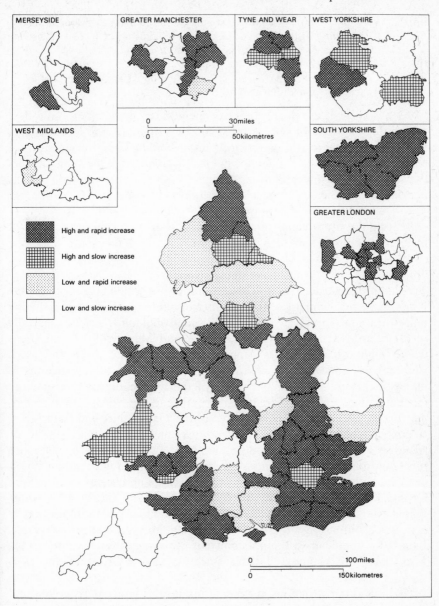

Fig. 7.7 Rate-borne expenditure per head 1978 and rate of change of rate-borne expenditure per head 1974–78 at constant prices relative to the median values.

(iii)   *Rich stinters* authorities with large tax base but small expenditure levels, reflecting either a choice of low tax rates allowing an increase in private incomes, or a low level of need to spend.

(iv)   *Poor stinters* authorities with a small tax base and low expenditure levels, reflecting either a choice to balance their books at a poor level of services, or a low level of need to spend.

Note that in each case the balance of expenditures and tax base can reflect very different levels of service provision depending on the local need to spend. Hence, there is a major problem in holding constant measures of service quantity, quality and scope in making comparison between jurisdictions. As a result interpretations of figure 7.8 must be made with care as interim judgements pending a more thorough analysis of the different factors underlying the determination of expenditure need.

Bearing these problems in mind, figure 7.8 shows a close association of high expenditures and high tax bases in most of the south east, the metropolitan Counties of Dorset, Warwickshire, Cheshire and Cleveland, in many London Boroughs and in the older city areas of Newcastle-upon-Tyne and Manchester. A close association of low expenditure and low tax bases characterises many of the non-metropolitan Counties, especially the more rural ones, and a small number of metropolitan Districts, especially in Greater Manchester, West Yorkshire and Merseyside. Those areas where high rateable values support low expenditure levels (the 'rich stinters') are concentrated in a remarkably small set of jurisdictions: a majority of London Boroughs, most of the west midlands metropolitan Districts, two inner city areas (Liverpool and Trafford) and the non-metropolitan Counties of Northamptonshire, Oxfordshire and Hampshire. In contrast the areas with high expenditures and poor tax base (the 'poor spenders') fall into three broad groups; first, a small number of the more rural non-metropolitan Counties; second, many of the more urban Counties especially in South Wales; and third, a large number of metropolitan Counties in South Yorkshire, West Yorkshire, Tyne and Wear, Greater Manchester and Merseyside. It can be expected that severe differences in tax rates resulting from these resource-expenditure differences will result. Especially significant is the possibility of achieving lower tax rates in the 'rich stinter' areas compared with the need for higher tax rates in the 'poor spender' areas.

The relation of variations in expenditure levels to local party colour and inner city status have been reported in Bennett (1982a). This analysis, using a three-way ANOVA, showed that expenditures bore no significant relation to degree of urban stress, but for most years differences in administrative area and party colour were statistically significant. The variations in expenditure for different areas and local party control are shown in table 7.5. This table shows the historically higher spending levels of London and Wales, but it also shows Labour Party control to have

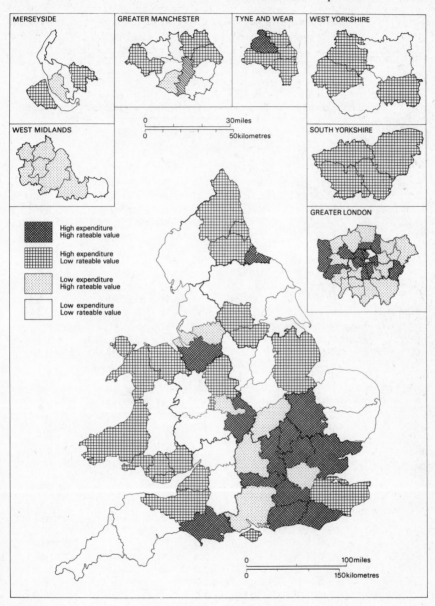

Fig. 7.8 Rate-borne expenditure per head in relation to rateable value per head in 1978 relative to median values.

Table 7.5. *Total expenditure per head at constant 1980 prices for different administrative areas and party control*

Fiscal year	Party colour	English Counties	Welsh Counties	Met. Districts and County Boroughs	Inner London	Outer London
1962/3	1	27.2(0.9)	–	33.1(1.6)	35.6(4.9)	N/A
	2	29.7(1.4)	32.5(1.7)	35.1(7.5)	32.4(9.9)	N/A
	3	29.3(6.4)	42.4(8.6)	30.6 –	–	N/A
	4	29.6(1.7)	36.9(7.4)	35.0(3.6)	–	N/A
1965/6	1	39.8(2.8)	–	38.4(1.7)	125.8(110.2)	34.1(4.8)
	2	40.6(2.2)	43.5(2.1)	41.8(14.6)	56.3(18.2)	40.5(2.1)
	3	42.1(4.1)	56.2(11.3)	38.8 –	–	36.9 –
	4	43.1(3.0)	51.9(13.9)	42.2(7.6)	–	39.0(1.6)
1967/8	1	47.7(3.3)	–	58.4(22.9)	89.8(65.1)	59.3(4.9)
	2	49.3(0.6)	50.2(1.6)	84.6(114.8)	63.5(16.4)	62.4(2.9)
	3	74.2(86.6)	64.2(13.6)	53.1(3.0)	–	–
	4	43.5(14.8)	53.6(10.1)	56.5(5.4)	60.9 –	–
1969/70	1	52.7(3.8)	–	64.0(5.5)	35.2(8.1)	55.7(3.9)
	2	55.3(0.2)	57.6(2.5)	66.3(6.9)	39.3(9.6)	62.6(0.5)
	3	52.7(2.6)	69.3(12.6)	62.8(4.7)	–	–
	4	52.8(3.9)	62.2(11.3)	59.4(10.6)	41.2 –	–
1971/2	1	67.4(4.4)	–	77.3(12.6)	279.1(208.8)	90.8(3.9)
	2	70.0 –	76.2 –	99.4(84.4)	98.9(36.5)	96.0(9.0)
	3	67.2(3.7)	89.1(16.1)	197.4 –	–	–
	4	69.3(5.6)	81.7(15.4)	80.2(5.8)	–	80.9 –
1973/4	1	92.5(5.5)	–	109.4(13.3)	82.8(9.7)	95.2(5.6)
	2	100.5 –	105.4 –	121.3(73.4)	87.8(23.0)	112.6(7.7)
	3	92.4(5.3)	123.4(20.7)	121.4 –	–	–
	4	94.5(7.7)	111.2(14.9)	107.1(8.9)	–	91.2 –
1974/5	1	131.4(8.2)	–	139.1(9.0)	412.5(256.8)	140.6(8.4)
	2	143.5 –	152.5(1.3)	152.1(18.8)	193.8(62.2)	162.1(14.9)
	3	123.7(6.3)	171.2(18.9)	181.8 –	–	–
	4	148.4(38.2)	142.9(4.2)	135.0 –	–	148.0(9.5)
1975/6	1	173.8(10.4)	–	186.6(16.2)	618.9(409.8)	197.9(13.1)
	2	191.5 –	210.7(12.3)	206.7(22.7)	256.4(89.2)	229.5(17.6)
	3	166.9(6.8)	224.3(22.2)	222.5 –	–	–
	4	197.7(35.3)	196.9(4.6)	189.2 –	–	197.6(0.7)
1976/7	1	189.8(11.2)	–	203.6(17.7)	690.1(474.3)	217.2(10.1)
	2	206.7 –	237.3(9.3)	227.9(20.8)	294.6(99.7)	256.5(18.9)
	3	176.9(4.9)	238.8(21.5)	238.0 –	–	–
	4	224.6(51.9)	221.5(3.0)	209.5 –	–	221.5(13.2)
1977/8	1	198.3(12.8)	–	217.9(17.9)	727.5(491.5)	226.8(10.9)
	2	211.6 –	247.8(1.5)	246.4(26.5)	319.0(104.6)	268.5(18.9)
	3	188.7(9.1)	245.2(20.4)	267.5 –	–	–
	4	234.1(53.8)	233.4(8.0)	221.4 –	–	239.4(4.2)
1978/9	1	227.7(14.2)	–	240.1(23.7)	781.3(497.8)	251.5(7.7)
	2	242.0 –	273.0(1.4)	365.3(35.1)	359.7(107.1)	300.6(23.0)
	3	214.9(11.2)	268.0(19.8)	288.1 –	–	–
	4	238.8(16.3)	256.7(1.5)	242.9 –	–	267.8(1.3)

Table 7.5 (*cont.*)

Fiscal year	Party colour	English Counties	Welsh Counties	Met. Districts and County Boroughs	Inner London	Outer London
1979/80	1	248.4(17.6)	–	271.5(20.0)	855.6(542.7)	282.6(9.1)
	2	279.7 –	313.2(7.1)	306.4(30.8)	415.9(136.4)	380.4(31.6)
	3	251.1(11.5)	305.9(23.7)	325.4 –	–	–
	4	262.9(19.1)	289.4(7.7)	264.9 –	–	298.6(2.5)
1980/1	1	284.2(14.7)	–	313.1(21.4)	1029.2(693.8)	314.4(14.3)
	2	307.9 –	345.8(12.8)	365.3(35.1)	489.4(173.6)	394.3(41.9)
	3	281.4(12.6)	336.0(23.8)	399.9 –	–	–
	4	297.1(34.6)	319.6(19.0)	321.2 –	–	335.5(5.7)

1, Conservative; 2, Labour; 3, Other; 4, No overall majority; N/A, not available.

maintained higher spending in almost all areas for a very long period of time. In Inner London this pattern is confused by the very low population levels of Westminster which give it high expenditure per head, even though it is Conservative. 'Other party' areas in Wales have generally high spending, whilst in England their spending levels are lower than the rest of authorities in their group. Clearly other factors, as well as party and administrative type, affect expenditure levels. But after controlling for need, grant and wage differences, the results in Bennett (1982a) confirm the statistical importance of the party effect.

7.4 The tax rate position of local authorities

The local rate poundage, or property tax rate, represents the ratio of the size of local expenditure requirements to the size of the local tax base (excepting the effect of intergovernmental transfers). As such, it represents what is frequently termed the tax *effort* of local government in providing for its expenditures from its own sources. Rate poundage in England and Wales increased by an average rate of over 9% per year between 1974/5 and 1978/9 in current price terms, but actually decreased in constant price terms. This reflects the relative inflexibility of the local tax base of the rates, the expansion of expenditures being increasingly made up by central grant transfers.

Table 7.6 shows the changing level of rate poundages exacted in different administrative areas since 1962. Comparisons are made difficult by revaluations of the tax base, but the table does show that, as a result of the combination of high expenditure levels with large tax base, London as a whole has tended to have lower rate poundages, but has experienced (with the exception of Outer London) some of the highest rates of increase

Table 7.6. *Changes in the mean and standard deviations of rate poundages (£ in £) for different administrative areas of England and Wales*

Fiscal year	English Counties	Welsh Counties	County Boroughs or metropolitan Districts	Inner London	Outer London	Total
1962/3	3.97	4.68	4.0	3.44	3.88	4.01
	(0.28)	(0.45)	(0.46)	(0.23)	(0.40)	(0.47)
1965/6	1.80	2.10	1.85	1.63	1.73	1.82
	(0.17)	(0.17)	(0.21)	(0.11)	(0.15)	(0.21)
1967/8	1.90	2.15	1.99	1.78	1.88	1.95
	(0.12)	(0.19)	(0.23)	(0.14)	(0.20)	(0.21)
1969/70	1.90	2.20	2.11	1.83	1.93	1.97
	(0.12)	(0.14)	(0.25)	(0.19)	(0.24)	(0.23)
1971/2	1.90	2.18	2.06	1.92	1.91	2.00
	(0.13)	(0.19)	(0.24)	(0.24)	(0.21)	(0.23)
1973/4	0.73	0.82	0.85	0.79	0.82	0.81
	(0.08)	(0.12)	(0.13)	(0.09)	(0.10)	(0.12)
1974/5	1.50	1.77	1.66	1.31	1.47	1.54
	(0.14)	(0.11)	(0.20)	(0.13)	(0.11)	(0.49)
1975/6	1.59	1.91	1.75	1.58	1.81	1.70
	(0.12)	(0.18)	(0.22)	(0.14)	(0.12)	(0.19)
1976/7	1.37	1.66	1.49	1.32	1.46	1.43
	(0.10)	(0.17)	(0.16)	(0.13)	(0.9)	(0.15)
1977/8	1.33	1.53	1.42	1.26	1.31	1.36
	(0.8)	(0.17)	(0.15)	(0.12)	(0.9)	(0.12)
1978/9	1.32	1.53	1.41	1.21	1.26	1.25
	(0.7)	(0.18)	(0.14)	(0.14)	(0.10)	(0.13)
1979/80	1.38	1.53	1.39	1.22	1.22	1.33
	(0.10)	(0.18)	(0.15)	(0.14)	(0.13)	(0.15)
1980/81	1.40	1.64	1.54	1.38	1.32	1.44
	(0.10)	(0.17)	(0.20)	(0.22)	(0.20)	(0.19)
1981/2	1.40	1.63	1.55	1.40	1.32	1.46
	(0.10)	(0.17)	(0.20)	(0.21)	(0.21)	(0.24)

Note that revaluations of rateable value occurred in 1963 and 1972. All figures are at constant 1980 prices and are before deduction of domestic element.
+estimate: CIPFA.

in poundage. The non-metropolitan Counties in contrast, with smaller tax bases and relatively high expenditure levels, have been forced to raise higher rate poundages, and also to increase these more rapidly. Comparing rate poundages and rate bases, it is clear that the rate yield has remained higher in English non-metropolitan Counties and Inner London Boroughs, whilst Welsh Counties, metropolitan Districts and Outer London Boroughs have been able (in constant price terms) to reduce the relative burdens they have placed on the local rates.

Relative differences in local tax rates, or tax effort, have been a

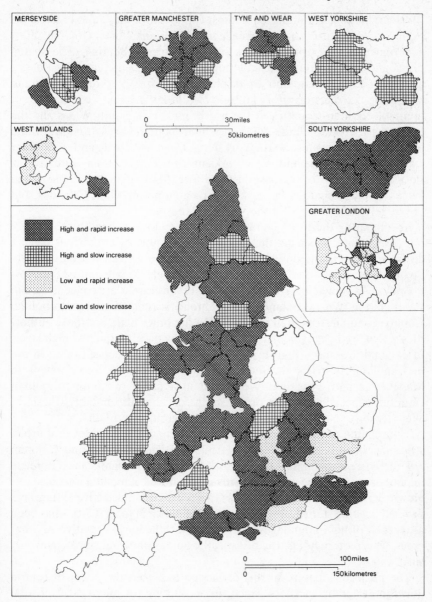

Fig. 7.9 Local tax rate (rate poundage) in 1978 and change in rate poundage 1974–78 at constant prices relative to median values.

considerable focus of attention in most countries. In the United States, for example, tax effort has been made an important element of US inter-governmental grant programmes such as Revenue-Sharing. The spatial distribution of differences in rate poundages for recent years are a major gauge of fiscal pressure and are displayed in figure 7.9. Areas of high and increasing tax rates or effort fall into two broad groups: first, a range of non-metropolitan Counties in the north, midlands, South Wales and the south east; second, a large number of metropolitan Districts concentrated in Greater Manchester, South Yorkshire, Tyne and Wear, and, to a lesser extent, in Merseyside and West Yorkshire. In contrast, areas with low tax rates and low rate of increase in tax effort (lower than the England and Wales median) characterise a scatter of non-metropolitan Counties, but are largely concentrated in the west midlands and London.

In comparing the three features of tax base, local expenditure, and tax rates, therefore, it seems that two overall groups of authorities can be distinguished. On the one hand are those authorities which are extremely favoured by a large tax base, and variable levels of expenditure, but low tax rates. As shown in figures 7.10 and 7.11 these areas are concentrated in London, the West Midlands and a broad scatter of non-metropolitan Counties and metropolitan Districts. On the other hand are those author-ities which are heavily penalised by a small tax base, frequently high levels of expenditure and a high tax rate. These areas are concentrated in the metropolitan Districts of South Yorkshire, and to a lesser extent in Manchester and Tyne and Wear, and in a range of non-metropolitan Counties especially in parts of the south east, but also notably Cheshire, Warwickshire, Northamptonshire, Cambridgeshire and Dorset.

The relation of rate poundages to local party control and level of urban stress has been reported by Bennett (1982a) and Harrison and Jackman (1978). These results show that up to 1980 there is no significant difference in poundages between areas once covariates have been taken into account. However, since 1980/1 urban stress, party and administrative differences have all become statistically significant. As shown in table 7.7 this has been as a result of high poundages in Labour-controlled areas relative to other areas. This pattern holds true historically, but is significant statistically only since 1980/1.

The neglected feature in this discussion has been the effect of central government grant transfers which, in accounting for about 50% of total local expenditures over the period considered, have enormous signifi-cance. However, it is already clear from the analyses above that high local expenditures in areas with low tax rates, although having variable tax bases, can be supported only with major redistributive effects deriving from the manner in which central grants are allocated. In contrast to the North American situation, greater or lesser fiscal pressure has resulted to a large extent from major involvement by central government overlain on

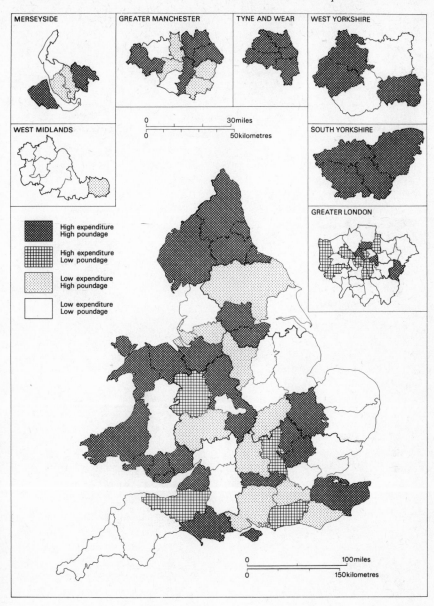

Fig. 7.10 Local rate poundage in relation to rate-borne expenditure in 1978 relative to median values.

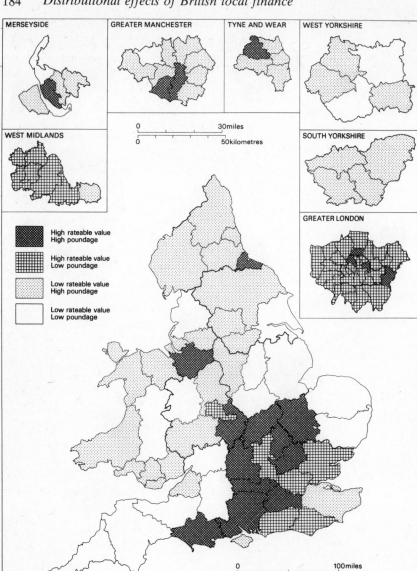

Fig. 7.11 Local rate poundage in relation to rateable value in 1978 relative to median values.

the tax base – local expenditure balance. More detailed discussion of the effects of these central grants is given in the next chapter.

7.5 Capital and debt position of local authorities

This book is concerned primarily with the revenue account of local authorities relating to current budgets and spending financed largely by taxation. The other main component of local finance, the capital account, is financed by various forms of borrowing. Although in theory independent of the revenue account, the capital account does in practice become interrelated in three main ways (see Bennett, 1980). First, local authorities frequently borrow during the year to even out the flows of tax receipts. This is borrowing which will be made good by revenue receipts at the year end. A second source of interrelation is the need for revenue income both to service the interest charges of debt burdens and to fund the revenue side of capital projects (e.g. salaries in schools are a revenue account burden whilst school building is a capital account burden). A third source of interrelation is the use of loans to finance current expenditure. This latter policy is illegal and largely impossible in Britain; however, it has been the cause of financial crises in North American cities like New York.

As a result of these interrelationships, capital finance cannot be ignored in studying the revenue position or grant position of local authorities. Capital finance represented about 20% of total local finance in 1980/1, or approximately £4000 m. The main sources of this income, as shown in figure 3.4, are loans, grants, and sales of assets. Loans represent about two-thirds of the total and derive from both central and local sources. Grants represent about one-quarter of the total and derive from central government through various schemes.

The overall level of local authority debt has steadily declined in Britain since a peak in 1973/4 (see figure 3.5). However, the position is strongly variable between local authorities. Figure 7.12 shows the variation in the level of total local debt for the relevant class of local authority but with non-metropolitan District debt not aggregated to the County level. Since housing represents the largest single component in the capital account it is not surprising that the highest levels of debt per head are recorded in those areas where public housing policy has been pursued most vigorously. Inner London, South Wales, the north east, the north west, most metropolitan Districts and the old County Boroughs stand out with the highest debt levels. The high debt levels also recorded in some non-metropolitan Counties are associated either with older, pre-reorganisation County Boroughs or are areas of planned overspill and new towns (see Smith 1983).

The total level of debt is not necessarily correlated with its total costs. However, in Britain the level and costs of debt are closely interrelated.

Table 7.7. *Average rate poundage (£ in £) at constant 1978 prices for different administrative areas and party colours*

Fiscal year	Party colour	English Counties	Welsh Counties	Met. Districts and County Boroughs	Inner London	Outer London
1962/3	1	1.12(0.06)	–	1.01(0.16)	0.93(0.05)	1.19(0.08)
	2	1.12(0.02)	1.32(0.09)	1.21(0.11)	1.01(0.06)	1.15(0.14)
	3	1.14(0.10)	1.37(0.11)	0.95 –	–	1.01 –
	4	1.16(0.06)	1.25(0.03)	1.14(0.13)	–	1.11(0.10)
1965/6	1	0.53(0.02)	–	0.50(0.06)	0.44(0.01)	0.52(0.03)
	2	0.59(0.13)	0.65(0.02)	0.57(0.06)	0.50(0.02)	0.53(0.05)
	3	0.53(0.03)	0.63(0.04)	0.47 –	–	0.50 –
	4	0.54(0.03)	0.64(0.13)	0.55(0.05)	–	0.52(0.05)
1967/8	1	0.60(0.63)	–	0.63(0.07)	0.56(0.05)	0.60(0.06)
	2	0.63(0.01)	0.76(0.69)	0.67(0.08)	0.60(0.02)	0.65(0.03)
	3	0.60(0.03)	0.68(0.05)	0.58(0.04)	–	–
	4	0.63(0.25)	0.67(0.01)	0.65(0.06)	0.61 –	–
1969/70	1	0.66(0.63)	–	0.70(0.09)	0.61(0.04)	0.67(0.08)
	2	0.70(0.02)	0.83(0.01)	0.72(0.08)	0.69(0.06)	0.77(0.03)
	3	0.65(0.04)	0.77(0.03)	0.68(0.06)	–	–
	4	0.69(0.04)	0.73(0.08)	0.71(0.07)	0.75 –	–
1971/2	1	0.82(0.04)	–	0.82(0.105)	0.70(0.024)	0.79(0.085)
	2	0.87 –	1.02 –	0.89(0.103)	0.83(0.096)	0.83(0.091)
	3	0.78(0.06)	0.90(0.07)	0.80 –	–	0.73
	4	0.81(0.01)	0.93(0.03)	0.83(0.053)	–	0.78 –
1973/4	1	0.40(0.03)	–	0.43(0.064)	0.34(0.005)	0.39(0.029)
	2	0.41 –	0.50 –	0.45(0.066)	0.42(0.033)	0.44(0.051)
	3	0.36(0.04)	0.40(0.047)	0.47 –	–	0.38 –
	4	0.37(0.03)	0.44(0.051)	0.38(0.053)	–	0.30 –
1974/5	1	0.52(0.03)	–	0.56(0.04)	0.39(0.01)	0.50(0.03)
	2	0.66 –	0.66(0.02)	0.62(0.08)	0.48(0.04)	0.53(0.04)
	3	0.63(0.04)	0.62(0.04)	0.65 –	–	–
	4	0.57(0.08)	0.59(0.01)	0.53 –	–	0.56(0.01)
1975/6	1	0.65(0.03)	–	0.68(0.07)	0.58(0.03)	0.72(0.05)
	2	0.76 –	0.85(0.08)	0.76(0.10)	0.67(0.05)	0.78(0.04)
	3	0.63(0.04)	0.75(0.06)	0.69 –	–	–
	4	0.67(0.06)	0.76(0.01)	0.70 –	–	0.74(0.00)
1976/7	1	0.70(0.04)	–	0.72(0.07)	0.62(0.08)	0.72(0.05)
	2	0.78 –	0.93(0.06)	0.81(0.08)	0.69(0.06)	0.78(0.04)
	3	0.64(0.03)	0.77(0.07)	0.71 –	–	–
	4	0.72(0.06)	0.84(0.02)	0.74 –	–	0.73(0.03)
1977/8	1	0.79(0.05)	–	0.80(0.07)	0.69(0.07)	0.74(0.03)
	2	0.81 –	1.00(0.02)	0.90(0.08)	0.76(0.07)	0.81(0.05)
	3	0.74(0.05)	0.81(0.08)	0.82 –	–	–
	4	0.80(0.05)	0.92(0.03)	0.79 –	–	0.82(0.06)
1978/9	1	0.87(0.05)	–	0.86(0.07)	0.72(0.04)	0.75(0.04)
	2	0.90 –	1.07(0.03)	0.97(0.10)	0.81(0.07)	0.82(0.05)
	3	0.82(0.07)	0.88(0.10)	0.86 –	–	–
	4	0.85(0.05)	0.99(0.07)	0.84 –	–	0.83(0.11)

Table 7.7 (*cont.*)

Fiscal year	Party colour	English Counties	Welsh Counties	Met. Districts and County Boroughs	Inner London	Outer London
1979/80	1	0.99(0.05)	–	0.99(0.07)	0.75(0.03)	0.85(0.07)
	2	1.08 –	1.25(0.08)	1.11(0.12)	0.93(0.09)	0.98(0.08)
	3	0.99(0.09)	1.05(0.12)	1.01 –	–	–
	4	0.94(0.06)	1.12(0.13)	0.92 –	–	0.95(0.12)
1980/1	1	1.04(0.06)	–	1.04(0.08)	0.81(0.03)	0.88(0.10)
	2	1.17 –	1.34(0.08)	1.25(0.13)	1.07(0.14)	1.10(0.12)
	3	1.04(0.07)	1.14(0.06)	1.25 –	–	–
	4	1.03(0.13)	1.20(0.19)	1.08 –	–	0.98(0.11)
1981/2[+]	1	1.40(0.07)	–	1.40(0.10)	1.08(0.05)	1.17(0.13)
	2	1.57 –	1.79(0.10)	1.67(0.18)	1.44(0.19)	1.47(0.16)
	3	1.39(0.09)	1.52(0.08)	1.68 –	–	–
	4	1.38(0.17)	1.60(0.25)	1.45 –	–	1.31(0.15)

1, Conservative; 2, Labour; 3, Other; 4, No overall majority.
[+], estimate: CIPFA, 1980 prices.

Figure 7.13 shows the total costs of local debt as measured by total interest charges. The highest levels of repayments are in the four main areas which also have the highest levels of total debt: first, Inner London and a few Outer London Boroughs; second, older industrial areas and inner cities in the Metropolitan areas; third, new and expanded areas in the non-metropolitan Counties; fourth, some areas of very low population, mainly in Wales, the north east and East Anglia. More detailed analysis of the spatial variation of level of debt costs for different service categories and for different sources of borrowing is given by Smith (1983).

When we turn to analysis of the interrelation of total local debt with party colour and other factors, Smith (1983) and Foster *et al.* (1980) have demonstrated that local party control is a very significant determinant of debt for almost all services. Of the services, housing and personal social service investments are the most closely related to party effects, being much higher in Labour-controlled areas. Indeed even the comparison of figure 7.12 with figure 1.2 for party control in 1978 shows a close relation between party colour and debt level.

7.6 Conclusion

This chapter has presented a preliminary analysis of the resource and total expenditure positions of local authorities in England and Wales. These components are inter-dependent with the influence of transfers of re-sources from central government, especially in the form of the Rate Support Grant, and with the overall debt position of local authorities. It

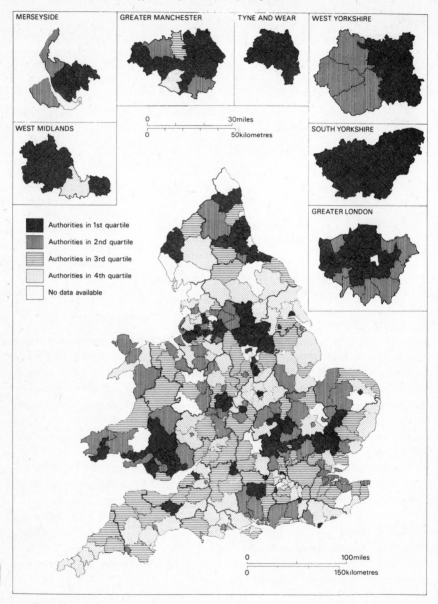

Fig. 7.12 Outstanding debt per head in 1978 in quartile groups. Debt is plotted for the relevant class of authority.

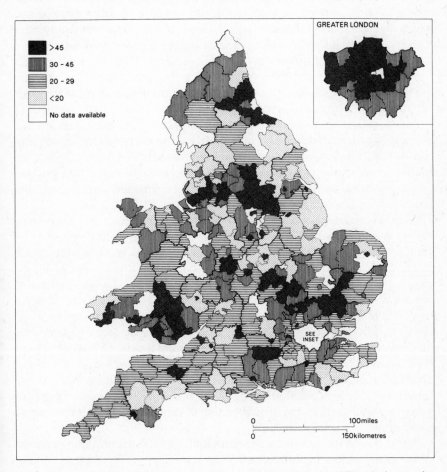

Fig. 7.13 Total interest charges on total outstanding debt in 1978 relative to the mean interest charge level of £29.8 per head.

has been attempted here to control for these features, but they are the more detailed subject of later chapters.

Despite the preliminary nature of the present results, it is possible here to reach a number of interim conclusions. First, there does seem to be a fairly clear distinction between, on the one hand, resource-rich, low tax rate areas in London, the West Midlands and some other areas, and on the other hand, resource-poor, high tax rate, high expenditure areas in the north east and most of the metropolitan Districts outside London and the west midlands. Second, it is clear that the concept of urban fiscal crisis, which has been formulated in the United States, does not apply very directly to England and Wales, at least on the level of analysis adopted here. The tax base of urban stress areas is relatively high, expenditures are high, and both debt and the average rate poundage, although higher, do not differ greatly or significantly from the non-metropolitan and other urban areas. Moreover, recent changes since 1974/5 have tended to decrease rather than increase these differences. This may indicate the success of urban programmes, particularly those deriving from manipulation of the Rate Support Grant. However, it is clear that *in fiscal terms*, the major urban stress areas which remain are probably not in the governments covered by the existing Inner Cities Policies, but are in the metropolitan Districts which do not receive specific urban aid.

A third conclusion derives from the analysis of the resource, expenditure and debt positions to discover the influence of party colour reported by Bennett (1982*a*) and Smith (1981). For these fiscal variables the major differences seem to be that Conservative-controlled areas have the lowest expenditure and debts and the highest rateable values. Labour-controlled areas have the highest expenditure and debt. However there is some evidence of convergence in expenditures and rateable values in the areas controlled by the two main parties. In addition those areas governed by other parties and no overall majority, do present a major constrast. These tend to have higher expenditure levels in Wales and lowest expenditures in England, together with lower and more slowly increasing tax rates. Since these can be only accounted for partly by their higher tax base, this result does suggest that there has been direction of central grants preferentially to these areas.

A fourth conclusion relates to rate poundages. Even after taking account of the major modifying influences of the Rate Support Grant, these do not differ significantly between administrative areas and areas of party control except for the most recent years studied, since 1980/1. These years were the first to be affected in any major way by the Conservative central government. It seems therefore that the effect of local finance on tax rates over the years up to 1979/80 has been one which has overall induced great similarities in rate poundages irrespective of location. In achieving this, the Rate Support Grant must be accorded a primary role.

Each of these conclusions is preliminary and tentative. However, they do indicate that considerable differences in the resource and expenditure position of local governments exist, and that local party factors may be important as one determinant of these differences.

CHAPTER 8

THE DISTRIBUTION OF THE RATE SUPPORT GRANT

8.1 Introduction

The two previous chapters have been concerned with evaluating the resource, expenditure and need positions of local authorities. These chapters have provided estimates of the main characteristics which give rise to variation between local authorities and hence which must be held constant when attempts are made to assess distributional properties of the Rate Support Grant. The present chapter uses the results of these previous chapters in order to differentiate those aspects of the distribution of the Rate Support Grant, on the one hand, which arise from the 'objective' components of need and expenditure variation, from those aspects, on the other hand, which arise from central political party and local political party effects.

The discussion of this chapter falls into three sections. First, in section 8.2, the historical pattern of allocation of the resources element to local authorities is discussed. This is followed in sections 8.3 and 8.4, respectively, by comparisons of allocations of the needs element and total RSG between classes of authorities.

8.2 The resource element position

The distribution of the resource-equalising grants remained as a fairly constant element of central grants in Britain since their inception with the Rate Deficiency Grant in 1958. Significant changes in grant allocation occurred in 1967 with the introduction of the RSG resource element, and again in 1981 with the inclusion of resource equalisation in the block grant to a constant level of need. However, the basic distribution criterion of allocating grants to local authorities in proportion to their deficit in rateable value per head below national standard has remained unchanged.

The distribution of the resources element per head in different administrative areas is shown in table 8.1. Three features are clearly evident. First, the shift from the Rate Deficiency Grant to the RSG was very significant in increasing the level of grant received by the urban areas outside London, such that by 1969 their relative ranking for per capita resource grants rose

Table 8.1. *Means and standard deviation (in parentheses) of level of per capita RSG resources element (or Rate Deficiency Grant up to 1966) in different administrative areas at constant 1980 prices*

Fiscal year	English Counties	Welsh Counties	Metropolitan Districts or County Boroughs	Inner London	Outer London	Total
1962/3	19.1	60.6	7.8	0.6	0.3	13.5
	(13.1)	(37.3)	(8.8)	(0.9)	(0.8)	(20.3)
1965/6	12.0	35.7	14.4	0.4	—	12.6
	(7.6)	(23.5)	(16.7)	(1.4)	—	(16.0)
1967/8	51.4	91.0	12.3	0.5	—	26.5
	(106.7)	(53.2)	(15.2)	(2.0)	—	(63.5)
1969/70	12.3	35.9	13.1	0.7	—	12.1
	(7.9)	(21.9)	(13.8)	(2.4)	—	(14.5)
1971/2	12.6	38.4	15.6	0.8	—	13.5
	(7.7)	(22.4)	(16.9)	(2.9)	—	(16.3)
1973/4	12.9	40.7	18.6	0.9	0.5	15.2
	(8.3)	(22.6)	(18.2)	(3.1)	(0.2)	(17.5)
1974/5	62.0	127.9	90.3	7.9	6.7	60.2
	(29.2)	(34.6)	(44.4)	(16.3)	(12.7)	(49.1)
1975/6	84.9	156.0	111.5	13.6	19.1	79.3
	(29.4)	(34.8)	(44.1)	(25.5)	(24.6)	(54.4)
1976/7	75.0	136.1	97.1	11.2	17.7	69.6
	(24.5)	(29.2)	(36.8)	(20.9)	(20.7)	(46.6)
1977/8	65.5	122.0	88.3	8.8	13.4	61.5
	(23.7)	(27.8)	(34.7)	(16.4)	(17.9)	(43.3)
1978/9	63.3	110.5	81.9	9.2	13.4	58.1
	(21.5)	(23.6)	(30.7)	(17.0)	(16.5)	(39.2)
1979/80	58.7	102.5	76.1	8.5	12.4	53.9
	(19.9)	(21.9)	(28.5)	(15.8)	(15.3)	(36.2)
1980/81	58.2	107.7	76.2	7.3	11.3	53.8
	(22.2)	(24.8)	(31.3)	(14.1)	(15.4)	(38.4)

above the rural areas for good. This is largely a result of abandoning the population density and road mileage weighting to resources applied in the Rate Deficiency Grant (chapter 3, q.v.) which aided the Counties. A second feature is the narrowing of the differences between urban and rural areas from 1975/6 onwards. But a third major feature is the entry of London into resource equalisation, receiving a significant component from 1974/5 onwards. This is as a result of the progressive raising of the standard rateable value per head (see Table 4.7) allowing a large number of authorities to partake in resource equalisation.

The more detailed pattern of the distribution of the resources grant can be seen from figure 8.1, which displays the level of resources element per head and the rate of increase in grant per head from 1974/5 to 1978/9 in relation to the England and Wales medians. It is clear that the resource

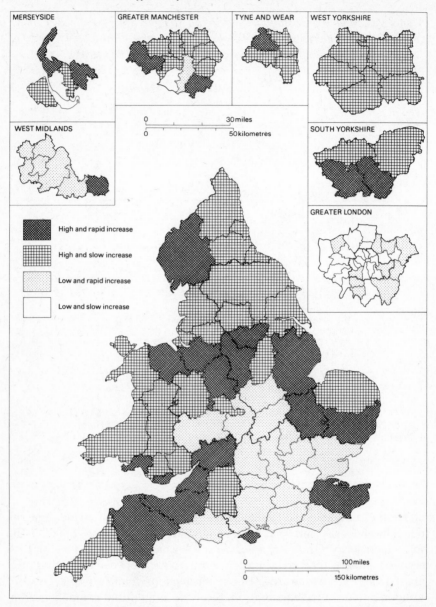

Fig. 8.1 Resources element of RSG in 1978 and rate of change in resources element 1974–78 at constant prices per capita relative to median values.

grant favours those areas with low tax bases: the rural and non-metropolitan Counties outside the midlands and south east, and also the metropolitan Districts outside London. The areas with the highest rates of increase in grants per head over this period are, however, in Outer London and the south east, together with a small number of metropolitan Districts and the non-metropolitan Counties bordering the midlands and south east. Whilst these differences between administrative areas are important, in an analysis of the relation of the resources element to party colour and inner city status, Bennett (1982c) found that neither were significant factors statistically for most years.

The significance of this distribution of resources equalisation can be assessed, in part, by its relation to expenditure per head, as shown in figure 8.2. This displays the very marked importance of the resources grant to the Welsh and north of England Counties, as well as the metropolitan Districts in South and West Yorkshire, Tyne and Wear and Greater Manchester. However, the resources grant is very significant to most of the non-metropolitan Counties outside the south east. The west midlands, south east and London stand out as areas which, because of their high rateable values, have a low level of support through the resources element.

The aggregate relation of the resources element to local rateable values, rate poundages and relevant expenditures is shown in table 8.2. The relationship of resources grant to rateable values is very close and inverse, as expected, i.e. as the tax base rises the resources grant diminishes. The elasticity of the resources element to local rate poundages is strongly positive and has increased steadily in recent years showing that grant is largely proportional to the rate poundage levied. In addition, the relationship of the resources element to local expenditures is also strongly positive. These two relationships may result from local governments exploiting the grant structure to obtain increased grant by raising higher levels of local expenditure at a high rate poundage. Alternatively, the reverse may be true: central grants may encourage a supply expansion of services because of their relative underpricing. Yet again, areas with low rateable values may correspond to areas with high spending need. Hence firm conclusions must await the results of simultaneous-equation estimation with the expenditure model (chapter 9). Table 8.2 also displays the relation of the resources grant to the level of average local wages. As in the case of rateable values, there is the desired inverse relationship, that grant levels decline with the levels of local incomes, but the degree of spatial progressivity has steadily declined over time. This decline in income redistribution effect is probably accounted for mainly by the Labour government of 1974–79 shifting grant allocations away from non-metropolitan to metropolitan Districts and London, which, despite frequently having severe environmental problems, are also characterised by high levels of personal income.

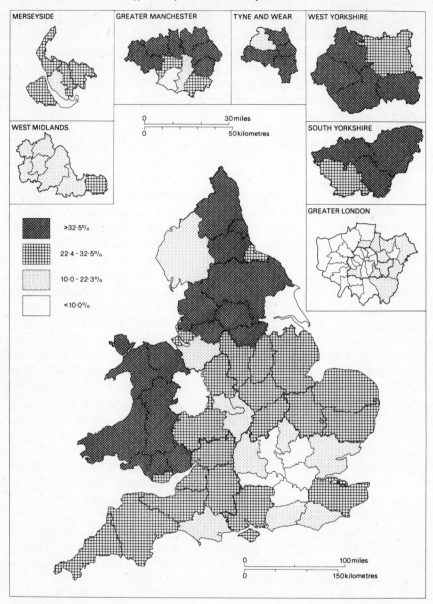

Fig. 8.2 Resources element per head in 1978 as a percentage of rate-borne expenditure per head in quartile groups.

Table 8.2. *Allocation of the resources element of the Rate Support Grant in relation to tax base, tax rate, expenditure and wages*

Fiscal year	Resource grant/head vs. RV/head	Resource grant/head vs. assessed rate poundage	Resource grant vs. relevant expenditure	Resource grant/head vs. average local wages
Rate Deficiency Grant				
1962/3	−0.208*	0.019*	0.193*	—
1965/6	−0.001	0.029	0.011	—
1967/8	−0.219	0.027*	−0.771	—
1969/70	−0.205*	0.205*	0.227*	—
1971/2	−0.251*	0.027*	0.219*	—
1973/4	−0.109*	0.016*	0.212*	—
1974/5	−0.640*	1.549*	0.258*	−2.697*
1975/6	−0.720*	0.732*	0.271*	−2.601*
1976/7	−0.738*	1.188*	0.200*	−2.544*
1977/8	−0.834*	1.550*	0.173*	−2.419*
1978/9	−0.846*	1.696*	0.147*	−2.153*
1978/80	−0.442*	1.507*	0.168*	−2.624*
1980/81	−0.524*	0.042*	0.169*	−3.205*

The result show the simple regression slope coefficient and exclude the City of London, and Westminster and all cases exclude areas receiving no grant.
** significant at 90% level or higher.

8.3 The needs element position

The distribution of the needs-equalising component of central grants in Britain has been subject to rapid changes, and great controversy. Prespecified needs indicators used in the General Grant up to 1966 and RSG needs element up to 1973 were replaced by regression indicators in 1974, only to be replaced by separate service categories of indicators in 1981. As a result of these major changes, and the host of minor changes discussed in previous chapters, it can be expected that the distribution of the needs element will show a relatively high degree of volatility from year to year.

This expectation is largely confirmed, as shown in table 8.3, for the distribution of needs element between different administrative areas. Year on year variations in needs allocations are small up to 1974. The relative rankings of different administrative areas are largely constant and carry over from the distribution of the General Grant. The major change is that Outer London markedly overtakes Inner London in its per capita allocation after 1971. From 1974 onwards, however, changes become more rapid. Outer London is overtaken by Inner London in 1974/5, a position only to be narrowed from 1978/9 onwards. Most significant, however, both

Table 8.3. *Mean and standard deviation (in parentheses) of levels of per capita RSG needs element (or General Grant up to 1966) in different administrative areas at constant 1980 prices*

Fiscal year	English Counties	Welsh Counties	Metropolitan Districts or County Boroughs	Inner London	Outer London	Total
1962/3	40.8	53.1	37.9	N/A	N/A	40.3
	(3.9)	(5.5)	(5.5)			(6.6)
1965/6	49.1	60.9	45.8	43.3	40.4	47.2
	(4.4)	(6.8)	(7.4)	(3.0)	(5.6)	(7.2)
1967/8	70.5	97.9	62.3	62.4	60.4	67.2
	(7.0)	(22.3)	(5.4)	(3.5)	(3.6)	(12.6)
1969/70	75.8	100.2	68.1	68.5	64.5	72.4
	(6.9)	(20.1)	(6.1)	(4.3)	(5.8)	(12.0)
1971/2	89.2	113.7	75.2	41.8	73.4	79.2
	(8.5)	(23.0)	(13.4)	(12.1)	(4.2)	(19.6)
1973/4	98.9	123.2	93.0	54.5	87.7	93.6
	(9.3)	(22.3)	(11.4)	(4.0)	(6.5)	(17.7)
1974/5	150.4	183.7	165.5	239.8	178.1	171.6
	(17.6)	(54.0)	(26.7)	(44.7)	(23.4)	(38.2)
1975/6	157.3	195.7	184.3	231.4	186.7	181.3
	(17.7)	(53.8)	(26.8)	(35.8)	(29.1)	(35.4)
1976/7	133.4	169.9	161.6	234.4	187.6	164.7
	(16.6)	(46.2)	(25.5)	(42.3)	(29.9)	(41.1)
1977/8	112.4	150.6	149.5	209.4	172.0	147.2
	(18.4)	(42.3)	(29.1)	(44,7)	(30,3)	(41.9)
1978/9	103.0	138.5	140.2	219.6	182.7	143.1
	(18.0)	(38.2)	(28.1)	(42.9)	(33.4)	(47.4)
1979/80	104.1	140.5	142.0	232.3	185.6	146.1
	(18.7)	(34.9)	(29.3)	(36.9)	(30.6)	(49.1)
1980/81	77.8	101.3	122.3	193.1	156.7	119.1
	(13.8)	(27.6)	(24.5)	(28.6)	(24.2)	(43.6)

the English and Welsh non-metropolitan Counties are overtaken by the London Boroughs, and for some years by the metropolitan Counties.

Comparisons of pre- and post-reorganisation local authorities are difficult, but whatever allowances are made for this factor, it seems clear that the Counties have been kept at relatively constant grant levels. The Districts and London Boroughs, on the other hand, have benefited from greatly increased per capita grant levels: some 200–300% for Inner London, 150–180% for Outer London, and 110–150% for the metropolitan Districts. In large measure this has been a result of the increased use of 'social need' indicators in the needs element allocation formula (see table 4.8). Variables such as overcrowded households, houses lacking basic amenities and one parent families have entered the needs formula from 1975/6 onwards, and were partially included in the 1974/5 formula through social service units. These social need variables all tend to indicate higher service requirements in city rather than rural County areas. Hence, as table 8.3 shows, the historic rise in RSG in 1973/4 and 1974/5 has largely gone to give additional funding to city needs whilst support for County needs has been reduced a little (to about 80% compared with 1973/4, and to about 90% compared with 1971/2).

One further feature of table 8.3 should also be pointed out; the rapid reduction in per capita receipts from needs elements for 1980/1. Coinciding with the first RSG settlement of the Conservative Government elected in 1979, this demonstrates the clear intention to reduce the level of central support for local services, and to reduce in aggregate the level of local spending. Comparison with table 8.1 demonstrates that cuts in the total level of grant support have been at the expense of need rather than resource equalisation. This in turn has tended to favour the Counties and metropolitan Districts when we look at aggregate RSG receipts.

The distribution of the needs element between areas, and the rate of increase in needs grant per head from 1974/5 to 1978/9 compared with the national median is displayed in figure 8.3. This figure displays the marked concentration of the needs grant in the London Boroughs, the metropolitan Districts (in Merseyside, Greater Manchester, Tyne and Wear, West Yorkshire and the West Midlands), and the non-metropolitan Counties of Nottinghamshire, Humberside and in Wales. Moreover, there is a very close correspondence between those areas with a high needs element and a high rate of increase in needs element, the Welsh Counties being the only major area with high levels of entitlement and relatively low rates of increase from 1974 to 1978. In an analysis of the relation of the distribution of the needs element to differences between administrative areas, party colour and inner city status, Bennett (1982c) found only administrative area and inner city status to be important for most years. This conclusion has particular significance in comparison with the USA. In Britain the needs element has been allocated on a massive scale towards cities in general, and inner city areas in particular. This result, combined with the high tax base of most of these areas, has prevented urban budget problems of the extent which have occurred in the USA. The shift in grant form after 1980, however, has reversed some of these characteristics, as we shall see below.

The relative importance of the needs grant to each authority can be judged by its relation to the total expenditure level of each area. As shown in figure 8.4 for those areas with high needs grants, the proportion of spending made up by the grant is in almost all cases over 50% of local relevant spending, and in many cases is over 90%.

The aggregate relation of the resources element to the tax base, tax rate and expenditure levels is shown in table 8.4. Since the needs element is allocated by taking account of the differing population levels, density,

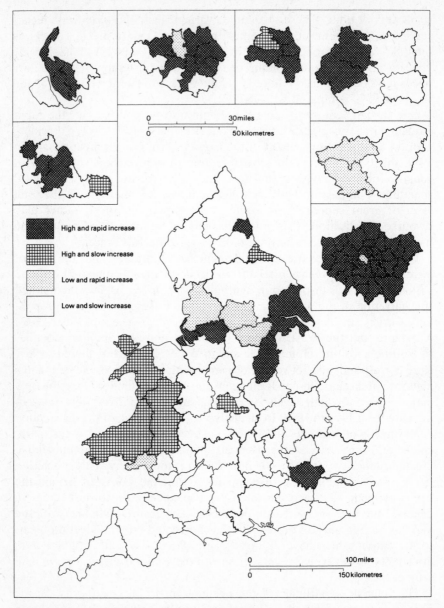

Fig. 8.3 Needs element of RSG in 1978 and rate of change in needs element 1974–78 at constant prices per head relative to median values.

Table 8.4. *Allocation of the needs element of the Rate Support Grant in relation to tax base, tax rate, expenditure and wages*

Fiscal year	Needs grant/ head vs. RV/head	Needs grant/ head vs. assessed rate poundage	Needs grant vs. resources expenditure	Needs grant/ head vs. average local wages
General grant				
1962/3	−0.389*	0.007*	0.069*	—
1965/6	0.227*	−0.415*	4.983*	—
1967/8	−0.109*	0.016*	0.012*	—
1969/70	−0.111*	0.014*	0.133*	—
1971/2	−0.241*	0.013*	0.170*	—
1973/4	−0.104*	−0.007	0.230*	—
1974/5	0.137*	−0.185	0.743*	0.853*
1975/6	0.109*	0.176	0.671*	0.830*
1976/7	0.214*	0.025	0.694*	1.639*
1977/8	0.226*	−0.261	0.695*	1.947*
1978/9	0.333*	−0.954*	0.737*	2.872*
1979/80	0.371*	−0.599*	0.726*	3.353*
1980/81	0.388*	0.062	0.684*	3.761*

The data exclude City of London, Westminster and Isles of Scilly in all cases and are based on determining the simple regression slope coefficient.
* significant at the 90% level or greater.

spending needs in education and old people's services, and differences in labour costs and other factors, it can be expected that its allocation will be closely related to aggregate levels of expenditure. However, it should not be closely related to the rateable value of a local government, except insofar that need and local tax resources are themselves inversely related. Table 8.4 displays the expected result that up to 1973/4, the needs element was *inversely* related to the tax base, but with the switch to the regression formula in 1974/5 it is clear that the needs grant has increasingly favoured areas with high tax bases. Hence, it has progressively worked against the aims of the resource equalisation and the resource element. However, the grant has generally favoured areas in relation to their expenditure. Up to 1973/4 there was an increasing tendency to distribute in favour of high expenditure levels. However, since 1974/5 the needs element has stayed roughly constant in the proportion of funding it supports at higher expenditure levels. In contrast, for rate poundages, it seems that the needs grant, although closely related to rate poundages up to 1971/2, has grown in the extent to which it is inversely related since then, but there is little general pattern. This feature is also reinforced by the relation of the needs

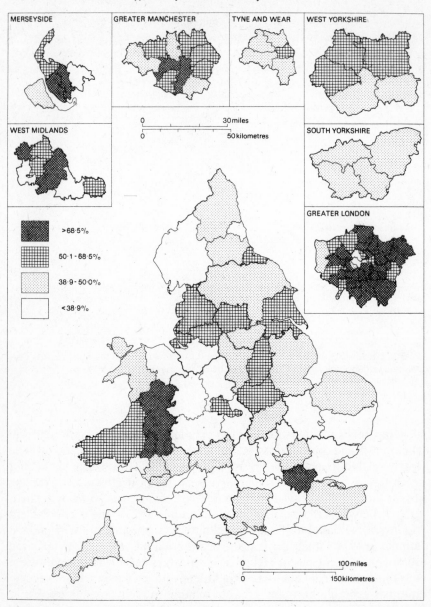

Fig. 8.4 Needs element per head in 1978 as a percentage of rate-borne expenditure per head in quartile groups.

Table 8.5. *Weights given to client, social and other need factors in the needs element 1975/6–1980/1*

Need factors	Fiscal year weight (£)					
	1975/6	1976/7	1977/8	1978/9	1979/80	1980/1
Client group	2.822729	2.185070	1.173993	1.044468	1.235263	1.692305
Social need	0.047238	0.479004	1.299382	1.466113	1.531454	0.964538
Other	0.011361	0.122432	0.040831	0.124870	0.307337	0.121366

The weights are undamped except 1980/1. Client factors include retired persons living alone and all education variables; other factors include all area factors, labour cost differential and housing starts; social need factors are all other variables including poor housing, one-parent families and social service indicators.
Source: ACC, 1980*b*, table 5.

grant to local average wages where the grant has steadily and markedly increased in the extent to which it has been distributed so as to favour areas containing high income earners.

The reasons behind these shifts lie in the choice of regression variables in the formulae between 1974 and 1980, and in changes in population distribution and their weights before 1974. In an attempt to rationalise the contributing factors to these changes in the post-1974 formulae, table 8.5 shows the changing role of three groups of factors: clients group size, social need, and other factors. These factors are taken from the RSG *Orders* for this period and are *undamped* except for 1980/1 (see chapter 4). This shows, up to 1979, the generally increasing weight placed on social need and other factors compared with the generally decreasing weight placed on client group size. This balance was markedly upset by the damped formula chosen in 1980/1 and has been completely reversed in the choice of client group variables for GRE in block grant allocations since 1981/2.

8.4 The combined needs and resource element position

Whilst the resource element of the RSG attempts to equalise between differences in local tax bases and the needs element attempts to equalise between differences in expenditure needs, the distribution of the combined RSG should incorporate the two equalising features. The resulting distribution of grant between different administrative areas is shown in figure 8.5 and table 8.6. For the period up to 1973/4 the relative positions of different areas remains roughly constant and the most significant features are the rapid increases in the share of grant of both Outer London and the County Boroughs. After 1974/5 there is a large jump in the level of support that each area receives. However, the most significant feature is that the London Boroughs (Inner and Outer) and metropolitan Districts each steadily outpace the English non-metropolitan Counties. They also very

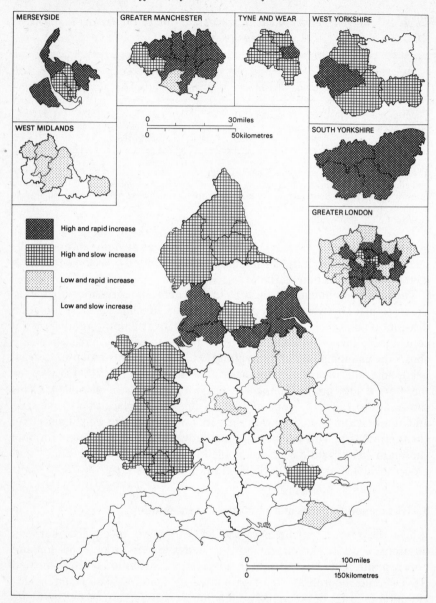

Figure 8.5 Total of resources and needs elements of the Rate Support Grant in 1978 relative to rate of change in total grant 1974–78 in relation to median values.

Table 8.6. *Mean and standard deviation (in parentheses) of level of per capita total RSG needs and resources elements (or Rate Deficiency and General Grant up to 1966, and block grant for 1981) in different administrative areas at constant 1980 prices*

Fiscal year	English Counties	Welsh Counties	Metropolitan Districts or County Boroughs	Inner London	Outer London	Total
1962/3	60.0	113.8	45.7	N/A	N/A	57.0
	(15.8)	(42.5)	(12.5)			(26.8)
1965/6	61.1	96.7	60.2	43.7	40.3	102.3
	(10.4)	(29.6)	(20.1)	(3.1)	(5.6)	(55.7)
1967/8	121.9	189.0	74.6	63.0	60.4	93.6
	(111.3)	(74.7)	(18.0)	(3.4)	(3.6)	(71.4)
1969/70	88.2	136.2	81.2	69.3	64.5	84.5
	(12.4)	(40.6)	(17.0)	(4.2)	(5.8)	(24.0)
1971/2	101.9	152.0	90.8	42.7	73.4	92.7
	(12.7)	(43.6)	(22.2)	(12.5)	(4.2)	(30,8)
1973/4	111.8	163.9	111.7	55.5	87.8	108.9
	(14.6)	(41.9)	(25.1)	(4.7)	(6.5)	(31.2)
1974/5	235.9	351.1	275.9	287.3	219.3	258.9
	(31.5)	(63.2)	(43.3)	(32.9)	(21.8)	(50.9)
1975/6	270.09	388.2	320.4	293.7	247.8	293.3
	(33.1)	(60.8)	(43.4)	(25.1)	(34.9)	(53.6)
1976/7	232.3	336.6	279.2	287.6	240.1	261.6
	(30.5)	(47.3)	(37.9)	(30.8)	(35.3)	(46.5)
1977/8	198.9	299.6	255.8	255.7	215.9	232.9
	(32.1)	(41.3)	(38.4)	(30.9)	(34.4)	(46.8)
1978/9	184.9	272.8	238.1	262.2	223.2	222.4
	(30.0)	(36.6)	(35.9)	(29.2)	(37.2)	(45.0)
1979/80	162.8	243.1	218.1	240.8	197.9	200.2
	(32.4)	(36.6)	(38.9)	(33.8)	(36.0)	(46.1)
1980/81	140.9	215.4	206.3	212.5	177.9	180.6
	(30.7)	(35.2)	(39.5)	(28.1)	(32.2)	(45.6)
1981/2†	164.5	249.1	226.0	178.9	180.8	185.6
	(43.7)	(46.7)	(42.8)	(265.4)	(43.2)	(137.3)

† estimate.

nearly match the grant levels of the Welsh Counties. Thus, in comparison with the late 1960s, the 1970s saw the English Counties' RSG fall steadily from having received the highest grant levels to receiving the lowest. By contrast London and the metropolitan Districts became the largest receivers of grant, with Inner London particularly favoured. The 1980s, however, have seen a renewed shift back to the Counties, initiated by the Conservative Government for 1981/2. It is clear that the block grant introduced at that date, although it has reduced grants to all areas, has reduced it most severely to the cities with the result that the per capita

Table 8.7. *Allocation of the Total Rate Support Grant in relation to tax base, tax rate, expenditure and wages*

Fiscal year	Total grant/ head vs. RV/head	Total grant/ head vs. assessed rate poundage	Total grant vs. relevant expenditure	Total grant/ head vs. average local wages[a] (× 100)
General Grant and Rate Deficiency Grant				
1962/3	−1.563*	0.026*	0.380*	−0.132
1965/6	−0.265*	0.043*	0.160*	−0.089*
1967/8	−0.554*	0.069*	0.233*	−0.055
1969/70	−0.315*	0.037*	0.359*	−0.046
1971/2	−0.492*	0.038*	0.004	−0.057
1973/4	−0.213*	0.005	0.442*	−0.036
1974/5	−0.106*	1.160*	0.281*	−0.479*
1975/6	−0.200*	0.821*	0.150*	−1.049*
1976/7	−0.106*	1.087*	0.211*	−0.332*
1977/8	−0.122*	1.110*	0.745*	−0.126*
1978/9	−0.037*	0.524*	0.325*	0.621
1979/80	−0.071	0.907*	0.277*	0.600
1980/1	−0.136*	1.005*	0.229*	0.371
1981/2†	1.310*	1.926*	0.435*	0.068

The results display the simple regression coefficient in each case and exclude the City of London and Westminster.
* significant at the 90% level.
[a] pre-1974 wage regressions are on a reduced data set using 1974 wage data (see Appendix 1).
† estimated.

allocations to London are closer to the English County allocations than at any time since 1973. One notable year is 1978/9. This was used as the base year for calculation of grant thresholds and penalties in 1981/2 (see chapter 5). Table 8.6 shows a marked relative shift in grant (within a fairly constant total grant) from the English and Welsh Counties towards the cities, especially London, as part of the Labour Government's inner city pro-gramme. The use of this base year for 1981 has thus had the effect of penalising the Counties relative to the cities.

A preliminary analysis of the effect of inner city status and party colour on total RSG allocation (Bennett, 1982c) shows that local party colour is seldom a major correlate of grant allocation, but inner city status, through the allocation of the needs element, has been a major correlate of RSG allocation at least since 1974. Thus, again the contrast to the USA should be noted: the British RSG has strongly favoured inner city problem areas over the 1974–80 period. Whether this favour has been sufficient to meet all of their higher spending needs relative to other areas must remain an

Table 8.8. *Multiple regression results of regressing the level of total RSG per head on tax and expenditure variables*

Fiscal year	Constant	Total rateable value/head	Relevant expenditure/ head	Assessed rate poundage
1962/3	25.66	−1.492*	0.319*	0.001
1965/6	16.97	−0.257*	0.178*	0.008
1967/8	27.34	−0.516*	0.234*	0.018
1969/70	24.87	−0.263*	0.189*	0.008
1971/2	47.70	−0.478*	0.018	0.016*
1973/4	60.22	−0.197*	0.619*	−0.109*
1974/5	116.34	−0.996*	0.805*	−0.399*
1975/6	195.39	−1.263*	0.792*	−0.694*
1976/7	146.72	−0.876*	0.675*	−0.402
1977/8	163.67	−0.958*	0.786*	−0.785*
1978/9	182.04	−1.021*	0.957*	−1.399*
1979/80	190.22	−1.121*	0.908*	−1.390*
1980/81	187.69	−1.165*	0.942*	−1.420*
1981/2†	48.08	−0.534*	0.517*	0.108

The results exclude the City of London and Westminster.
* significant at 90% level of greater.
† estimated.

open question until chapter 9 below. But it is clear that the extent of aid, combined with the large rateable value tax base of inner cities, has prevented major urban budget problems emerging in Britain.

The distribution of the combined grant in relation to tax base, tax rates, expenditures and wages is shown in table 8.7. Since 1974/5 the total level of grant payments has steadily become less redistributive with respect both to total rateable values and to the level of local incomes. However, it has since 1973/4 been relatively stable in redistribution towards areas with high expenditures. In general terms, therefore, it might be claimed that the RSG has achieved its aims of redistribution in favour of high expenditure areas, but declined in redistribution away from high resource areas until 1980. The pattern for rate poundages is inconsistent but generally shows an increasing effect in the post-1974 period. These results, however, do not control variations in tax base when assessing expenditure equalisation, or vice versa; hence it is necessary to recalculate elasticities as a multiple regression taking each factor into account simultaneously. This is undertaken in full in chapter 9, but preliminary estimates are given in table 8.8. These calculations show a broad pattern of decreasing total grant with tax base and rate poundage, and increasing total grant with expenditure level. The redistribution towards high expenditure areas and away from high tax base areas has increased markedly over time, particularly in 1974/5, but has markedly decreased since 1981/2. Redistribution away from areas with high rate poundages has steadily increased, again markedly since 1974/5, but with a reversal after 1981/2.

Taking these results together it can be concluded that, in the very broadest and aggregate terms, the RSG has achieved a major redistribution towards resource-poor areas and high expenditure areas; over the 1974/5 period the redistribution in favour of both was much more marked than either before or since.

8.5 Conclusion

This chapter has reviewed briefly the major distributional patterns of the RSG from 1967 to 1981 and has also compared this with the preceding methods of grant allocation. From this discussion it is clear that in the early days the needs element reflected the pattern of distribution of the previous General Grant and favoured Wales and the areas outside London. The resources element reflected the distribution pattern of the previous Rate Deficiency Grant and mainly favoured the Welsh Counties and metropolitan Districts. From 1974 to 1980 there was a massive increase in the total level of grant received by all local authorities. The main distributional consequence of this change was considerable receipts of resources grants by outer London Boroughs, but also a major change in the relative rankings of the areas receiving needs grant: from being the smallest receiver up to 1973, London, especially Inner London, became the largest receiver whilst the non-metropolitan areas suffered some losses in absolute grants (in real terms).

The reasons for this massive shift in support towards London lie mainly in the use, for needs element assessment, of the regression formula. Attempts to explain the shift by local political party factors show no statistically significant patterns, but inner city status is a significant statistical correlate. This suggests that it is central government party ideology, of the Labour Government from 1974 to 1979, to shift aid to the cities which has been a more important factor than local party effects on taxing and spending levels which subsequently affect grant allocation.

This feature is largely confirmed for the post-1981 period in which a Conservative central government has reduced grants overall, and shifted their allocation to be more equal between all areas in per capita terms with the result that the cities and London are returning towards their 1965 positions. The correlates and consequences of these shifts are explained in more detail in the next chapter where expenditure modelling allows many background determining factors to be controlled.

CHAPTER 9

MODELLING EXPENDITURE AND GRANTS

9.1 Introduction

The examination of the impact of grants on local finances cannot be undertaken in isolation from the rest of the local financial system in which they are enmeshed. Grants are only one element in a complex set of relationships which determine local decisions on budgets, service provision and need satisfaction. In order to isolate the effect of grants, and the effect of changing grant allocation programmes, it is first necessary to understand the broad structure of these relationships. The approach to this problem developed in this chapter employs a *model* of these relationships, the central component of which is an expenditure model.

Expenditure modelling is an essential step in the assessment of the economic and political impacts of central grants. Only by building a model of the interrelationships of expenditure level with other variables such as needs, debt, and level of grants is it possible to determine the separate impact of grants. For example, grants are usually allocated on the basis of past expenditure levels, indicators of need, and so forth. Only by removing the feedback effect of past expenditure levels on both grant allocation and need is it possible to determine the separate effect of grants on expenditure levels. Again, only by removing the effect of different choice of needs indicators on grant allocation is it possible to assess the extent to which grants satisfy different types of need. And only by removing both expenditure and need effects, and by controlling for other factors, is it possible to isolate any political aspects of grant allocations.

The interrelations between the elements of grants, expenditures, and needs suggest that in expenditure modelling at least three main model equations are required: one expressing expenditure levels in terms of grants, needs, resource base, other social and economic factors, and local political variables; a second equation expressing needs as a function of past expenditure, the resource base, taste and preference factors, social and economic indicators, grant levels, and perhaps local political factors; and finally, a third equation expressing grant levels in terms of past expenditures, tax base and other equalisation criteria, central political priorities, and measures of need. In addition, as we shall see, it is usually also necessary to develop subsidiary equations for service supply and tax rate determination. Each of the equations developed contains on the right-hand side variables which are also on the left-hand side. This suggests that a simultaneous equation approach is required to expenditure modelling which enables estimation of the separate expenditure, need and grant effects after controlling for each of the other components.

In an ideal world the interactions of grants with expenditure, tax rates and need would involve a complete analysis which included the impacts of the central and local government on the private sector in the local and national economy, the impact this has on both supply and demand for public goods, and the relation of local authority decision making to national economic health and central government economic policy. Such an approach would be one in a general rather than a partial equilibrium framework. In the present analysis these wider and longer-term effects are ignored and attention is concentrated on first-round impact. Such an approach is not altogether satisfactory, as noted by many authors in incidence studies.[1] However, it has the advantage over general equilibrium models, that it is tractable and, provided the background economic conditions are evolving only slowly, should provide a good approximation against which to assess at least the *relative* impacts of grants.

Bearing these aims and constraints in mind, this chapter is concerned with developing a consistent definition for each of these equations which allows the four equations to be linked in a single simultaneous equation model. The resulting model is then applied both to the examination of the effects of the Rate Support Grant on expenditure levels and needs, and to the assessment of distributional consequences of changed grant allocation formulae.

9.2 The expenditure equation

A considerable body of research has been built up which attempts to explain variation in expenditure levels between local authorities. Early work by Key (1951), Brazer (1959) and Dye (1966) has spawned a large range of studies in both the USA and Britain which emphasise the importance of the physical, social and economic environment in determining expenditure levels.[2] More recently this essentially economic literature has been supplemented and expanded by the inclusion of political variables. Two main forms of political variables have been used: either an indication of central-local political interrelationships reflecting the interplay of party ideologies (such as the number of local representatives corresponding to the national government party), or a measure of local party control reflecting the degree of marginality of a council (such as the number of local council seats held by a particular party).

Analyses of local government spending often approach the determination of the extent of political influence from the point of view of two models (see, e.g., Johnston, 1979):

(1)   *The adversary model*   This suggests that local governments introduce policies in line with their party ideologies; perhaps with extent of policy development related to the margin of their electoral majority.

(2)   *The vote-buying model*   This suggests that the more marginal an electoral situation, the more a local government will spend to benefit those it hopes will vote for it.

In the British context, the adversary model seems the most likely. Some evidence of this behaviour has indeed been found. For example, local party control seems to affect the expenditure level on some services, especially housing and education, with Labour-controlled councils having higher expenditure on these services.[3] In addition it has been suggested with considerably less certainty that local party colour also influences other services; for example, higher police and roads expenditure can reflect Conservative Party control (Boaden, 1971; Alt, 1971; Moore and Rhodes, 1971), higher personal social services expenditure can reflect Labour Party control for old peoples' services (Davies, 1968), and local health services (Boaden, 1971; Alt, 1971).

The most satisfactory political correlates of local spending are for single service categories. When the range of services and the social and economic environmental explanatory variables are aggregated, party effects seem to become of lesser significance. Thus, for total expenditure there is sometimes no discernible party political effect at all.[4] Despite these conflicting findings using the results of statistical analyses, most commentators concerned with explaining the budgeting decisions of individual local authorities do seem agreed that party control is significant in making expenditure decisions, especially at the margin.[5] On this basis Alt (1977), Wilensky (1975), and others have concluded that the problem is, first, to isolate the party effect from its multicollinearity with other social and economic variables (e.g., the strong correlation between voting and income), and second, to measure the party variable in a more satisfactory way. Particularly important seems to be the need to look at the continuity and incremental effect of political decisions which suggests that a time lag, or a time averaging, of political control is required (as discussed below).

There are also other reasons why the effect of local party control can be expected to be muted in Britain. There is frequently a lack of information about local parties or candidates, turnouts at local elections are low and have tended to decline, and the control of national issues and party allegiances is very significant at local level. As a result, it might be expected that the adversary and vote-buying models of political behaviour perhaps extend more fruitfully to central–local relations. The *adversary model*, for example, suggests that central government will allocate more to areas and programmes corresponding to its own party ideology. In Britain, this would tend to correspond, with a Labour central government, to increasing grant aid to education and housing, freeing local authorities of expenditure controls, and allocating grants to city rather than rural areas (from which not only the poorer people but also Labour supporters are perceived to come). A Conservative central government, in contrast, would tend to reduce expenditures on social programmes, increase central controls of total local spending, and provide grants preferentially to rural as opposed to urban areas. The *vote-buying model*, on the other hand, would tend to suggest the opposite pattern of grant aid: with Labour central governments preferentially aiding rural areas in order to woo Conservative voters, and Conservative central governments distributing grants preferentially to cities.

Such evidence as there is in Britain is relatively ambiguous, but does suggest that the adversary model has most relevance to explaining central government behaviour with grants and local government behaviour with expenditure.[6] This conclusion is reinforced by the results reported in earlier chapters of this book.

Clearly, however, the adversary and vote-buying models are great simplifications. Hence most workers now accept Newton and Sharpe's (1977) argument based on the marked social segregation of residences at local authority level, that it is not politics in isolation, but the combination of politics with social and economic variables which will produce particular expenditure patterns. Thus, for example, Labour control of local authorities is often associated with inner city areas and high expenditure need. Analysis of party effects alone will, therefore, find Labour-controlled areas to be higher spenders merely because of higher needs. As a result, political determinants of grant and expenditure patterns can be sought only whilst simultaneously controlling for social and economic determinants, and needs.

In addition to political and economic variables relating to the environment within which each local authority makes its expenditure decisions, the level of grants itself has major effects on expenditure levels. Grants adjust the price and distribution of public goods within local authorities by modifying both the supply of revenue and demand for goods through price. The classical public finance literature suggests that the result can be perverse resource allocation and hence inefficiency in the economy as a whole.[7] This suggestion has been questioned by Musgrave and Musgrave (1980) and others[8] who suggest that local authority finance can be economically efficient *only* by the employment of central grants. Grants improve efficiency in two respects: first, grants permit compensation for benefit spillovers along the lines of the Pigou compensation principle; second, grants permit equalisation of revenue burdens and hence diminish the impact of the differential propensity to produce or consume which would otherwise result from unequal tax bases.

Grants have three major impacts on local authority finance

(1)   *New spending effects*   Initiation of new services or expansion of existing operations.

(2)   *Service maintenance effects*   Continuation of existing programmes which would otherwise be terminated.

(3)   *Substitution (income) effects*   Tax reduction releasing income for consumption elsewhere, or tax stabilisation to avoid increases in burdens that would otherwise occur.

The degree to which grants have impacts in each category depends on the local revenue base, local incomes, the elasticity of demand for services with price, political and other factors. The impact of grants also varies with the type of grant. Musgrave and Musgrave (1980) and Wilde (1968) argue that general grants have a large substitution effect and only small effects on new spending; specific grants, however, have a much smaller substitution effect and produce most impact on new spending or maintain old spending. This expenditure expansion effect is heightened if the specific grants require local matching. These effects are shown in figure 9.1.

Analysis of various grant programmes in a large range of literature confirms these generalisations, although it is often difficult to isolate the various effects of service expansion, maintenance or substitution.[9] There has been very little analysis of elasticity of expenditure in response to allocation of the RSG in Britain. The main results available suggest an expenditure elasticity of almost unity: expenditure rises almost in line with needs element grants, i.e., there are new spending and maintenance effects. However, there is some evidence that, for the need grant and certainly for the resources element grant, there are substitution effects; i.e. there are expenditure elasticities of less than unity. Whilst the bulk of the literature has employed statistical methods (usually regression analysis), the comparatively recent approach of scientists using field workers to monitor new public service programmes has generally confirmed the broad major findings using statistical methods,[10] but there has as yet been no use of the field worker method in Britain.

The definition and measurement of the dependent variable, expenditure level, in such analyses, also presents difficulties. First, it is necessary to define precisely what it is sought to measure. Expenditure levels are only one measure of the benefits received from local government in any area. In addition to money spent, it is also necessary to allow for service extensiveness (the proportion of eligibles receiving a service, or satisfied need in comparison to latent need), intensity (the inputs per client need group), and quality (the outputs per client). Expenditure level is the main variable used in the ensuing analysis, but it must be borne in mind that this alone is incapable of differentiating differences in extensiveness, intensity and quality. Hence these factors must be incorporated in some other way; for example by controlling for them by the method of need measurement. A second difficulty presented by the expenditure variable relates to the type

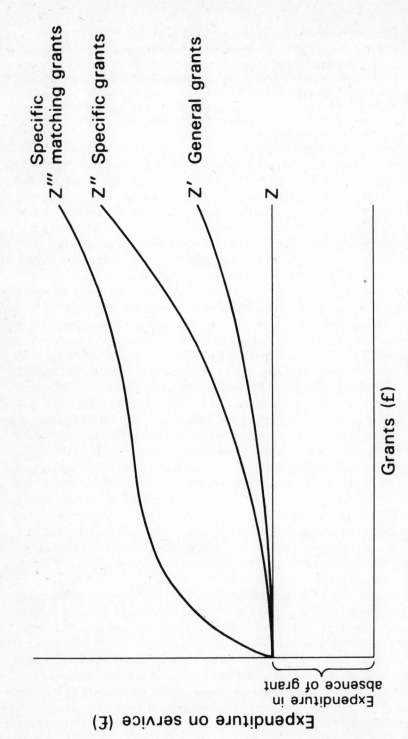

Fig. 9.1 The effect of different categories of grant on levels of expenditure as hypothesised by Wilde (1968) and mainly based on conclusions for the USA.

of service being considered. Aggregate expenditure analyses miss not only the detail, but also the differences in intent between services. Some local services are, in effect, providing private goods; as in the case of government contracts to private industry (although there will be additional public multipliers). Other services are public, but are intended to be provided equally for all, so-called beneficial services; as in the case of police, fire, refuse disposal, hospitals, infrastructure, etc. Many services, however, have a deliberate redistributional intent between client groups, so-called onerous services; for example, social services, welfare, old peoples' services, education, public housing. The relative balance of expenditure levels in any local authority between these three categories of service (private, public non-distributive (beneficial), and public redistributive) will depend partly on need and economic variables; but this is also an important area where political choice can be exercised, especially as to the level of redistributive services provided. Despite these difficulties, in many of the ensuing analyses it is not possible to differentiate between different service functions, although some such differentiation is built into needs measurement.

A further problem in defining the dependent variable (expenditure) is whether grant levels should be included within it. A number of authors have questioned the correctness of using grants as an independent variable in explaining expenditure levels.[11] This questioning has been based on three supposed problems: first, that much grant aid is distributed uniformly and thus becomes part of the regression intercept term (e.g., the population component of the RSG up to 1974/5; see table 4.9); second, that grants are allocated using criteria which derive from many of the other variables, especially tax base and need, and hence there is a high degree of multicollinearity between variables; and, third, that grants are not a true independent variable since, if they are matching grants, they *must* be spent, hence they appear as an identity on both sides of the equation. As a result of these problems, in a single-equation analysis of grant effects on expenditure, it has become normal to subtract the grant level from expenditure before carrying out regression estimation.[12] However, a preferable approach, to be followed here, is to estimate grant effects on expenditure allowing for feedback of expenditure on grants using simultaneous equation and reduced form methods.

This discussion suggests that the expenditure equation can be constituted in the following general form

$$E_i = c_0 + c_1B_i + c_2I_i + c_3X_{ij} + c_4G_i + c_5N_i + P_i + c_6D_i \quad (9.1)$$

where, for local authority i and coefficients c_j,

E_i = expenditure per head
B_i = tax base per head
I_i = local wages per head

X_{ij} = other social and economic environmental variables
G_i = grants per head
N_i = expenditure needs per head
P_i = political disposition of variable
D_i = debt costs per head

This is a form of equation very similar to that specified in the economic and political literature cited above. The tax base and local wage variables give two differing income measures reflecting the respective local authority's ability to produce and the individual's propensity to consume local services; the social and economic environment and expenditure need variables measure demands and costs of spending; and the political disposition variable measures the preferences of local government for spending on given services. The expenditure model given in equation (9.1) is essentially an extension of the economic model of rational choice originally applied to consumer behaviour, here expanded to describe local authority behaviour. Income in the consumer model is replaced in part by the local tax base and grants, preferences are replaced by local needs, and prices are replaced, in part, by political and other factors.

However, a major problem with the form of equation (9.1) is that it confuses two separate elements: on the one hand, supply, measured by the ability and preference of local government to produce; and on the other hand, the demand from local residents to consume. This confusion of supply and demand can be overcome by expanding expenditure into its various components derived from both supply and demand, i.e. write

demand equation

$$E_{ik}^d = \sum_j b_{jk} X_{ijk} + \alpha p_k + c_k G_{ik} \tag{9.2}$$

supply equation

$$E_{ik}^s = \sum_j c_{jk} X_{ijk} + \gamma p_k \tag{9.3}$$

price equation

$$p_k = \sum_j a_j X_{ijk} \tag{9.4}$$

Linear demand and supply functions are assumed, and the variables are defined for local authority i, service k and indicator j as

E_{ik}^d = expenditure demand

E_{ik}^s = expenditure supply

p_k = price

$$\tag{9.5}$$

X_{ijk} = all other economic, social, political and demographic indicators and a, b, c, α and γ are constant coefficients. The major difficulty with these equations is that the separate demand and supply effects cannot be separately estimated since there are no independent measures of E_{ik}^d, E_{ik}^s or p_k available in most countries. This difficulty can be overcome by two main approaches. Some authors have employed socio-economic variables as surrogates for price. This approach, developed mainly by Ohls and Wales (1972),[13] then employs the price surrogates to produce a reduced form for total expenditure, assuming supply and demand to be in equilibrium, which is estimated using nonlinear methods. The major drawback in this approach is that it is by no means clear which socio-economic variables should be used as surrogates for price. Ohls and Wales used a combination of population density, wages, population change and proportion of city/ suburb population as surrogate variables. This has the drawback that many of these variables are collinear with other variables in the demand equation and their relation to price is by no means clear. In a rather simpler approach Foster *et al.* (1980) employed rateable value as a price surrogate; this has the advantage of being a truly independent variable (in the short term), but it is very unlikely to have any simple relation to price and it confuses expenditure effects with grant effects since rateable values are used as the tax base in grant allocation.

A second approach to price estimation is to assume that local tax rates act as a form of price. For the local authority local tax rates act jointly with expenditure as simultaneous determinants of a given local budgetary decisions: a given expenditure change has tax rate implications (*ceteris paribus*) which represent a form of price constraint on supply. Similarly, for the local population, the local tax rates act as a price constraint on demand (*ceteris paribus*). This approach, therefore, requires introducing tax rates into the expenditure equations to determine a final reduced form.

The use of tax rates as price surrogates has been pioneered by Strauss (1974) and Ashford *et al.* (1976) and has been extended to Britain by Gibson (1980).[14] The Strauss–Ashford approach is that followed below, but a number of important modifications are made to the resulting equations to make them fully compatible with British data for the two periods of RSG: separate resources and needs elements up to 1980, and block grant after 1980. The definition of the resulting equations in each case is given in section 9.6.[15]

A further problem of expenditure modelling, which will not be tackled here, is that of imputing to expenditures levels of benefits different from their costs. In all the following analyses a cost basis of benefits is used. Hence it is assumed that benefits received are provided at equal costs to all consumers in a given area and can be aggregated as a simple linear addition. The alternative is to weight the level of benefits received by each client group by a term reflecting the marginal utility of the benefits to the recipient group. This alternative welfare basis has been used elsewhere,[16] but introduces a large measure of complexity and arbitrariness in the choice of the weighting coefficients adopted.

9.3 The need equation

The accurate assessment of expenditure need is an essential prerequisite in expenditure modelling since it allows description of how far expenditures meet similar service requirements in different areas, or how far expenditures differ because of differing need positions. This then allows isolation of how far need variation is reflected in the distribution of grants, and how far grants affect expenditures once need variation has been taken into account.

The approach to estimation of need adopted here is identical with that developed in chapter 6. Thus a representative index of expenditure need is employed. This, it will be recalled, is defined as the product of the client group size (or other relevant criteria) with the unit cost of servicing that need. As defined in chapter 6, unit costs are derived for each major type of administrative area as an attempt to control for the effects of discretionary spending. Of course, the unit cost figures will still be distorted by discretionary effects if large numbers of authorities within a given group exercise relatively high degrees of discretion to expend expenditures (this may particularly affect inner London); but the administrative group average should considerably damp these discretionary effects whilst maintaining the capacity to incorporate major areal differences in unit costs. The index employed is thus more susceptible than GRE to including discretionary effects but probably gives a truer measure of unit cost differences.

Defined in this way the representative index is a form of unit pricing for given services in different localities. As such it can be readily incorporated into either the Ohls and Wales or the Strauss–Ashford models of expenditure determination. In the former case, the representative need index can be interpreted as a direct price estimate which, because it is a deterministic index (not a stochastic measure of price as in Ohls and Wales' case) can be used directly to modify the expenditure equation to transform it from an expenditure demand measure to an expenditure outturn measure. This then obviates the need for nonlinear estimation techniques. In the case of the Strauss–Ashford model the representative need index enters both the expenditure demand equation and the grant equation and hence converts both of these from expenditure demand to expenditure outturn estimates.

The use of the representative need index also endows further advantages: it allows total expenditure outturns to some extent to be scaled by extensiveness and intensiveness. However, output quality has not been allowed for. *Extensiveness* of services measures the proportion of eligibles receiving a service and hence is a ratio of satisfied to latent need. The representative index, as far as it controls for discretionary variations, translates variations in extensiveness to other controlling variables; in most of the ensuing analysis such variations are ascribed to differences in local party control. *Intensiveness* of services measures the inputs per client (e.g. worker-hours per social service case, pupil/teacher ratios, etc.). Again, the representative index, as far as it controls for discretionary differences in labour-output ratios will ascribe differences in intensiveness to political and other variables. *Output quality* is measured by such criteria as examination passes, housing and environmental conditions, crime levels, etc. Such quality variations can be controlled for in the definition of the needs index; e.g., police expenditures can be scaled by population multiplied by a crime level factor. The representative index does not scale for output quality in the way that GRE has attempted for some services. Hence, in this case it is assumed (or is implicit in the needs index) that expenditures and needs translate into the same quality measure of final service. Although a drawback, the level of error introduced into the need index by this omission is likely to be small. In any case, the present state of research on service outputs in most cases is not yet sufficiently accurate to allow use of quality variation controls to be more than arbitrary (see Newton and Sharpe, 1976; Newton, 1981), although clearly future research should be devoted to this task.

9.4 The grant equation

Central grants are allocated using a variety of criteria, but all include elements which will reflect variation in local need, local tax base and local expenditure. They can have important effects on both the demand and supply side of local authority behaviour. On the supply side, local authorities can view grants, to some extent, as 'new' income and hence can expand services, or substitute for previous income sources (this is, of course, in main part a fiscal illusion (Strauss, 1974) since part of the grants is financed by local residents; but this is less visible than local taxation). On the demand side, block grants provide the possibility of reducing the price of local services as a given tax rate. It is necessary, therefore, in determining the effect of grants on expenditure levels to determine how far they are intercorrelated with other terms on the right-hand side of the expenditure (demand and supply) equations.

The Rate Support Grant in England and Wales has been allocated, as discussed in chapters 4 and 5, using a wide variety of needs and resources criteria. These criteria and the weights applied to them are well documented and can be used to define the grant equation for the simultaneous equation model. However, the resulting equation would possess a very large number of terms (40 or 50 for some fiscal years), many of which have variable thresholds, and the intercorrelation of these is exceedingly complex.[17] Hence, instead of employing the known equations used in allocating the RSG, in the following analyses the equations for allocating the RSG will be reduced to a relatively small number of terms which represent the major factors determining grant allocation. This simpler equation will then allow the ready development of a reduced form equation which can be more easily interpreted.

Since the resources element of RSG is a pure revenue equaliser, attention is concentrated here on the needs element. This is allocated on the basis of a complex set of need indicators, variation in tax bases, tax rates, and past expenditure levels. The equation used to capture the pattern of grant allocation is specified as follows

$$G_i^N = b_1 N_i + b_2 E_i^{t-1} + b_3 I_i \qquad (9.6)$$

where for local authority i, E_i^{t-1} = the local expenditure level (expenditure per head) in the previous year and other variables are defined as before. The first terms on the right hand side represent those which were found to be the best general descriptors of the needs element in the preliminary analysis by Bennett (1982*d*).

9.5 Debt

The primary concern of this book is with central grants to local governments to fund *current* account expenditure. However, it is inevitable that the current and capital accounts can never be kept totally distinct. For example, capital spending makes impacts on general administrative costs and is, for some local authorities, funded to a significant extent from internal advances and revenue balances, especially for interest charges on past capital expenditures. Hence, the capital account makes direct impact on current expenditure levels and affects local liquidity. Similarly, the level of current spending may affect the capital account. Changes in maintenance schedules of roads and buildings, heavy current account spending, and other factors, will have long-term impacts on the needs for capital funding. As a result of this two-way relationship, in a complete analysis it is not possible to ignore the impact of the capital account on current expenditure levels.

A full analysis of local authority debt would require separate equations to be developed to describe total debt for each local service category, the costs of that debt, the impacts of variable maturity date of the debt, and analysis of the form of debt employed (central loans, temporary loans, revenue balances, internal advances, stock issues, bonds, mortgages, etc.).[18] In the present study such a complete analysis is neither possible nor warranted. Indeed it is only necessary to concentrate attention on the impact of debt on the current account. For this purpose service breakdowns will not be employed, and all analysis proceeds using total debt levels.

9.6 Estimation of expenditure model

The discussion of the previous sections has introduced a set of equations which describe, respectively, expenditures, tax rates, expenditure need, level of grants and debt. Each of these equations contains, on the left-hand side, variables which are also on the right-hand side. Hence, there is a problem of *simultaneous* dependence which induces biased parameter estimates when least-squares regression is used. Thus special statistical estimation techniques are required. There is a large literature on statistical techniques which can be applied to simultaneous equation estimation. However, in this chapter one approach will be consistently followed: that of reduced form and simultaneous equation estimation. These methods have the properties that their estimates are consistent and normally distributed. However, simultaneous equation methods have the disadvantage that they have large variance for small data sets, arising from the fact that an error in any one equation is transmitted to all equations. As a result, attention is given below both to reduced form results deriving from single equation estimates which take simultaneous effects into account, and to simultaneous equation estimates. In most of the analysis the Strauss–Ashford approach to defining the expenditure equation is employed. For this model we have the following five equations defined, from the previous discussion, in their most simple form as

$$E_i^d = c_0 + c_i f(B_i) + c_2 t_i + c_3 G_i + c_4 N_i + c_5 D_i^t + c_6 I_i + P_i \qquad (9.7)$$

$$G_i = b_0 + b_1 N_i + b_2 E_i^{t-1} + b_3 I_i \qquad (9.8)$$

$$t_i = d_1 (E_i - G_i)/B_i \qquad (9.9)$$

$$N_i = \sum_k a_{ik} X_{ik} \qquad (9.10)$$

$$D_i^t = C_i(a(1 - \alpha) + \alpha D_i^{t-1} + d_2 B_i + d_3 G_i^N + d_4 N_i). \qquad (9.11)$$

The approach followed below differs from that of Strauss, Ashford *et al.* and Gibson in that a tax base term enters the expenditure demand equation and there are also terms for need, income, debt and political factors. A second difference is that the tax rate equation is stochastic since the effects of other income, debt service etc., are either excluded, or appear in other equations. A third feature is that the need equation (9.5) is in all cases *not* estimated but is specified from the deterministic representative needs index developed in chapter 6. Fourth, the results in each case are calculated separately for each administrative area. Fifth, the political party factor (as discussed in chapter 1) is treated as a categorical variable.

Sixth, the grant equation for 1981/2 has to be modified to take into account the change in RSG to the block grant. With these changes the definition of the expenditure and tax equations for the two periods (before and after 1980/1) is given below.

(1) *Expenditure modelling up to 1980/1*

For the period up to 1980/1 the RSG was distributed in three separate elements: domestic, resources and needs. The domestic element is excluded from the analysis. The incorporation of the resources and needs elements into the five equation model (9.7) to (9.11) results in the following substitutions: first, write the expenditure demand equation, for simplicity, as a shortened version

$$E_i^d = c_o + c_1f(B_i) + c_2t_i + c_3G_i^N + t_4N_i \qquad (9.12)$$

in which need is the main demand variable but is modified by local tax base, needs grant G_i^N, etc. The related supply equation is

$$E_i^s = d_1(t_iB_i + t_i(\bar{B} - B_i) + G_i^N) \qquad (9.13)$$

which is stochastic because it excludes other income, debt charges, etc. The term $t_i(\bar{B} - B_i)$ represents the 1974–80 resources element grant. The placement of the needs grant in the supply and demand equations and resource grant only in the supply equation reflects their status in expenditure determination: needs grant may have an effect on expenditure demand, but resources grant affects only supply of expenditure to authorities below the standard tax base $\bar{B}$. However, to the extent that the resources element is not fully equalising of tax base differences, expenditure demand will be affected by tax base. Hence, in equation (9.12) tax base effects have a two-fold functional structure for $f(B_i)$; one for areas in which $B_i > \bar{B}$ and tax base excess can be used to decrease tax rate, increase expenditure or both; and a second for $B_i \leq \bar{B}$ where imperfections in resource equalisation give local authorities differential ability to spend and tax (arising from 'closed-ending', 'clawback' and other factors discussed in chapter 4).

For estimation purposes the reduced form for expenditure is gained by substituting equation (9.13) into (9.12). First, rearrange (9.13) to give

$$t_i = (E_i^s - d_1G_i^N)/d_1\bar{B}. \qquad (9.14)$$

Substitute this into (9.12) to give the equilibrium expenditure when supply of expenditure equals demand for expenditure

$$E_i = c_0 + c_1f(B_i) + c_2(E_i - d_1G_i^N)/d_1\bar{B} + c_3G_i^N + c_4N_i. \qquad (9.15)$$

Taking E_i to the left hand side and rearranging yields[19]

$$E_i = \frac{d_1 \bar{B}}{d_1 \bar{B} - c_2} \left[c_0 + c_1 f(B_i) - \left(c_3 - \frac{c_2}{\bar{B}} \right) G_i^N + c_4 N_i \right]. \quad (9.16)$$

Notice that since $\bar{B}$ is a constant for any year and since the original demand and supply coefficients cannot be estimated, this equation can be simplified to give a new set of coefficients

$$E_i = g_0 + g_1 f(B_i) - g_2 G_i^N + g_3 N_i. \quad (9.17)$$

Other terms can also be added to this equation (for debt, income, etc.) in terms of new coefficient g_i.

The equilibrium tax rate is obtained by substituting the right-hand side of equation (9.16) into (9.14) to give[20]

$$t_i = d_1 \left[\frac{d_1 \bar{B}}{d_1 \bar{B} - c_2} [c_0 + c_1 f(B_i) + \left(\frac{c_3 - c_2}{\bar{B}} + 1 - \frac{d_1 \bar{B}}{d_1 \bar{B} - c_2} \right) G_i^N + c_4 N_i] \right]$$

$$(9.18)$$

which can be simplified to give

$$t_i = h_0 + h_1 f(B_i) + h_2 G_i^N + h_4 N_i \quad (9.19)$$

and other terms can be added if required.

(2) *Expenditure modelling after 1980/1*

For the period since 1981/2 the RSG has been distributed using a block grant combining resource and need equalisation, and domestic rate relief grant. As in the preceding discussion, the domestic rate subsidy is excluded. For this period the expenditure demand equation is written

$$E_i^d = c_0 + c_1 t_i + c_4 N_i. \quad (9.20)$$

Now grants become purely supply equation terms

$$E_i^s = d_1 (t_i B_i + G_i). \quad (9.21)$$

The block grant is defined (from equation 2.16) as

$$G_i = E_i - K B_i \bar{t} \quad (9.22)$$

where $K = (E_i - \bar{E}_i)$ in England and $K = (E_i/\bar{E}_i)$ in Wales above certain thresholds; $K = 1$ below these thresholds (see chapter 5). Substituting this into equation (9.21) and rearranging we obtain

$$t_i = \frac{1}{d_1 \bar{B}_i} (E_i - K\bar{t}). \quad (9.23)$$

Substituting this into (9.20) gives the expenditure equation when supply equals demand as

$$E_i = \frac{(d_1 B_i - c_2)}{d_1 B_i}[c_4 N_i - Kt].$$

(9.24)

Other terms can again be added inside the square brackets of this equation, but note that, in contrast, with the RSG up to 1980/1, this equation presents a nonlinear estimation problem.

(3) *A new model of expenditure*

A development of the Strauss–Ashford model, which can be applied without nonlinear estimation to the post-1980/1 period, is to replace the tax base term in the supply equation by a revenue estimate $\hat{E}_i$, i.e.

$$E_i^s = (\hat{E}_i + G_i).$$

(9.25)

This is a realistic assumption since local authorities set rate poundages based on an estimate of expenditure and grants. If, then, expenditure demand is given in a form similar to that used before, i.e.

$$E_i^d = c_0 + c_1 B_i + c_2 t_i + c_3 N_i + \ldots$$

(9.26)

we can find the equilibrium solution when supply equals demand by defining

$$t_i = d_1(\hat{E}_i - G_i)/B_i$$

(9.27)

where the stochastic relation reflects the ignored effects of balances and other factors. Substitute this tax equation into (9.26) to give

$$E_i = c_0 + c_1 B_i + \frac{c_2 d_1}{B_i}\hat{E}_i + \frac{c_2 d_1}{B_i}G_i + c_3 N_i + \ldots$$

(9.28)

If we now make the assumption that local authorities estimate expenditure requirements on the basis of past expenditures, i.e.

$$\hat{E}_i = gE_i^{t-1}$$

(9.29)

we obtain a final equation

$$E_i = g_0 + g_1 B_i + g_2 E_i^{t-1}/B_i + g_3 G_i/B_i + g_4 N_i + \ldots$$

(9.30)

In this equation both per capita past expenditure and present grant elasticities are rendered per pound of rateable value per head. This is a form of multiplicative nonlinear equation, but one which can be more easily estimated than the nonlinear Strauss–Ashford model (9.24).

Similarly the equilibrium tax rate equation is given by direct substitution of (9.29) into (9.27) as

$$t_i = d_1 g E_i^{t-1}/B_i + d_1 G_i/B_i$$

(9.31)

The assumption (9.29) that a deflated lagged expenditure term can be substituted for the expenditures estimate is supported by a large amount of empirical evidence (see, e.g., G.B. Committee of Enquiry, 1976; Lynch and Perlman, 1977; and Foster *et al.*, 1980) which suggests that although authorities systematically overestimate their expenditure requirements, their estimates in large part derive from a multiple of the previous year's expenditure.

Estimation results

In the estimation results that follow two sets of procedures are employed. First, single-equation estimates are given for expenditure and tax equations using the reduced form estimators developed above. These results are presented separately for the pre- and post-1981 periods. The second estimation procedure is to estimate the expenditure model using simultaneous equation methods. This latter approach is theoretically preferable to the former single-equation approach because it allows all the simultaneous interdependence between taxes, expenditures and grants present in reality to be taken into account. With small data sets, such as those available here, the variance of simultaneous equation estimators may be high, and the number of categorical factors that can be included is reduced. For this reason the results presented for simultaneous equation estimates concentrate only on the main categorical factors of administrative types. However, all categorical factors are included in the single-equation estimates.

Despite the possible limitations of simultaneous equations methods with small data sets, those developed below are highly significant statistically and offer improvement over the single-equation results in terms of their relation to reality. All results tabulated in this chapter relate to data which have been expressed in terms of constant 1980 prices and which have been standardised (by subtracting the mean and dividing by the standard deviation); as a result comparison between coefficients illustrates the relative strength of different variables as correlates of expenditures, tax rates or grants.

For both periods of analysis the differences between authorities dependent upon their administrative type, party control and level of urban stress are controlled for. The administrative type within which an area falls is controlled in five groups: (1) English non-metropolitan Counties (NMCE); (2) Welsh non-metropolitan Counties (NMCW); (3) the metropolitan Districts (MD); (4) Inner London (IL) and (5) Outer London (OL). The local party variable gives four categories of authorities for time averages of party control coinciding with periods of central government party control: (1) Conservative; (2) Labour; (3) Other party; and (4) no overall majority or changes in party control. The urban stress variable also produces four groups of authorities deriving from the 1978 legislation: (1) major stress (with city partnerships); (2) stress (with inner city programmes and special powers); (3) other urban areas; and (4) non-metropolitan Counties. The definitions and difficulties in both of these classifications are outlined in chapter 1.

The consequence of adopting the nominal scale of measurement of party and urban stress effects is that, for expenditure modelling, categorical regression methods must be adopted. The statistical estimates method used to provide estimates for these models derives from the *G*eneralised *L*inear *M*odelling (GLIM) computer package. This approach represents a new method of analysis of local government finance, but it is one which, because of its inclusion of both interval and nominal data variables, is better suited than most previous approaches to analysis of local authority financial behaviour.

The interpretation of categorical regression methods is identical to that of dummy variables in normal regression methods. Interval scale data are related, and tested by statistical significance tests, in the same way as in normal regression analyses, but now these estimates and tests are for different categories of authority. In the ensuing analysis the three categorical variables are administrative type (County, metropolitan Districts, London Borough etc.), local party colour, and level of urban stress. The interpretation of the resulting regression coefficients can be understood from figure 9.2 for the example of political party effects. The example shows the combination of this categorical variable with interval variables in four possible forms

A.   No categorical party effect
B.   Categorical party intercept effect
C.   Categorical party slope effect
D.   Categorical party intercept and slope effect.

In the first case (figure 9.2A) all parties have similar effects and the categorical variable can be dropped from the analysis. In the other three cases, however, the categorical variable produces different regression interrelationships depending upon the categorical value of the party variable; in effect two regression relations are estimated jointly. In case fig. 9.2B, the categorical variable modifies the level of the expenditure-grant relationship, but the relationship remains the same. In case fig. 9.2C, the categorical variable modifies the relationship between expenditure and grants (indeed they could also differ in sign, being positively related for one party grouping and negatively related for another). In the case of fig. 9.2D, both the level and form of the relationship are modified. As we shall see, examples of each of these categorical variable effects are necessary to capture administrative, party, and urban stress effects in British local authorities. Using this technique, the two approaches to estimation of the

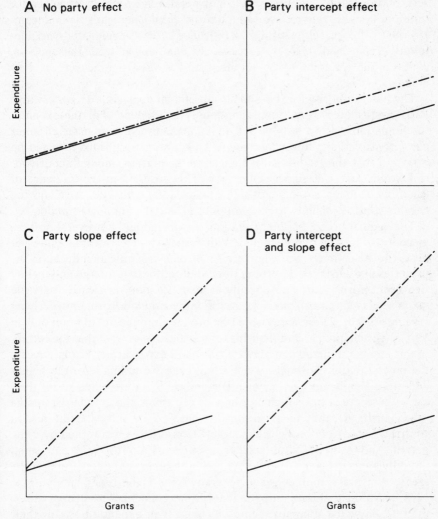

Fig. 9.2 Possible effects of party variable interacting with other variables to modify local financial structures. The example shown is of modifications to expenditure consequent upon different levels of grant for two different types of local party control (shown by solid and broken lines).

expenditure model listed above are each followed in turn below; namely, reduced form categorical regression; and simultaneous equation estimates.

(1) *Modified Strauss–Ashford model 1974/5–1980/1*

Estimates of the modified Strauss–Ashford model for both expenditure and tax equations (9.17) and (9.19) are given in tables 9.1 and 9.2. In both sets of equations, a number of individual regression coefficients for different administrative classes of authority are not statistically significant. However, the overall statistical significance of each aggregated variable, tested by partial F tests by successively removing it from the analysis, confirms all variables to be required at the 95% significance level or greater.

The base authority for this and all subsequent analysis is a Conservative Party-controlled non-metropolitan County in England. For the expenditure equations shown in table 9.1, in comparison with this base, all other party controls have lower intercepts. This is surprising since it would be expected that the traditionally higher spending areas under Labour control would have the highest intercepts. This arises from the influence of generally steeper slope coefficients associated with the areas having Labour control—mainly the metropolitan Districts and inner London.

The main burden of interpretation of the model, however, must be placed on the slope coefficients. With respect to the rateable value per head tax base there is a relatively small and insignificant effect for the Counties and Districts. However, for London, there is a large and mainly significant positive effect especially in inner London. For early years the tax base elasticity to spend is highest in inner London Boroughs receiving resources grant. These areas were first brought into receipt of grant only in 1974/5 (see chapter 4) and their higher coefficients suggest that the receipt of this grant stimulated a modest expansion of expenditure. For later years the inner London Boroughs which do not receive grants have the higher and more significant coefficients. This indicates the expected result that, since these areas have high tax base (i.e., above the standard rateable value per head), they have 'surplus tax base' to expand expenditures. In contrast, outer London, which is largely Conservative-controlled over the period, shows similar but smaller coefficients up to 1978 and small insignificant coefficients thereafter. For both inner and outer London the areas receiving resources grant generally have higher coefficients up to 1978, but after that date only the inner London high tax base areas above the rateable value standard seem to increase their expenditure using their tax base advantage. Overall the elasticity to tax base shows a major transition at the point where London is brought into the England and Wales machinery for distribution of RSG and where the resources element is subjected to clawback for London, i.e. after 1976.

The relation of the RSG needs element to expenditure levels shows the expected positive effect for almost all coefficients. For early years it is generally the English Counties which have the highest coefficients and which are greater than unity suggesting an expenditure multiplier effect of grants. For later years, however, it is outer London which has the highest coefficients. Relating this to various periods of central government control, it seems that there was a major once-and-for-all expansion effect of grants on expenditures in 1974/5 following local government reorganisation. However, the period from 1975 to 1979 was characterised by a stable pattern of high expenditure in relation to grant in the metropolitan Districts, whilst in the English Counties expenditures, although highest, steadily declined. In this period outer London expenditures stay fairly high in relation to needs grant, but inner London expenditures show only minor stimulus from needs grant (being more affected by resources grant, as noted above). After 1979 grant effects grow steadily for all areas despite the attempts by the Conservative government to use grants to limit expenditures.

The measurement of need used in the distribution of the RSG needs element was rather different from the representative needs index used to define need in the present analysis. In relation to this measure of need few coefficients are significant; most of the need effects are probably already accounted for in RSG needs element allocation (despite differences in measurement) and in administrative type. The surprising result, however, is that the only individually significant coefficients indicate that expenditure declines as need increases in inner London over most of the period. Taken with the other coefficients this suggests that the inner London areas with highest need are spending least and those with lowest need are spending most; this in turn suggests significant over- or under-provision in relation to need as measured by the representative index.

With respect to the other variables, as expected, debt charges are generally a positive correlate of expenditure and are most significant in London and the English Counties. Wages should be expected to be negative correlates of expenditure; as local income rises demand for public service provision should decrease according to most of the results of classical public finance theory (see e.g. Musgrave and Musgrave, 1980; Oates, 1972). This is the general pattern for many areas, especially in early years in London and the metropolitan Districts.

Turning to the estimates of the tax equation (9.19) shown in table 9.2, these conform to most of the expectations of the Strauss–Ashford model. Relative to the Conservative English County base, Labour and other party areas are seen to have very small overall intercepts. The intercepts for administrative type are generally positive but insignificant except occasionally in the Counties.

The relation of tax rates to rateable value would be expected to be negative, as it is in general for the Welsh Counties and in outer London. The generally positive coefficients in inner London conform to the hypothesis that these areas use higher tax base to expand expenditures which can be achieved at lower marginal tax rates than other areas. The surprising result is that the English Counties also show this effect,

Table 9.1. Estimates of the coefficients of expenditure equation (9.17) for years 1974/5–1981/2 and equation (9.30) for 1981/2

Variables	Fiscal year							
	74/5	75/6	76/7	77/8	78/9	79/80	80/1	81/2
Base: Conservative, non-metropolitan County England (NMCE)	1.55	1.61	1.98	0.09	2.19	1.12	0.44	-4.41
Labour	0.44*	0.46*	0.22+	0.43*	0.65*	0.61*	0.64*	-0.01
Other party	0.07	-0.07	-0.26	-0.10	-0.05	0.18	0.19	0.32*
Change of party	0.35*	0.24+	0.35*	0.28*	0.08	0.14	0.08	-0.004
Non metropolitan County Wales (NMCW)	4.59	18.10+	18.90	12.76	5.05	9.71	6.13	16.64
Metropolitan District (MD)	6.16+	6.93*	1.44	0.31	0.17	3.73	0.39	2.17
Inner London (IL)	-2.30	4.59	0.77	7.63*	5.76	25.36*	28.67*	-6.69
Outer London (OL)	-3.14	-1.93	-2.22	0.82	1.09	-0.23	-1.51	2.04
B. NMCE ⎫	0.07	0.10	0.03	-0.01	0.23	0.17	0.001	0.92*
B. NMCW ⎬ $(B < \bar{B})$	-0.28	0.32	0.48	0.58	0.34	0.38	-0.08	0.48
B. MD ⎭	-0.22	-0.17	-0.16	-0.18	-0.23+	-0.23+	-0.14	3.60*
B. IL $(B > \bar{B})$	1.89*	1.47*	0.42	1.36*	1.32*	0.29	0.03	⎱ 0.48*
B. OL $(B < \bar{B})$	1.38*	1.37*	0.33	1.34*	1.37+	1.29*	1.20*	⎰
B. OL $(B < \bar{B})$	1.16*	1.18*	0.95*	0.54*	0.53+	0.33	0.01	⎱ 1.53
B. OL $(B > \bar{B})$	1.00*	1.05*	0.83*	0.47*	0.49*	0.37	0.08	⎰
[a]G^N. NMCE	1.29*	1.27*	1.42*	0.92*	0.85*	0.49*	0.65*	0.12
G^N. NMCW	2.28*	-0.77	0.13	0.56	0.02	0.03	0.27	-5.33
G^N. MD	1.26*	1.22*	1.15*	1.10*	1.12*	0.96*	0.99*	3.16*
G^N. IL	1.33*	0.92*	0.46	0.87*	0.44	-0.56	-0.60	1.67*
G^N. OL	0.89*	0.93*	0.96*	0.87*	0.71*	0.86*	1.12*	2.24
N. NMCE	0.11	-0.05	-0.16	0.16	-0.04	0.15	0.12	0.29+
N. NMCW	-2.21	2.10	0.01	-0.18	0.33	0.06	0.18	0.82
N. MD	0.08	0.02	0.14	-0.02	0.16	0.09	-0.03	-0.23*
N. IL	-0.19	-0.25+	0.23	-0.63*	-0.30*	-0.91*	-1.07*	1.17*
N. OL	0.15	-0.23	-0.06	0.26	0.19	0.19	0.17	-0.08

E. NMCE	—	—	—	—	—	—	—	—
E. NMCE								
E. MD								
E. IL								
E. OL								
D. NMCE	0.02	0.02	0.31	0.004	0.10^+	0.14^*	0.13^*	0.03
D. NMCW	0.04	−0.29	−0.04	−0.02	0.04	0.07	−0.07	0.47
D. MD	0.10^*	0.10^*	0.43^*	−0.004	−0.03	−0.02	0.01	−0.02
D. IL	0.11	0.30^+	0.79^*	0.97^*	0.52^*	1.02^*	1.17^*	$−0.33^*$
D. OL	0.04	−0.03	0.36^*	0.06	0.53	0.07	0.06	−0.01
I. NMCE	−0.001	−0.01	−0.01	0.21^+	0.10	0.13	0.17^+	0.07
I. NMCW	0.34	−1.28	−0.94	−0.47	−0.21	−0.38	−0.14	−0.72
I. MD	$−0.38^+$	$−0.43^*$	−0.17	0.22	0.06	−0.08	0.20	0.23
I. IL	−0.01	−0.35	−0.13	$−0.43^*$	−0.35	$−0.94^*$	$−1.06^*$	0.36
I. OL	0.13	0.16	0.06	0.01	−0.08	0.03	0.15	0.05^*
Total deviance	10.92	11.33	12.87	15.77	10.50	9.99	9.80	11.11
R (correlation coefficient)	0.67^*	0.80^*	0.79^*	0.73^*	0.84^*	0.82^*	0.80^*	0.94^*

* denotes coefficient at the 95% level and + at the 90% level; [a] G^N is replaced by G for 1981/2;

B = Rateable value per head;
G^N = RSG needs element;
N = expenditure need;
D = debt costs on current account;
I = local average wages;
E = expenditure for previous year;
NMCE = non-metropolitan County in England;
NMCW = non-metropolitan County in Wales;
MD = metropolitan District;
IL = inner London Borough;
OL = outer London Borough.

particularly in the 1976–79 period; although this must also be accounted for by the shift of grant aid away from the Counties over this period towards the urban areas (see Foster *et al.*, 1980; Bennett, 1982*c*).

It should also be expected that tax rates would be negatively related to needs grant. However, although this is true for the Counties, other areas show a tendency for tax rates to increase with needs grant. This again tends to confirm 'over-provision'. Similar conclusions relate to the relation of taxes to need. As far as the needs element of RSG is imperfectly equalising, we would expect a positive relationship with need; where it is perfectly equalising a zero relationship; and where equalising, but tax rates subsidised or reduced, a negative relationship. Table 9.2 demonstrates that some under-provision of need occurred in both London and the English Counties resulting in tax subsidies, while Wales and the metropolitan Districts either over-provided for need, or were subject to poor equalisation under the RSG.

(2) *Lagged expenditure model 1981/2*

This second approach to expenditure modelling is similar in structure to the Strauss–Ashford approach and can be expected to yield similar results. However, the entry of the two modified variables of lagged expenditure and grants, both per £ of rateable value, represents a more direct method of estimation which also yields results for the pre- and post-1980/1 period.

The estimates of the expenditure equation (9.30) are given in table 9.1. Most of the results are similar to those shown in the earlier columns of the table, and bear similar interpretations. The most important difference is that tax base rateable values consistently show a significant positive effect on expenditure which is highest in London especially among those pre-1980 authorities receiving resources element.

The lagged expenditure variable, which is additional to the variables included up to 1980, shows a consistent high, statistically significant, and positive effect on expenditures. This positive serial correlation is to be expected, and, taken as per £ of rateable value, shows that expenditures have the highest positive elasticities in the urban and resource-rich areas (those above the standard $\bar{B}$).

The coefficients for grants per £ of rateable value show, for significant coefficients, a small positive relationship in the English Counties, and strong positive coefficients in the metropolitan areas, with the strongest positive, expansion effect on expenditures in outer London.

The coefficients of the tax rate equation (9.31) are given in table 9.2. Again these show much in common between the Strauss–Ashford model estimates and the lagged expenditure estimates although, because of redefinition of variables, comparisons cannot be direct. Lagged expendi-

Table 9.2. *Estimates of coefficients of tax rate equation (9.19) for years 1974/5—1980/1 and tax rate equation (9.31) for 1981/2*

	Fiscal year							
	74/5	75/6	76/7	77/8	78/9	79/80	80/1	81/2
Base: Conservative, non-metropolitan								
County, England (NMCE)	2.81	3.98	6.47	7.47	8.53	8.13	6.84	4.73
Labour	0.49*	0.86*	0.91*	1.01*	1.13*	1.13*	1.36*	0.31*
Other Party	−0.16	−0.24	−0.39	−0.32	−0.24	0.03	0.34	−0.17
Change of party control	4.23*	0.07	0.20	0.16	0.16	−0.10	0.22	−0.15
Non-metropolitan County Wales (NMCW)	1917*	5.85	9.53*	6.57	1.12	3.07	0.73	1.96
Metropolitan District (MD)	2.49	2.65	0.20	3.40	1.25	1.48	1.08	−4.31
Inner London (IL)	−0.42	−1.05	−1.26	−0.58	−4.14	−2.88	−0.87	1.14
Outer London (OL)	3.92	5.11	1.76	3.46	1.29	3.63	2.81	−6.52
[a] B. NMCE	0.12	0.44	0.81*	0.91*	0.64*	0.24	0.49+	0.30
B. NMCW $\}$ $(B < \hat{B})$	−0.10	−1.11	−0.56	−0.11	−0.55	−0.48	−0.66	2.63*
B. MD	−0.89*	−0.81*	−0.65*	−0.44+	−0.39	−0.69*	−0.22	4.06*
B. IL	0.53	0.49	0.60	0.25*	0.69+	0.45	−0.004	0.61*
B. IL $(B > \hat{B})$	0.08	0.16	0.32*	0.18	0.29*	0.31*	0.20+	
B. OL $(B < \hat{B})$	0.17	0.59	0.19	−0.81*	−0.80	−0.85	−1.14*	4.52*
B. OL $(B > \hat{B})$	0.02	0.37	0.03	−0.95*	−0.83*	−0.53*	−1.03*	
G^N. NMCE	−0.09	0.21−	−0.53+	0.32	0.11	−0.07	0.08	−0.27
G^N. NMCW	1.78+	−0.68	−0.006	−0.60	−2.05+	−1.85*	−1.99*	−6.05*
G^N. MD	0.65*	0.61*	0.51*	0.39*	0.29*	0.09+	0.14	−10.24*
G^N. IL	0.59+	0.57	0.58	0.05	0.26	0.86*	0.89*	−1.12*
G^N. OL	0.32	0.11	0.16	−0.19	−0.39*	0.05	0.42+	−9.17*
N. NMCE	1.09*	0.51	−0.14	−0.11	10.18	0.05	−0.12	–
N. NMCW	−3.57*	0.82	−0.64	−0.14	1.07	0.59	0.75	–
N. MD	0.35	0.05	0.31	−0.30	−0.15	0.02	−0.09	–
N. IL	−0.06	−0.02	−0.27	0.03	0.05	−0.42*	−0.53*	–
N. OL	−0.22	−0.38	−0.01	0.27	0.34	−0.31	−0.29	–
Total deviance	43.12	45.84	39.54	44.04	42.52	40.38	38.39	23.08
R	0.55*	0.53*	0.59*	0.50*	0.53*	0.59*	0.61*	0.77*

Notes: [a] Same as in Table 9.1 except for 1981/2 where variable is E and not rateable value B.

ture per £ of rateable value is a consistent positive correlate of tax rates and is highest in Wales, the metropolitan Districts and outer London. The relation of tax rates to grant levels is, for almost all areas, significantly negative indicating the general effect in reducing tax rates; however, the relation is most strongly negative in outer London and the metropolitan Districts. Taking these results with those in table 9.1, grants seem to have had a more important effect in reducing tax rates than in increasing expenditures, with this effect being most marked in London and the metropolitan Districts. However, this conclusion does not take the simultaneous and feedback effects between expenditure and taxes directly into account during estimation. Hence, as we shall see below, this conclusion needs to be modified when these effects have been jointly estimated.

Simultaneous equation estimation

The single-equation reduced form estimates discussed above have the advantage that a wide range of categorical variables can be included in the estimation stage of the analysis. However, they do not allow for the totality of interdependence between equations: the estimates of the coefficients and error covariance in other equations are not taken into account when estimating the coefficients in each single equation. A full simultaneous equation approach is, therefore, preferable since it allows estimation of the joint interdependence between expenditure, tax rate and grant decisions which occurs in reality. This is what is undertaken below. The simultaneous equation approach does suffer from the disadvantage, however that, because of instability of estimators and loss of degrees of freedom, not all the categorical variables used in the single-equation approach can be included. It is particularly unfortunate that the political variable is one of those which has to be excluded. Estimation was undertaken in three blocks: (1) the England and Wales Counties, (2) the metropolitan Districts, and (3) the London Boroughs. In the case of London categorical variables are introduced to control for the cases (up to 1980/1) where areas are or are not in receipt of resources element. This depends upon whether or not $B > \bar{B}$. A further categorical variable controls for whether an area is an outer or inner London Borough. In all cases estimation is carried out using the full-information maximum likelihood (FIML) estimation method. This is a standard simultaneous equation estimation technique which has the advantages that it is very robust, has few limiting assumptions which apply to the present data sets, and with careful choice of starting values will converge to stable consistent estimates even for the relatively small data sets of the present problem. Full details are given by Johnston (1972) and Bennett (1979). For simplicity of presentation results are given here only for alternate years up to 1980, and for 1981/2.

(1) *Simultaneous equation model 1974/5–1980/1*

Estimates of the simultaneous equation model for this period are shown in table 9.3. Various approaches to specification of the model are possible; however, in the development below the form of equations (9.7), (9.8) and (9.9) which is estimated is as follows

$$E_i = g_0 + g_1 B_i q_1 + g_2 B_i q_2 + g_3 t_i + g_4 G_i^N q_3 + g_5 G_i^N q_4 + g_6 N_i + g_7 D_i + g_8 I_i$$

$$t_i = h_0 + h_1 N_i + h_2 I_i + h_3 B_i q_5 + h_4 B_i q_6 + h_5 G_i^N q_7 + h_6 G_i^N q_8 + h_7 E_i^{t-i}$$

$$G_i = e_0 + e_1 E_i^{t-1} + e_2 N_i + e_3 I_i + e_4 (\bar{B} - B_i) q_8 + e_5 (\bar{B} - B_i) q_9$$

Notice that the reduced form structures are not used here. Although these could be estimated simultaneously, the simultaneous equation approach in any case allows for joint interdependencies without the need for substitutions, and this also allows a more direct interpretation. The q_i terms represent dummy variables controlling for whether or not an area is in receipt of resources element. They are set to one or zero.

The results of the FIML estimates shown in table 9.3 are the first major application of simultaneous equation estimation to British local authority finance modelling. They are also extremely satisfactory in terms of stability, speed of convergence, levels of significance obtained, and relationship to expectations based upon other analyses of financial patterns both in sign and relative magnitude. The response of expenditure to tax base is almost always positive and is generally highest in London, especially in inner London for areas above the rateable value standard. Although significant for the Counties in early years, however, tax base becomes of minor significance in expenditure effects until 1981/2. In general, larger tax base encourages larger expenditures particularly for tax-rich areas not receiving resources grant; but for many years this conclusion also holds true for London and metropolitan Districts with tax base below the standard.

The response of expenditures to tax rates is, as expected, rising with taxes, usually in the range of 40–70% of tax rates. This elasticity is highest in London, particularly inner London (except 1978/9). The overall effect of tax rates is consistently larger than that of tax base in the areas outside London, and smaller within London.

Of major interest is the response of expenditure to grant levels. For almost all areas and times the elasticity is close to unity, or greater than unity. This suggests that grants give a direct throughput into expenditures rather than reducing taxes, whilst for some areas they stimulate an increase in expenditure. In general, the metropolitan Districts show the smallest tendency to expand expenditures based on grants, and they are closely followed by the Counties (except in 1978/9). However, for London there is a more complex pattern. Early and later years show high levels of

Table 9.3. *Full-information maximum likelihood estimates of expenditure, tax and grant parameters*

Equation and variables	1974/5	1976/7	1978/9	1980/1	1981/2
(1) Expenditure					
Intercepts: NMC	−1.11	3.11	−0.23	0.53	1.68*
MD	0.01	−0.01	−0.02	0.02	0.01
L	−2.32*	−3.36*	−4.99	−4.01	2.53*
B. NMC	1.87*	−0.05	0.13*	0.18	0.85*
B. MD	0.33	0.06*	−0.01	0.17*	0.30*
B. IL $(B > B)$	1.26*	1.16*	1.18*	1.13* ⎫	
B. IL $(B < B)$	0.01	−0.26	0.63+	0.68+ ⎭	0.48*
B. OL $(B > B)$	1.11*	0.99*	0.95*	1.09* ⎫	
B. OL $(B < B)$	1.57*	0.89*	1.08*	1.10* ⎭	0.01
t. NMC	2.11	0.68*	0.46*	0.45	0.32*
t. MD	0.49*	0.38*	0.52*	0.51*	0.09*
t. IL	0.49*	0.68*	0.69*	0.53*	0.34*
t. OL	0.25+	0.52*	0.74*	0.46*	0.04
G^N. NMC	1.18	2.64	1.06*	1.06+	1.07*
G^N	0.84	0.94	1.03*	0.89*	0.45*
G^N. IL $(B > B)$	0.85*	0.62*	0.94*	1.15* ⎫	
G^N. IL $(B < B)$	1.38*	0.86*	1.03*	1.32* ⎭	0.98*
G^N. OL $(B > B)$	1.27*	0.86*	1.03*	1.32* ⎫	
G^N. OL $(B < B)$	1.12*	0.96*	0.91*	1.31* ⎭	−0.05
N. NMC	−1.46	−1.43	−0.07*	−0.06	−0.17*
N. MD	0.07	0.06	0.16+	0.08	0.01
D. NMC	−7.32	−0.01	0.001	0.07	0.02
D. MD	0.08	0.07	−0.01	0.01	0.02*
I. NMC	−0.96	−0.22	0.003	−0.04	−0.06*
I. MD	−0.07	−0.03	−0.11*	−0.16+	−0.04
(2) Tax rate					
Intercepts: NMC	6.22	−5.01	0.51	3.88	8.11*
MD	−0.04	0.03	0.16	−0.04	0.01
L	4.47*	7.25	6.07	2.67+	5.88*
N. NMC	−6.28*	2.23	0.14*	2.68	0.76*
N. MD	−1.15	−0.10	0.81	−0.17	−0.07
I. NMC	−1.11	0.37	−0.004	2.00	0.02
I. MD	0.22	0.07	0.23	0.08	0.02
G^N. NMC	0.32	−3.93+	−2.31*	−3.78*	−3.28*
G^N. MD	−0.08	−2.14*	−4.71+	−1.91*	−0.11
G^N. IL $(B > B)$	0.38*	0.25	0.51*	1.25* ⎫	
G^N. IL $(B < B)$	0.48	0.07	0.36	2.06* ⎭	0.14
G^N. OL $(B > B)$	0.61*	0.57*	0.36	1.39 ⎫	
G^N. OL $(B < B)$	0.84*	0.62*	0.54*	1.75* ⎭	−0.49
B. NMC	0.01	−0.02	−0.28*	−2.28*	−2.47*
B. MD	−0.26	−0.02	0.09	0.05	0.13
B. IL $(B > B)$	−0.03	0.05	0.02	−0.14 ⎫	
B. IL $(B < B)$	−0.14	0.33	0.26	−1.89 ⎭	1.14*
B. OL $(B > B)$	0.003	−0.17	0.28	−0.25 ⎫	
B. OL $(B < B)$	−0.23	−0.21	0.07	−0.89+ ⎭	2.01*

Table 9.3. (*cont.*)

Equation and variables		Fiscal year				
		1974/5	1976/7	1978/9	1980/1	1981/2
E. NMC		5.31	1.37*	2.16*	−4.26	0.46*
E. MD		1.89	2.26*	1.93*	2.13*	1.71*
E. IL		–	–	–	–	0.93*
E. OL		–	–	–	–	1.80*
(3) Grants						
Intercepts:	NMC	3.63	7.98	1.94	−3.87[+]	1.20*
	MD	−6.53*	−4.77*	−4.03	4.26*	0.17
	L	−0.26	−0.74	−0.95	−1.11[+]	−4.71*
E. NMC		3.76*	2.38*	1.56*	0.03	–
E. MD		1.58	*0.62*	0.17	0.77*	–
E. L		0.42*	0.67*	0.69*	0.62*	–
N. NMC		−3.05*	−1.73	−0.23	0.37	0.22*
N. MD		−1.53*	0.12	0.41	0.02	0.25*
N. L		0.23*	0.11*	0.12*	0.07[+]	0.80*
I. NMC		−0.96	−0.65	−0.42*	0.19	–
I. MD		0.29[+]	0.23*	0.30	0.13[+]	–
I. L		−0.003	−0.04	−0.03	−0.01	–
($\bar{B}$ − B). NMC		0.51	0.07	0.62[+]	1.97	−0.14*
($\bar{B}$ − B). MD		2.68*	1.67*	1.40	1.49	0.02
($\bar{B}$ − B). L	($B > B$)	0.45*	0.71*	0.69*	0.74*	–
($\bar{B}$ − B). L	($B < B$)	2.14*	1.99*	2.64*	1.98*	–
B.t. NMC						−0.13*
B.t. MD		–	–	–	–	0.02[+]
B.t. L						0.05*

* indicates a coefficient significant at the 99% level; [+] at the 95% level. Estimates for each year are undertaken using modifications to equations (9.7), (9.8) and (9.9) in three blocks for non-metropolitan Counties in England and Wales (NMC), metropolitan Districts (MD) and London as a whole (L) with dummy variables for Inner London (IL) and Outer London (OL). All other abbreviations are as in table 9.1. The City of London, Westminster and Isles of Scilly are excluded from the analysis.

expenditure in relation to grant. For inner London there is an expenditure increase of up to 145% for areas below the rateable value standard, and expenditures of at least 129% of grant are maintained over the whole period of analysis suggesting that in these areas needs grant facilitated a marked increase in expenditures. For other inner London areas, above the rateable value standard, grants do not become completely throughput into expenditures; some reductions in ·tax rate are also achieved. For outer London expenditure elasticities to grants are high but generally lower than in inner London; they also bear a close relation between areas above and below the rateable value standard.

Need, debt and wage variables show comparable pattern to table 9.1 and because of reduced degrees of freedom were not estimated for London. In each case, although only small effects are present (except in 1974/5), these variables are significant to the overall significance level of the model.

For the tax equation the relation of tax rates to grant levels shows a consistent and marked division between London and non-London authorities. Especially for later years the needs grant has a marked effect on reducing tax rates for non-London authorities; but within London the needs element is strongly and positively related to tax rates for areas both below and above the rateable value standard. This conforms with the expenditure effects noted above. For 1980/1 the London relationships are very strong suggesting that, despite grants rising in line with tax rates and hence with fiscal effort, London areas have used the needs element as a price support and have been encouraged to raise taxes generally at a rate of a further 20–30% above needs support. For non-London authorities the reverse is true with grants leading to reductions in tax rates.

The relation of tax rates to tax base shows only small but relatively consistent results. Whilst generally negative, most coefficients are positive in 1978/9, and few attain statistical significance until the most recent years. In contrast the lagged expenditure variable is a consistent and strong positive correlate of tax rates over the whole period. Need and wages, although rarely statistically significant, do increase the overall fit of the model significantly.

The grant equation is intended to provide a simple description which captures the main determining structure so as to include feedback from expenditure and other variables when estimating the other two equations. The results tabulated are generally as expected, with past expenditures and tax base having the most significant and positive effects. Need, as measured by the representative index, and wages, show only weak relations, although once again they are important to overall model fit.

(2) *Simultaneous equation model 1981/2 onwards*

For the post-1981/2 period there are, as yet, much fewer data to use in estimation, but the last column in table 9.3 does allow a preliminary estimate of the effects of the RSG block grant. Interpretation of this period is also limited by the fact that, at the time of the analysis, only estimates of expenditure and not final outturns were available. These limitations must be borne in mind in interpreting the results. The form of the estimation equations for this period are slightly different than for the pre-1981 period, and are as follows

$$E_i = g_0 + g_1 B_i + g_2 t_i + g_3 G_i + g_4 N_i + g_5 D_i + g_6 I_i$$

$$t_i = h_0 + h_1 E_i^{t-1} + h_2 B_i + h_3 G_i + h_4 N_i + h_5 I_i$$

$$G_i = e_0 + e_1 N_i + e_2 B_i t_i$$

The major change from the earlier period is the definition of the grant equation which takes account of the block grant structure. The terms g_i, h_i and e_i are a reduced parameter set and again for London, debt, need and wages are omitted. The results of the FIML estimates are given in the last column of table 9.3.

The relation of expenditure to tax base shows an important reduction in London, but an increase in the Counties, from the pre-1981 levels. Similarly, expenditures decline in relation to tax base in London, but they increase significantly in the Counties. Expenditures stay fairly high in relation to grant levels, but are reduced to below 100% in inner London. In outer London grants have little relation to expenditures, being used largely for tax reductions in real terms (compare the tax rate equations).

The tax rate equation shows a marked decline from pre-1981 levels in London in the relation of tax rates to grants, changing previous tax rate expansion to contraction in outer London. Tax base and past expenditures become even more important correlates of tax rates and suggest that it is tax base which is allowing the maintenance of expenditures in tax-rich areas. Grant equation coefficients are broadly comparable with the results for the earlier period, but many more coefficients are statistically significant. The more important role played by the needs index is particularly noteworthy since the index used in these estimates relates so closely to GRE used in RSG allocation in this year (see chapter 6).

9.7 Assessment of estimation results

The estimation results presented above suggest that in general, for pre-1981, tax base, needs grant and past expenditure levels are the prime determinants of both expenditures and tax rates. London, especially inner London, has frequently taken advantage of its high tax base to increase expenditure levels; it is also in London that the greatest tendency is shown to increase expenditures based on both needs and resources grant. London up to 1981 has also, as a result, shown the strongest tendency to increase tax rates in relation to needs grant, using this as a subsidy and price support to expand services. After 1981 the pattern changes somewhat, with grant levels becoming the prime determinant of expenditures, and tax base the main correlate of tax rates. Most important are the results for London which show that, for inner London, grant levels have been used for a less than 100% throughput into expenditures and have hence induced a lower rate of taxation than in earlier years; for outer London grant levels have led to reductions in overall levels of real spending and taxing. Both effects correspond to the objectives of central government in this period, of trying to reduce local expenditure, especially in high-spending areas.

The results of modelling expenditures and taxes in the manner developed above are, of course, only aggregate generalisations of statistical regularities over groups of authorities. They cannot necessarily be taken as typical of any one area within each group. However, this aggregate relationship does give an overall average or statistical yardstick against which individual areas can be gauged in order to see how far they deviate with respect to each equation: expenditure, tax rates and grants. In the broadest possible terms it is then possible to identify categories of 'high spenders', 'low spenders', 'high taxers', 'low taxers', 'grant favoured', and 'grant disfavoured' areas. This is undertaken in figures 9.3, 9.4 and 9.5 for each of the three equations. The comparison is, in effect, a plot of the residuals from the model estimates, and the local authorities are divided into groups depending upon whether their estimated values from the FIML estimates are more than plus or minus two standard deviations (2σ) of the actual values. Primary emphasis can be placed only on these large residuals since estimator variability and the overall confidence levels of the equations do not justify looking in detail at the smaller residuals. In each case residuals are calculated relative to other areas in their groups, i.e. (1) non-metropolitan Counties in England and Wales, (2) metropolitan Districts, and (3) London. Hence comparison of residuals between groups can be made only in the most cautious of terms. Results are mapped for the 1978/9 year which is fairly representative of the pre-1981 results. As in previous figures the metropolitan Counties are averages for their districts, but in the cases of the GLC as a whole, Westminster and the City of London, these are shown blank since they are omitted from the analysis. In interpreting these figures it must be borne in mind that each represents residuals in one equation *after controlling for other equations*; hence each of the three figures should be looked at together.

In the case of expenditures, shown in figure 9.3, the high spending areas form a fairly predictable group with many notable metropolitan Districts and London Boroughs included, frequently corresponding to areas controlled by Labour councils: Birmingham, Liverpool, Humberside, West Glamorgan, Lambeth, Lewisham and Camden. The low spending areas conform to many well-known examples of authorities practising relatively frugal levels of service provision, frequently controlled by Conservative councils: Gloucester, Hereford and Worcester, E. Sussex, Norfolk, Suffolk, S. Tyneside, Bradford, Tameside, Bury, Knowsley, Harrow and Bexley.

Figure 9.4 shows the pattern of high and low taxing areas in relation to model estimates, after controlling for grant and expenditure effects. High taxing areas are thus high in relation to their levels of expenditures and grants. In this case most of the low spenders are also high taxers in that they tax relatively highly for the level of expenditures they make. Also differing levels of efficiency will affect this relationship. The high taxers are: Hereford and Worcester, Gloucester, E. Sussex, Norfolk, Suffolk,

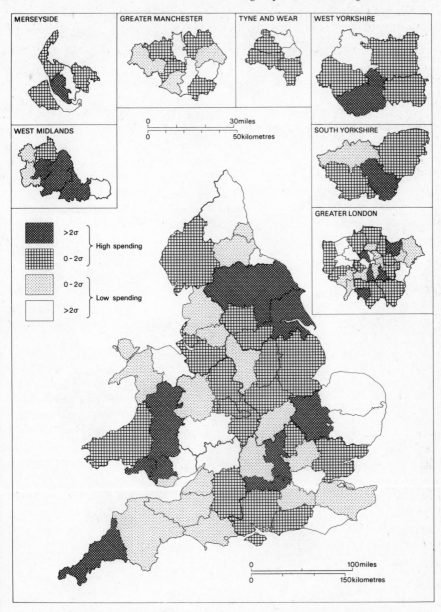

Fig. 9.3 Deviations from the aggregate relations of the expenditure equation in the FIML model in terms of the standard deviations (σ) of the residuals in 1978.

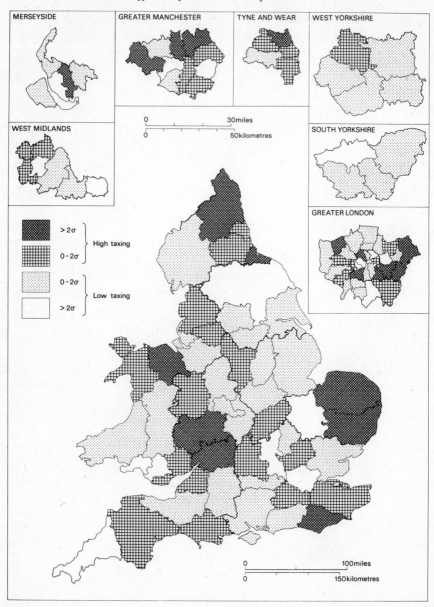

Fig. 9.4 Deviations from the aggregate relations of the tax rate equation in the FIML model in terms of the standard deviations (σ) of the residuals in 1978.

Clwyd, Cleveland, Northumberland, Bury, Knowsley, Newcastle and Harrow: almost all of these are Conservative controlled over the period of analysis. Most are also relatively grant favoured (figure 9.5). In contrast, the low taxers are in some cases the high spenders, e.g. N. Yorkshire, Cambridgeshire, Buckinghamshire, Cornwall, W. Glamorgan, Mid Glamorgan, Liverpool and Sandwell. These are a mixed group of authorities, many of which are relatively disfavoured in grant allocation. Interestingly two experienced change of party control just before 1978: to Liberal (Liverpool) and to Conservative from Labour (Sandwell). Many of the others are Liberal or Independent controlled (Wales and Cornwall) and this may help to explain what may appear to be a relatively high level of efficiency. Other low taxers are Coventry, Salford, Tameside, Barnsley, Newham, Islington, Kensington and Chelsea, and Croydon. In these a mixture of high tax base, low spending levels, and grant favouring offer explanation of the pattern.

Figure 9.5 shows the pattern of areas which are relatively favoured or disfavoured in grant allocation. The strongly favoured areas are Clwyd, Knowsley, Wigan, Bury, Rochdale, Newcastle, N. and S. Tyneside, Harringey, Hackney, Islington, Newham, Southwark and Merton. The strongly disfavoured areas are E. Sussex, Buckinghamshire, Cumbria, Sandwell, Dudley, Tameside, Calderdale, Havering, Bexley, Greenwich, Lewisham, Camden and Hillingdon. Obviously the grant equation used in this analysis is not in one-to-one correspondence with the complex group of factors used in actual RSG allocation. However, the inclusion of the major elements does give a good aggregate description of grant patterns. Hence, from this aggregate, major deviations display the effect of RSG favouring particular authorities. This pattern, and its subsequent effects on expenditures and taxes, is one of the crucial interests of the analysis. The resulting pattern is complex but the analysis of grant favouring and disfavouring in relation to expenditures and grants does bring out some important and highly expected underlying features. A tabular approach to describing these relationships is given in table 9.4.

The typology in table 9.4 shows that for all areas relatively favoured by grants, London is in marked contrast to other areas. Each of the favoured London areas has low relative tax rates; of these one half have low relative expenditure and one half high relative expenditure. In contrast for the (all non-London) areas having high taxes, almost all have low expenditures. The relatively disfavoured areas have the common feature that all high spenders and high taxers are Labour-controlled in London. In table 9.4 the groups of high spenders and high taxers, or low spenders and low taxers are marked normal for their level of service provision. Within each grant category they have the expected pattern of tax and expenditure relations. The other categories produce the more controversial results. Areas such as Newham and Southwark are highly advantaged in that they gain high

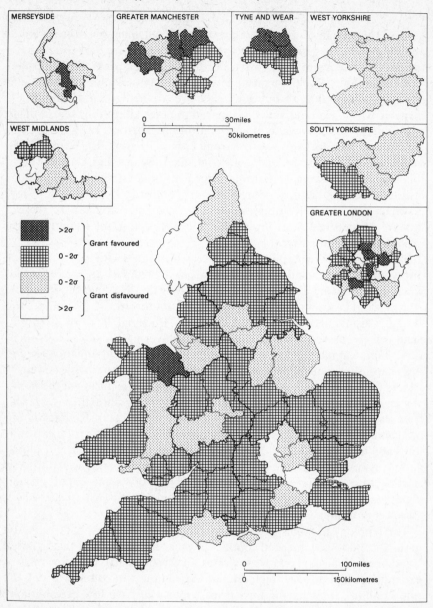

Fig. 9.5 Deviations from the aggregate relations of the grant equation in the FIML model in terms of the standard deviations (σ) of the residuals in 1978.

Table 9.4. *Typology of local authorities in 1978/9*

Grant state				Major examples	Possible examples
Grant favoured	High expenditure	high taxes	Normal high provider	–	Rochdale
		Low taxes	Highly advantaged	Newham Southwark	–
	Low expenditures	High taxes	Insufficiently favoured: stressed	Clwyd Bury Knowsley S. Tyneside N. Tynside	Wigan
		Low taxes	Advantaged normal low provider	Islington Harringey	Hackney
Grant disfavoured	High expenditures	High taxes	Normal high provider	Camden Greenwich Lewisham	–
		Low taxes	Insufficiently disfavoured	Bucks. Sandwell Calderdale	Cumbria
	Low expenditures	High taxes	Highly disadvantaged: stressed	Bexley E. Sussex	Havering
		Low taxes	Disadvantaged normal low provider	Tameside	–

The typology is based on the degree of grant favouring or disfavouring, and expenditure and tax rate decisions in 1978/9 using the results of the FIML estimates in table 9.3. Major examples are areas for which grant estimates deviate by more than 2σ, and estimates of the other two variables deviate by at least 1σ from actual values. Possible examples are cases in which grant estimates deviate by at least 2σ, and at least one of the other variables deviates by more than 2σ from actual values.

expenditures with low relative tax rates; by implication they are 'over-advantaged' relative to their tax base and relative to other areas with similar characteristics. In contrast, areas such as Tyneside, Knowsley, Bury, etc. are favoured by grants, but have only low relative expenditures at high tax rates in comparison with other areas in their groups. They are termed 'stressed' in the sense that, even after grant favouring, they still have to make large fiscal effort to support expenditures.

The corresponding groups in the grant disfavoured category are highly disadvantaged or insufficiently disfavoured in the terms discussed here. The highly stressed areas have low expenditures at high tax rates and low level of grant. In contrast areas such as Buckinghamshire, Sandwell and Calderdale have high expenditures at low relative tax rates, mainly as a result of large tax base. The stressed and advantaged areas exhibit one of the difficulties of the pre-1981 grant in that it gave insufficient weight to the extent to which areas used tax base to satisfy expenditure requirements.

The typology in table 9.4 is a relatively crude and contentious one, but it does allow comparison between areas and time periods. Table 9.4 relates to 1978, one of the last years affected by the Labour government of the 1974–79 period, and it is fairly typical of the FIML estimates for this period. We would expect a marked contrast of these results with later years in which local authorities have had to cope with the changed priorities and changed RSG grant structure of the Conservative central government holding office from 1979. Table 9.5 displays the equivalent typology to that of table 9.4 but for 1981/2, the first year of the RSG block grant. As in table 9.4 only the residuals exceeding the 2σ limits on the grant variable are used.

There are few similarities between tables 9.4 and 9.5. The highly advantaged category in 1981 has now been reduced, but the stressed category has increased. The favoured high providers contain a fairly clear and expected group who decide to spend and tax to provide high service levels and also gain grant advantage. Amongst the disfavoured areas Islington moves from being favoured in 1978 and Greenwich changes from being a low taxer. Tameside maintains its position as a disfavoured low provider. In general terms grant favouring seems to have shifted highest benefits from many London and central city metropolitan Districts to a number of non-metropolitan Counties and some of the more suburban metropolitan Districts, often those with high tax base, e.g. Kensington and Chelsea and Bury.

It is clearly too early to say what the long term effects of the post-1981 RSG block grant will be, but the model estimates given in table 9.3 and the residuals discussed above give grounds for believing that considerable shifts in benefits and burdens have resulted. In aggregate this has produced a marked reduction in per capita grant levels particularly in inner London

Table 9.5. *Typology of local authorities in 1981/2*

Grant state				Major examples	Possible examples
Grant favoured	High expenditure	High taxes	Normal high provider	Bury Bedfordshire Kensington & Chelsea	Rotherham Newcastle
		Low taxes	Highly advantaged	–	Kent Wandsworth Hounslow
	Low expenditure	High taxes	Insufficiently favoured: stressed	Durham Gateshead Cleveland Mid Glamorgan	Berkshire
		Low taxes	Advantaged normal low provider	Knowsley	Hillingdon
Grant disfavoured	High expenditure	High taxes	Normal high provider	Hackney	Hammersmith
		Low taxes	Insufficiently disfavoured	Humberside Islington Greenwich	Salford St. Helens Bradford
	Low expenditure	High taxes	Highly disadvantated: stressed	Leicestershire Nottinghamshire Brent	Devon
		Low taxes	Disadvantaged normal low provider	Tameside	Wiltshire

The typology is based on degree of grant favouring or disfavouring, and expenditure and tax rate decisions in 1981/2 using the results of the FIML estimates in table 9.3 and the same criteria for groups as in table 9.4.

and some metropolitan areas. This has produced an aggregate tendency to lower levels of expenditure and lower levels of taxes in real terms.

9.8 Conclusion

This chapter has been concerned with defining a model to capture the main interrelationships which determine the level of local expenditures, needs and grants. This model is obviously a simplification of the complex reality which determines local budgetary decisions and central–local political processes. However, it does provide a fairly sophisticated tool for assessing the impact of central grants, and its definition has incorporated most of the features which have been received from the literature of economics, economic geography and political science. The model developed in this chapter has been applied to assessing the impact of the Rate Support Grant on local expenditure and tax rate decisions. The conclusions of this analysis have important implications for both expenditure modelling and assessment of the impacts of the RSG in Britain.

The aggregate statistical results which have been developed have considerable importance because they allow the finance of local authorities in England and Wales since the 1974 reorganisation of government to be related to overall theories of public finance. This relation of the practice of British local authorities to aggregate theory has been notably lacking until recent years in Britain. The developments of this chapter, giving the first major simultaneous equation estimates of the British grant, expenditure and tax relationships, should be an important stimulus both to further theory and to practical decisions in grant design in the future. Although there are variations between estimators, the simultaneous equation results offer a high level of stability and statistical significance. They also relate to most of the prior ideas we hold of reality. For the pre-1981 period tax base, needs grant and past expenditure levels are the main determinants of both expenditure and tax rates. It is suggested that inner London has used its advantage of high tax base to support high expenditure levels: grants have a direct throughput to expenditures at around 130% of the level of grants. For most areas in this period the level is at about 100% of grants going directly into expenditures, but in the Counties quite high levels find their way into tax rate reductions. All of these effects change after 1981; the central government attempts to control local expenditure in this period have produced a reduction in the level of grant throughput with a greater impact on local tax rates, and this effect is particularly marked in London and some metropolitan Districts.

Analysis of the residuals from the simultaneous equation model for both 1978 and 1981 shows important patterns of stress not answered, or not answered sufficiently, by response of grant allocation. It also shows benefits going disproportionately to some areas despite low expenditures or high tax bases. Although the results are highly aggregated and preliminary, the pattern of residuals does suggest unequal and shifting burdens and benefits of the aggregate system of financial relations in local authorities in England and Wales.

Results based upon aggregate statistical relationships must always be treated with caution, especially in trying to make inferences relating to individual cases. However, this chapter has developed a relatively straightforward methodology for estimating expenditure, taxing and grant effects based upon a set of objective and aggregate characteristics which allows comparisons to be made between areas. Although no account has been taken of local variations and idiosyncrasies it does offer a good approach to understanding the generalities. As an approach it may well commend itself to central governments concerned with aggregate comparisons between authorities. However, the general approach and analysis of deviant cases is not intended to do more than, at this stage, highlight aggregate relationships and structures and allow the overall effects of the RSG to be assessed. In the same spirit, the next chapter uses the implications of the results of this and earlier chapters to suggest modifications to the whole system of local finance in which grants are more firmly enmeshed in the processes of decision-making on tax and expenditure policies at both central and local levels of government.

PART III

TOWARDS A MORE RATIONAL PROCEDURE

CHAPTER 10

AN APPRAISAL OF ALTERNATIVES

10.1 Introduction

It is clear from the discussion of the preceding chapters that the allocation of the Rate Support Grant has possessed a number of inadequacies which have induced significant inequities into its allocation between areas resulting in important differences in total rate bills and, hence, in an unequal geographical pattern of fiscal incidence. Moreover, the reform of the RSG in 1981/2, although mitigating some of these effects, has still left many problems untouched, and in some cases has introduced important new inequities.

Because of the problems that still remain in the method of distributing RSG it is necessary to appraise alternative methods that might be employed. In order to undertake this, the present chapter starts from the premise of how, for the existing structure of British local government and the services it provides, a grant system should be designed which allows full fiscal equity to be achieved. The distribution of central grants should be critically dependent upon the two features: first, the methods of local raising of revenue, for example, the tax sources, distribution and equity of tax bases, extent of user charges, and so forth: second, grants should also depend on the form of local expenditure, especially the extent of discretionary components, the variation in costs, levels, efficiency and output between areas, and the extent of redistributional intent built into service areas. As a result of the interaction between these two factors of local revenues and expenditures, it is not possible to discuss new forms of central grants in isolation from broader questions relating to the local financial system as a whole.

The interlinkage between central grants, local taxes, and local expenditures has been recognised most recently in the Green paper, G.B. Government (1981b), which presages the shift of domestic rating to a capital valuation principle and appraises alternative and additional local tax sources which can be used to replace the domestic rates. This appraisal by the Conservative Government was restricted largely to alternatives which could be implemented in the short term. The present discussion takes a somewhat longer-term view and assesses the major means of co-ordinating central and local finances through taxes, grants and expenditures. Then, in the next chapter, a new form of central–local financial structure is proposed. The present chapter is divived into three main sections. In section 10.2 the present structure of local finance in Britain is critically assessed in order to determine where improvements are required. Then, in section 10.3, five main methods of improving central–local co-ordination are discussed. This is followed by an assessment of each of these alternatives in section 10.4. The conclusion is that a mixture of solutions rather than one simple solution is required. The mixture suggested is of improved rating valuation, increased user charges, tax-base sharing (involving tax credits), extended specific grants and improved unitary block grants. This conclusion is discussed further in the next chapter where its practical implications are assessed.

10.2 Difficulties in the present structure of central–local relations

The present structure of finance of public services in Britain derives from the interplay of the two sides of the public fisc: revenues and expenditure. In Britain, as in all western countries, the two responsibilities of making expenditures and levying taxation are divided, to a greater or lesser extent, between different levels of government. In Britain this division is between central and local government, with additional complexity added by the two-tier structure of local government (Counties and Districts in England and Wales; Regions and Districts in Scotland). Within such patterns of multi-level government there is a major problem of balancing the taxation raised at each level with the expenditure required at each level.

In Britain, under existing financial provisions, local government undertakes about 27% of total public spending using revenues which provide only 22% of total government revenue. As a consequence, considerable transfers from central government are required in order to match locally-controlled expenditure with the revenues required to support them. As a consequence of this heavy dependence on transfers, a number of problems of local finance have increasingly pressed themselves to attention in Britain. The five most important of these have been: (1) deficiency in independent local revenues; (2) the undermining of local accountability; (3) the loss of local discretion; (4) encouragement of economic inefficiency; and (5) perverse effects on income distribution. Each of these has been raised in earlier discussion but is briefly reviewed below.

The problem of deficiency in local revenues is a major underlying problem of multi-level government in Britain. The major taxes are mostly held at national level: the three sources of personal income and corporate taxes, and VAT alone account for 65% of total taxation in 1977/8 (see table 10.1). Local government has only one major tax source and as a result it has frequently been argued that the local authorities have been seriously

Table 10.1. *Sources of revenue in England and Wales by level of government*

Revenue source	£m (1979 prices)				Percentage of total revenue at each level			
	1967/8	70/1	74/5	77/8	67/8	70/1	74/5	77/8
1 CENTRAL REVENUE								
Personal income tax (and Surtax)	14982	20019	24878	24499	27.4	28.5	44.1	45.0
Corporation tax	4463	5299	6801	4693	8.2	7.5	12.1	8.6
Capital gains (+ DLT) tax	58	465	908	487	0.1	0.7	1.6	0.9
Death and Estate duty (+ CTT)	1214	1794	806	558	2.2	1.7	1.4	1.0
Stamp duties	362	392	471	528	0.7	0.6	0.8	0.9
Petroleum revenue tax	—	—	—	1	—	—	—	—
Alcohol taxes	2751	3116	2695	2893	5.0	4.4	4.8	5.3
Tobacco taxes	3839	3820	3182	2884	7.0	5.4	5.6	5.3
Purchase tax and VAT	2751	4252	5965	5939	5.0	6.1	10.6	10.9
Betting taxes	250	437	567	450	0.4	0.6	1.0	0.8
Fuel taxes	3566	4664	3882	3450	6.5	6.6	6.9	6.3
Customs duties	13690	15766	1250	1206	25.1	22.4	2.2	2.2
Car tax and Licence fees	990	1409	1503	1882	1.8	2.0	2.7	3.5
Employment taxes (SET and National insurance)	3914	6662	5700	1633	7.2	9.5	0.1	3.0
Others	1705	2767	3444	3304	3.1	3.9	6.0	6.0
Total central income	**54481**	**70262**	**56357**	**54407**	**100**	**100**	**100**	**100**
2 LOCAL REVENUE (ENGLAND AND WALES)								
(i) Locally derived								
Property tax (rates)	4869	5492	6964	6580	25.5	24.5	20.9	29.2
Licences	4	4	1	0.3	—	—	—	—
User Charges	2687	1026	1070	1218	14.1	4.6	3.2	5.4
Other Revenue a/c income		1833	3323	3334		8.2	10.0	14.8
Loans	4531	5348	8107	67	23.7	23.9	24.3	0.3
Other Capital a/c income	884	1239	1265	87	4.6	5.5	3.8	0.4
Total	**12975**	**14942**	**20730**	**11286**	**68.0**	**66.7**	**62.2**	**50.1**
(ii) Transfers from central government								
Rate Support Grant	4722	5915	10264	9084	24.7	26.4	30.8	40.4
Specific grants	1087	807	1833	1960	5.7	3.6	5.5	8.7
Grants to Capital a/c	271	376	332	700	1.4	1.7	1.0	0.1
Other transfers	35	371	180	176	0.2	1.6	0.5	0.8
Total	**6115**	**7469**	**12609**	**11227**	**32.0**	**33.3**	**37.8**	**49.9**
Total local income	**19090**	**22411**	**33339**	**22513**	**100**	**100**	**100**	**100**

Central revenue refers to Great Britain totals (Source: Central Statistical Office, *Financial Statistics*), Local revenue refers to England and Wales (Source: Department of Environment, *Local Government Financial Statistics*).

weakened, and this has often restricted the accountability and discretion of local finances. Whilst the property tax has the advantages of cheap administration, difficulty of evasion, its undeniable local character, its high yield and its amenability to multi-level use, it does nevertheless possess important disadvangages: that it is not buoyant,[1] it is not redistributive, it tends to discourage improvement of property, and there are difficulties in maintaining up-to-date valuations (see chapter 3). As a result of the balance of these relative advantages and disadvantages, it has seldom been proposed that the property tax should be abolished altogether. However, it has become increasingly clear that the rates should be supplemented and the need for additional revenues has been recognised in a series of government statements (G.B. Government 1966, 1971, 1977, 1981*b*; G.B. Royal Commission on the Constitution, 1973; G.B. Committee of Enquiry, 1976). Various alternative sources of revenue suggested in these statements are examined in section 10.3 of this chapter.

The undermining of local accountability has resulted from inadequate locally-raised revenue and from the increasing dependence of local government on central transfers especially the RSG (see table 10.1). These two factors were sufficient to make the Layfield Report (G.B. Committee of Enquiry, 1976) suggest the need for alternative revenue sources, the preferred alternative being a local income tax. Accountability, therefore, overlaps to a great extent with the difficulties arising from the dependence on the property tax as the major source of locally-raised revenue.

The loss of local discretion has been associated with the increasing role of central grants which have brought with them increased scrutiny, increased subjection to the uncertainties of national economic management policies (including monetary controls and policies on pay and prices), and increased direction through departmental circulars and legislation on the manner in which expenditures should be made (especially important in housing, education and general capital spending). Especially since 1981/2, strings have been attached to the RSG by setting a standard expenditure (GRE) and standard rate poundage (GRP) against which to assess local authorities and, since 1980/1, authorities that 'overspend' relative to the expenditure and poundages chosen have been penalised.

The encouragement of economic efficiency has also been a side effect of the increased level of central grants and relative reduction in the importance of local taxation. On average, about 60% of total local relevant expenditure has been funded centrally, and for some areas the proportion is over 90%. As a consequence it is often claimed that services to the consumer are severely underpriced in terms of the perceived cost borne in taxation. As a result demand has tended to increase, politicians and councils have been encouraged to oversupply services, and inefficiency, especially as a result of overmanning, has been encouraged. It was argued, for example, by the Department of the Environment (G.B. Committee of Enquiry, 1976, Appendix 1, pp. 91–3), that a high grant percentage in itself does not act as an incentive for local authorities to increase expenditure, and moreover, marginal expenditure above grant settlement has greater impact on the local tax rates as grant rises. However, since 1980 large grant percentages have been increasingly regarded as stimulating expenditure. Three main factors are important in this regard. First, historic patterns of service supply and demand affect the levels of need that can be recognised in many services categories with the consequence that previously high expenditures can attract high grants and this was reinforced especially by the RSG regression formulae and damping used up to 1980/1. Second, neither central nor local forecasts of future price changes in local authorities are independent of local behaviour, especially in setting pay and manning levels; changes which have usually been passed on in grant increase orders. Third, the Rate Support Grant has not equalised perfectly between spending needs and tax base resources but has been subjected to annual relatively volatile shifts, to feedback and deficiency payment of the resources element, and other difficulties (see G.B. Committee of Enquiry, 1976) which up to 1980/1 may have encouraged some local authorities to overspend, if not others. Certainly by the late 1970s the RSG system was seen as stimulative of overspending to the extent that its reform to encourage greater control seemed essential (G.B. Government, 1980*a*, *b*).

The perverse effect of local finance on income distribution arises not, like the previous four factors, from too large a role being played by central grants and too small a role by local taxation, but instead derives from three main problems in the manner of allocation of central grants and a lack of relation of local finance to total taxation policies. One major problem has been that the RSG has been insufficiently equalising and has encouraged either poverty traps or cumulative cycles of local overspending. A second problem has been that the taxation and expenditure sides of local finance have not been sufficiently interrelated. For example, as shown in chapter 9, people living in areas with large tax bases in relation to need have benefited from reduced tax bills, whilst people in areas with small tax bases in relation to need have been penalised by unnecessarily high tax bills. However, a third problem, which grants alone cannot solve, is that deriving from unequal income distribution within local government units. Equalisation of criteria such as average domestic rate bills (ADRB) between local authorities, which is a subsidiary aim of the Rate Support Grant, leaves income distribution *within* areas unaffected and it will frequently be the case that equalised ADRB will be perverse to other income distribution criteria. For example, a local authority may, relative to its population, have a large number of old people needing home-helps, meals on wheels, and other services, hence qualifying it for a large payment of RSG. However, these large numbers of old people may either be able to pay for the services themselves, or it might be that the tax base of the local authority is sufficient to support their needs. What is required, therefore, is a means of relating tax resources and need not only at the spatial level (between local authorities) but also at the individual level; and this in turn requires considerations of local finances to be integrated with consideration of the system of total taxation and benefits in the nation as a whole.

Each of these problems with the present structure of local finance in Britain draws us back to two general issues: first, the inadequacy of independent local tax sources, and second, the high level of dependence of local government on central grant transfers. Additional issues arise because the grant system may be seen as not the most appropriate and because central finance of local government has often been perverse with respect to income distribution criteria. A central question, therefore, is to determine if alternative methods of local finance, including alternative tax and grant structures, would offer improvements. The rest of this chapter discusses the major alternative methods of co-ordinating central and local finances.

10.3 Alternative forms of central–local relations

Five main methods are available for co-ordinating central and local finances to provide local revenue and expenditure balance (see Hunter, 1977; Bennett, 1980). First, additional local taxes can be used to provide many of the advantages of local rates without having their disadvantages. This alternative has so far been rejected by the central government for immediate implementation. The second alternative is tax base sharing. This was never seriously considered in any detail by the central government until the 1981 Green Paper (G.B. Government, 1981b). The third alternative, of tax credits and deductions, whilst considered in Appendices to the G.B. Committee of Enquiry into Local Government Finance (1976), has never received a great deal of support from the central government. A fourth method of increasing local revenues is to make increased use of charges for a larger number of local services. A fifth method of coordination is to use new forms of grant from central government. This was the alternative proposed by the central government in 1966 (G.B. Government, 1966) and employed since that date. Each of these methods is discussed in turn below.

Alternative local tax sources

Local tax sources which are alternatives or additions to the rates have been considered in a number of government statements, and the most frequent proposal has been for the introduction of the local sales tax, poll tax, or income tax (see G.B. Government, 1971, 1977, 1981b; G.B. Committee of Enquiry, 1976). However, proposals for major new local taxes have so far been firmly rejected by the central government.

A major issue related to the degree to which additional local tax sources can be employed derives from the extent to which revenue separation or competition is induced. Revenue separation is represented by the present financial system in Britain with totally separate revenue sources used by central and local government. Revenue competition results when two levels of government tax the same tax base, as would occur if a local income tax was introduced. Separate revenue sources have the advantage that they make it clear, to the taxpayer, to whom and for what a particular tax is paid, they are easier to administer, and allow income distribution effects to be more clearly assessed (Newcomer, 1917). However, separate tax sources are often expensive to administer, the sources assigned exclusively to local level usually provide inadequate yields, and separation does not dispose of the need for revenue transfers (such as grants) to equalise between areas with rich and poor tax bases. Moreover, the problems arising from the alternative of overlapping and competition for taxes can often be overcome to a great extent by grants, credits and deductions.

Six main arguments have been made against the case for providing additional local tax sources in Britain. First, most of the taxes which it would be feasible to implement at local level in Britain are too low-yielding to add significantly to local revenue as, for example, with betting, dog licensing, vehicle licensing and similar taxes. As a consequence the Layfield Report saw LIT as the only viable alternative to the rates; but sales tax is another alternative (see G.B. Government, 1981*b*). A second argument has been that the major high-yielding taxes which are available, such as personal and corporate income taxes, payroll taxes, and sales taxes, have major difficulties in local application; either as a result of high costs of collection and administration or of identifying the base of the appropriate local taxing authority. For example, with income taxes there is the question whether a local income tax should be apportioned to living or working unit. In addition, for corporate taxes there is the problem of assigning the appropriate level of tax between production units in multi-plant enterprises. The administrative costs and other difficulties are particularly quoted by the Treasury and Inland Revenue (see G.B. Committee of Enquiry, 1976, Appendix 1). A third argument with the high-yielding local taxes, especially those on personal or corporate incomes, is that the tax base is spatially variable to a very significant extent. Hence, local use of these taxes would still require central government equalisation. A fourth argument, also emphasised by the Treasury, is that the basis for variation in tax bases between areas would in many senses be 'irrational'. For example, for fuel tax it would depend on the level of vehicle ownership and use; for sales tax variations would derive from different levels of retail sales or turnover; a payroll tax would vary with employment/population ratios; and a betting tax would vary with the incidence of gambling. A fifth argument has been that of central government managing the economy if some of its fiscal instruments are being used also by local governments. The sixth argument relates to the stimulus major differences in personal and corporate income taxes might give to 'fiscal migration' if the rate of local tax was at all significantly variable between areas: the so-called horizontal inefficiency argument (Oates, 1972; Foster *et al.*, 1980). In the USA, for example, where local taxes have been implemented and where considerable variations in tax rates occur, differences in local tax burdens have been identified as a major stimulus to migration (see e.g. Tiebout, 1956; Mills and Oates, 1973; Oates, 1969; Margolis, 1961). As a result they form part of a model of so-called fiscal migration.

Up to 1974, many of these difficulties in the way of diversifying the local tax base in Britain seemed to arise from the fiscal migration which might result from tax rate variation in a large number of small government units within the same journey-to-work area. For example, G.B. Government (1966, p. 4) recognised that 'In a country as heavily populated as this local authority areas can never be self-contained, but *larger areas* would possess a better balance between different types of development and activity, while a reduction in the *number* of authorities would greatly ease the administrative problems connected with the collection of possible taxes' (italics added). However, since the reorganisation of local government in 1974 many of these difficulties have been removed: in particular local units have become larger for most major services.

Most of the objections to diversifying local revenues seem to arise not, however, from objective difficulties in implementing changes, but from the massive inertia to change built into the outlook of central government deriving in part from the economic philosophy of its principal fiscal agency the Treasury. This has produced what Prest (1978) terms 'an everlasting theme song' of central government rejection of new local taxes. As Prest notes, this is perhaps nowhere more clearly stated than for LIT in the G.B. Committee of Enquiry (1976, Appendix 8, p. 269) evidence by the Treasury: 'We need income tax as a fiscal instrument; we need to have it unimpaired, and we need to maintain its flexibility in the hands of central government'.

This is not the place to use a great deal of space in assessing and rejecting each of the arguments against providing additional local tax sources. For the major feasible additional local tax, the LIT, a full dismissal of the major arguments has already been given (G.B. Committee of Enquiry, 1976; Bennett, 1981*b*). These rejections can be briefly summarised as follows. First, the practical issue of assessing LIT on a place-of-residence basis is quite feasible for personal income tax (Prest, 1978), but may involve high administrative costs. However, these costs are not as excessive as is often claimed. The Layfield Report estimated these as £50 m in direct collection costs plus a similar amount for the additional work required by the private sector (at 1975 prices). Certainly, as Prest (1978) and Foster *et al.* (1980) conclude, the implementation of LIT would be considerably simplified and the costs reduced if Britain had a self-assessment system of income tax, such as that proposed by Barr, James and Prest (1977). However, since it is likely that the collection costs of LIT will be more or less fixed as the level of the tax rises, then the costs are not excessive provided the tax is used to a reasonably large extent. For example, if LIT were used to the same extent as the rates in providing about 30% each of local revenue needs, then the relative administrative costs would not be excessive (about 2.3% of revenue yield and roughly comparable with the rates of about 2%). Moreover, the relatively low collection costs of the rates has often been achieved at the expense of infrequent updating of the valuation tax base. Originally proposed to be on a uniform five-year basis of revaluation, in fact valuations have been undertaken much less frequently, as evidenced by the recent postponement, on grounds of cost, of the valuation due for 1982. If these revaluation costs are taken into account the collection and administration costs rise considerably above 1.5–2% of costs in relation to tax yield.

With respect to the other objections to new local taxes, those most worthy of attention relate to undermining of central economic management and inducements to fiscal migration through horizontal inefficiency. Regarding central management of the economy an allocation of a new tax to local use is equivalent to assigning about 25% of the personal income tax (thus providing about 30% of local resources). Since this is the highest level which might be considered, and since this level is also approximately equivalent to the total share of local spending in total public expenditure, the assignment does not seem altogether unreasonable. Since it is also true that local spending is much more fixed and inflexible than central spending, and has in recent years been largely stable or counter-cyclical during the economic cycle (Neild and Ward, 1976; Foster *et al*. 1980) then it is not clear how central ability to manage is undermined.

With respect to inducement to fiscal migration, the arguments are far more serious, but again can be dismissed provided that the right structure of new local taxes is chosen. First, it must be stated that no scheme for providing additional local revenue can be seriously proposed in isolation from national taxing and spending policies. Hence a new local tax requires a system of tax credits and deductions as discussed further on page 263 below. When this set of procedures is taken into account, then almost all of the objections to new local taxes collapse.

Taking these arguments together it is clear that it is not economic to implement LIT or any other additional local tax for long-term use at a comparatively low level of utilisation. It would be worthwhile having a low rate of LIT, however, for a short transition period. When LIT is compared with rates *at the same level of tax yield* the differences in collection and administration costs tend to become very small and perhaps disappear altogether. Hence, from this discussion it must be concluded that a major alternative or addition to the rates exists in the form of LIT although, as we shall see, this needs to be connected with tax credits, index-linked tax definitions of thresholds and marginal rates, and to local collection.

Tax-base sharing

It certainly seems clear from the preceding discussion that no new *separate* local tax source is available for local use which is of sufficient size. Hence, diversification of local revenues must rely on implementing a sharing of tax bases between central and local government; and the main candidate for such sharing is the personal income tax.

Tax-base sharing (or revenue-sharing) involves the allocation of fixed proportions of selected revenue sources derived from central government to local governments on a permanent basis. There are two forms of tax-base sharing. First, special revenue-sharing assigns fixed percentages of central revenues to specific spending programmes. Second, general revenue-sharing assigns a fixed proportion of central revenues with minimum strings attached. Most federal countries employ a significant proportion of tax-base sharing in their intergovernmental programmes. This is, for example, a major component of revenue transfers to intermediate governments in Canada, Germany, Austria, Switzerland and, since, 1976, in Australia (see Bennett, 1980).[2]

Tax-base sharing has never been given serious consideration in Britain. Its use was not even considered in the devolution debate, and few people have been aware of its advantages over a purely local income tax. There was, for a short period after 1887, following Goschen's (1872) suggestion, a form of 'assigned revenues'. However, their basic form of allocation made them more akin to grants. They can now be seen, in retrospect, as a transition form associated with placing British local authorities on a uniform basis following the 1888 administrative reforms.

The aim of tax-base sharing is to give to local authorities the advantage of a wider revenue base, which is also more flexible and buoyant, but without the disadvantages of attempting complete separation of revenues between levels of government. In addition, tax-base sharing allows the greater use of taxes which can be progressive, and hence redistributive, at central level, but which if applied at local level have to be proportional or even regressive. Hence, it relieves the burden on the more regressive property tax, and does not conflict with national fiscal policies which aim at total income redistribution.

However, tax-sharing, like grants, has the disadvantages of diffusing lines of accountability and underpricing local services. But tax-base sharing differs from grants in three important respects. First, it represents a more certain and stable source of revenue than grants. In Germany, Canada, Switzerland and Austria, for example, shared tax bases are guaranteed by constitutional amendments and requirements. Second, equalisation cannot usually be combined with tax-base sharing since this would involve the employment of criteria for equalisation which usually must derive from the central government, and certainly require a degree of central intervention to assess local variations in needs and resources. Hence, tax-base sharing cannot substitute for equalisation grants where these are required. Third, shared tax bases can be made relatively buoyant, and this will normally represent a considerable advantage over central grants. Since shared taxes usually derive as a fixed percentage of central taxes, they rise naturally with the growth of central taxes, national income, or prices. This has the additional feature of providing better guarantees of support for local services.

With the advantages of buoyancy and guarantees of total sums available, tax-based sharing is most appropriate for providing general unhypothe- cated revenue support on a per capita or similar basis. It cannot be used for major equalisation without running into the same difficulties as grants, in which case the difference between tax-base sharing and grants becomes what has been sometimes termed merely 'presentational'.

The implementation of tax-base sharing in Britain can be taken with, or separately from, the issue of providing new local taxes; i.e. LIT could be administered as a form of assigned or shared revenue, or in addition to shared revenues. Implementation of tax-sharing, however, would mean that the central government would have to accept a permanent, or at least a long-term, commitment to release to local authorities a share of income tax or some other source. In the British context a constitutional guarantee of the inviolability of this arrangement is probably impossible. However, strong and significant commitments via a new *Local Government Act* would be possible and for general, unhypothecated aid would be a significant change increasing the discretionary independence of British local government.

Tax credits and deductions

Tax credits and deductions allow an individual a rebate from central government taxes equal to, or in proportion to, the levy of a specified local tax or taxes. As such, credits and deductions can only be considered jointly with a particular form of local and central taxation. In practice, in Britain, this means the use of credits or deductions for local use of income tax (although they could also be applied to the rates).

Credits and deductions both have the advantage that central govern- ment, in effect, supports a proportion of local revenue burdens (equal to the sum of deductions on credits allowable) and thus, like grants and revenue-sharing, permit the wider, more flexible and more redistributive national revenue base to be used instead of the local tax base. However, deductions and credits have two main disadvantages. First, they induce a

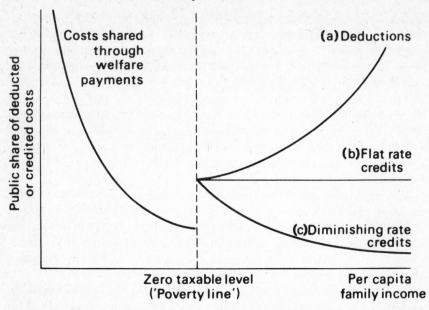

Fig. 10.1 The effect of tax deductions and tax credits (of local tax against central tax bills) on per capita income (after Pogue, 1974; and Bennett, 1980).

form of central government involvement into new areas of local action. Second, both deductions and credits erode the central tax base which ultimately may require new tax sources to be found.

Credits and deductions have always been rejected in Britain, but are used extensively in federal countries. For example, in the USA deductions are a major source of intergovernmental co-ordination, whilst in Canada and West Germany, tax credits are used. Applying the definitions usually employed, tax credit and tax deduction can be expressed by the following equations

(1) Tax deduction

$$T = (Y\text{-}K\text{-}D)t_1 \tag{10.1}$$

(2) Tax credit

$$T = (Y\text{-}K)t_1 - (Y\text{-}K)t_2$$
$$= (Y\text{-}K)(t_1\text{-}t_2). \tag{10.2}$$

In these equations T is the total tax paid by an individual, t_1 is the central tax rate and t_2 the local tax rate (both of which may vary with income), Y the total income of an individual, K the tax threshold, D the tax deduction, and $(Y\text{-}K)t_2$ the tax credit. The impact of these two methods of financial co-ordination is shown in figure 10.1. Tax deductions allow an individual to deduct local taxes paid from income liable to tax *prior* to assessment of central tax. Tax credits, in contrast, allow reductions in central tax liability for local tax actually paid *after* central tax has been assessed. Since deductions reduce the income liable to taxation they give greater benefits to taxpayers liable to higher rates of taxation. Hence, deductions are distributionally perverse in that they affect the marginal rate of tax on different income levels. Credits, on the other hand, leave marginal tax rates unaltered, provided that local government employs the same tax definitions as central government (see Maxwell, 1962; Pogue, 1974; Sunley, 1977; Bennett, 1980). As a result credits are usually deemed more socially equitable and play a major role in Canada.

Clearly the use of tax credits in the British system of local finance would allow the achievement of similar properties to that found in other countries, of eliminating horizontal inefficiency. Indeed every country which employs an LIT also employs some system of tax credits or deductions; and no recent proposal for the introduction of LIT into Britain has been made without considering some form of deduction or credit to be essential.

The Layfield Report (G.B. Committee of Enquiry, 1976, pp. 448–51) proposed one form of tax credit which required that both tax base assessment and tax collection be operated by the Inland Revenue, with the tax rate being set by both the local authorities and central government. As noted by Foster *et al.* (1980), this proposal creates great problems for perceptibility since the tax is collected by employers through PAYE, paid to the Inland Revenue, and paid back to local authorities without the necessity of the taxpayer knowing the tax rate or form of LIT at all. The Layfield Report suggests overcoming this by the Inland Revenue providing to each taxpayer a statement at least annually of the rate and level of LIT paid. But this is both cumbersome and costly, and also undermines perceptibility and hence accountability.

The only other major proposals for tax deductions and credits in Britain derive from the Inland Revenue evidence to the G.B. Committee of Enquiry (1976, Appendix 8, pp. 90–2). In this evidence, the Inland Revenue correctly considered tax deductions to be out of the question because of both their perverse distributional effects and the difficulties of relating information on local ratepayers, held by local authorities, to information on taxpayers held by the Inland Revenue (estimated to require at least an addition 12 million IR tax returns each year). The Inland Revenue have also considered a flat rate tax credit system to be out of the question. Running at a suggested level of about £40 per year (1976 prices), this method was deemed administratively difficult since the credit would involve payment to many non-taxpayers; or if the credit was itself regarded as taxable income, then it would increase the total number of taxpayers in the country as a whole by about 200 000. In addition it would affect the marginal benefit position of special groups such as pensioners. To simplify the scheme by exempting non-taxpayers from credits, and exacting tax at basic rate on higher income tax earners, would again be perverse in distributional terms.

The only scheme which the Inland Revenue have considered would be administratively easy and cheap is a deduction of basic rate income tax from rate bills. This would mean that instead of central government giving deductions for local tax paid, local government would have to give deductions for central income tax paid, the difference being made good by central government payments. This proposal has attractive features since it would mean that each domestic ratepayer would receive the same relief. The outlay on income tax and rates combined would leave each ratepayer in the same income tax position as if no rates were paid, provided he was a basic rate income taxpayer. If he was a higher rate income taxpayer, relief would be restricted to the basic rate alone. Whilst attractive from the total tax point of view, the proposal has the disadvantage that it relies on central government making good the central tax allowed as deductions by local government and hence leaves it in a relatively weaker position than with a deduction or credit given by local government.

The Layfield and Inland Revenue proposals, however, are not the only possible form of tax credits for LIT. Whilst there is insufficient space here to explore in detail the extent and practicalities of other schemes by which credits for LIT could be administered, it is possible to sketch briefly one alternative form of credit for LIT which overcomes most of the criticisms of perceptibility, and other problems in the Layfield scheme as well. This can best be understood by considering three separate components of central–local taxation: assessment of the tax base, setting the tax rate, and tax collection.

A form of credit for LIT which gives better properties of perception is to remove the collection stage of the tax process from the Inland Revenue and give it instead to the local authorities. This has the advantage that the costly and contentious reallocation of LIT from the Inland Revenue to the local authorities is no longer required, nor is it necessary to inform taxpayers of what they pay in terms of LIT. This suggestion would still require the Inland Revenue to assess income liable to tax. However, instead of being collected through employers via PAYE, notices of personal income tax liability could be sent to local authorities who would issue to taxpayers, in accordance with place of residence, the tax demand for LIT. Tax announcement and collection would, therefore, be conducted in a similar manner to that of the present rating system, but with the tax payer identified rather than taxable property. Tax credits against central tax liability for LIT levied would be collated by the Inland Revenue (as suggested by Layfield) and coded out in the coding forms sent to employers for collection of PAYE. This scheme is a hybrid of local and national tax collection, but is an integrated national scheme for tax assessment purposes, and is rather distinct from the schemes discussed by the Layfield Report or

in the evidence to it. To be sure there would be difficulties to be ironed out
in operating such a scheme. Particularly important amongst these would be
income levels at year-end which differ from initial estimates of tax
liabilities rendered to local authorities. However, these difficulties could
be overcome on a following year basis, or some other criterion, if there was
a desire to make a system of LIT operate. That such a scheme is practical is
now accepted in G.B. Government (1981*b*).

The point to note here is that an adjustment of the Layfield proposal to
allow tax collection by local authorities provides a very satisfactory form of
tax credits for LIT. It would be only a little more costly (in addition to the
Layfield estimates of the costs of LIT) because demands could be sent with
rating demands (or instead of them) through the existing local government
administrative machinery. Moreover, perceptibility would be retained and
improved as desired in the Layfield Report.

Central grants

A further means of modifying the balance of local revenues and expenditure is by use of new central government grants. This is the alternative
which has been adopted by every Westminster government since 1929. On
the one hand, this approach has the advantage of allowing taxes to be
raised from the (largely) progressive tax base available to the national
government, thus maintaining central control of the social distribution
effects of taxation. On the other hand, central grants allow the considerable differences between local expenditure needs and tax bases to be taken
into account, and hence allow a significant degree of equalisation. Both
features accord well with the theory of fiscal federalism in public finance
(see Musgrave and Musgrave, 1980; Oates, 1972, 1977).

The disadvantages of such an approach, to the local authorities, are that
such a source need not be particularly buoyant; it undermines local
discretion; and it carries with it a greater degree of central control.
Moreover, it also confuses the lines of accountability to the electorate (on
which the Layfield Report placed so much emphasis) since it is not clear
who is paying for what. In addition, the level of central support is
extremely imperfectly perceived by local tax payers such that the local
services they receive are, in effect, underpriced and this can lead to
inefficient allocation of goods (both public and private) in the economy as
a whole. In addition, for the local councils, since a proportion of
expenditure increases can be passed on to national taxpayers, central
support can act as a stimulus to supply-expansion of services. Hence, if a
significant proportion of local finance is supported by grants, this tends to
lead to poor lines of accountability, to inefficient revenue allocation, and
to an overall increase in public spending.

Since 1979, it has been suggested by the central Government that many of these problems can be overcome by employing a block grant, and this has been explicitly employed to allocate the RSG since 1981/2. As we have seen in earlier chapters, this method has the benefits that, first, advantage can be taken of coincidences of high needs and high resources on the one hand, and low needs and low resources on the other hand; second, arising from such coincidences, tax-rich low-need Authorities can be penalised by exacting negative payments from the resources element by the device of reducing their needs element, hence permitting resource transfers from resource-rich to resource-poor areas; third, the unitary grant allows full equalisation, but at a lower grant level than separate grant elements; and fourth, a simpler and more equitable system can be implemented. However, there remain the disadvantages that grants confuse local accountability, services can be underpriced in perceptual terms, and local discretion is undermined by a high degree of central control.

A further difficulty with central grants which also is sometimes raised is that, when multiple taxes are used by local authorities, it is impossible to implement central grants without imposing very severe constraints on local government setting of tax rates. This argument has been propounded mainly by the G.B. Treasury (G.B. Government Committee of Enquiry, 1976, Appendix 1) and Foster *et al.* (1980, pp. 505–7). It is based on a particular premise as to the way in which central grants are allocated to local governments and Bennett (1981*c*) demonstrates it to be without general validity. The criticism is overcome by appropriate choice of tax rates (see G.B. Government, 1981*b*; US, ACIR 1962, 1971; Reischauer, 1974; Halstead, 1971). This literature suggests that fiscal capacity and fiscal effort of localities can be assessed only over average or representative systems which require that instead of separate standard tax rates that there is only one *representative* tax rate $\bar{t}$. This suggests a new grant formula in terms of the block grant as:

$$G_i = \bar{E}_i - k\bar{t}\,(B_{1i} + B_{2i} + \ldots + B_{ni}) \qquad (10.3)$$

where B_{ji} is the tax base of authority i for tax sample j (j = 1, 2, . . ., *n*), k is a constant, $\bar{t}$ is the representative tax rate and the other terms are defined as before. The representative tax rate is then given by the average term of equation (10.3) where $\bar{E}_i$ is the standard expenditure assessed by central government for authority i. This block grant structure operates with a representative tax rate and an average tax base derived by aggregating each of the individual tax bases, each expressed in per capita terms. The representative tax rate is derived by either (a) dividing total local expenditure by centrally-assessed standard expenditure, or (b), subtracting total local expenditure from centrally-assessed standard expenditure. Using this grant the central choice of representative tax rate has no impact on the choice of local tax rates. Moreover the choice of local authorities to tax one base higher than another has no impact on grant levels. This demonstrates that the difficulties of achieving equalisation by central grants in a multiple tax local financial system can be overcome by a suitable choice of central grant structure.

The block grant implemented in Britain since 1981/2 has improved some aspects of the financial system considerably, but there are still important difficulties arising from imperfect equalisation and reductions in the level of discretion, accountability and efficiency which it allows at local level (see chapter 9). As a result, it is necessary to find an improved method of allocation which overcomes these difficulties, and can be integrated with multiple local tax sources, tax-base sharing and credits. One such method, which is that proposed in this book, is to attempt better to differentiate discretionary local spending (deriving from local preferences, politics and choice) from non-discretionary local spending (where local authorities act as devolved agencies of central government). For the first, discretionary, category of spending the most appropriate form of central grant is a general block grant, unhypothecated, based on equalising local tax bases and hence ability to exercise discretion, but with very restricted total magnitude. Such a grant might be related to tax-base sharing as discussed earlier. For the second, non-discretionary, category of spending the most appropriate form of central grant is a set of specific grants hypothecated to achieve national norms of service provision in each expenditure category. However, the difficulties of accurately measuring local needs and costs suggest that a barrage of specific grants cannot be used, but that instead a unitary grant assessment on the basis of the standard expenditures could be defined for norms in each non-discretionary service category. This unitary grant is then hypothecated to achieving these service norms. The most obvious service to be accorded such treatment is that of education. This is a service to which national minimum standards already apply to an extent which severely limits local discretion such that there is relatively little variation in standards or expenditures between areas for the education of five to sixteen year olds, although there is considerable variation in provision for nursery, secondary, post-16 secondary, further education, and special education. Education is also a service on which central government has traditionally imposed national criteria, and for which there is probably the political will to use a specific grant status (see G.B. Government, 1981*b*); indeed, education was a major form of specific grant until 1958.

The division of block grant support into two categories, one hypothecated and one unhypothecated, is a central proposal of this book and is developed in detail in the next chapter.

Table 10.2. *The use of fees and charges as a percentage of gross expenditure on selected services*

Service	Year				
	1969/70	1971/2	1973/4	1975/6	1977/8
Education	9	9	8	6	6.8
of which: school meals and milk	36	41	34	29	29
Personal Social Services	N/A	N/A	15	14	14.4
Rate fund services[a]	10	8	7	6	6.5
Trading services (total)[b]	69	70	56	56	54
of which: passenger transport	76	88	82	85	89.5
Housing revenue account					
Rents	69.3	72.0	57.7	45.1	45.3
Other local income	3.1	3.9	4.3	3.2	3.9
Central Government subsidies	18.7	22.5	19.6	40.4	44.1
Rate fund contribution	8.4	4.8	18.2	11.8	9.3
Total charges	**4.0**	**4.6**	**3.2**	**3.3**	**5.4**

[a] includes: libraries, museums, art galleries, refuse collection and disposal, vehicle parking, cemeteries, crematoria and other minor services.
[b] includes: passenger transport, harbours, ports, piers, markets, slaughterhouses, aerodromes, civic restaurants, and other services.
Source: G.B. Consultative Council, 1979, Annex H, table 1.

Charges for local services

The final major alternative method for improving the local revenue tax base is to charge an increasing proportion of service costs directly to the users of the service. User charges form a relatively small proportion of total revenues in Britain, 5% in 1977/8 (see table 9.1). Moreover, they are a source of revenue which has tended to decline in extent used during the early 1970s, although their use increased in the later 1970s, as shown in table 10.1 and figure 3.1.

In more detail, table 10.2 demonstrates the extent to which fees are used to fund particular service categories. It is clear that the two main groups of redistributive services for which local authorities are responsible, education and personal social services, are funded to the least extent by charges. In contrast the diverse group of trading services, of which the most important is usually passenger transport, have been funded by a very significant use of charges. A special case of such trading services is public housing and here again user charges form a very significant proportion of the total.

Table 10.2 makes clear a number of important features which have characterised the use of charges in Britain. First, they have tended to decline when there has been a Labour Government at Westminster and to increase when there has been a Conservative one. This has particularly characterised the provision of school meals and milk (included in education), and also the use of central government subsidies on the Housing Revenue Account. Second, the pattern of decline in level of user charges, especially for the 1973/4–1975/6 period, has been made possible by increased use of both central grants and increased rate contributions. Third, there is usually a major distinction made between the employment of user charges for two classes of services: (i) beneficial (or non-redistributive) services, and (ii) redistributive (or onerous) services. For the former, charges can be extensively employed, but for the latter the use of charges undermines the object of providing the service at all.

Charges have two main advantages. First, they allow the introduction of market forces into the allocation of public goods so that demand, and hence supply, can be regulated to some extent by price. This should encourage economic efficiency overall and also accords with the benefit principle of taxation (Buchanan, 1966). Second, charges are a revenue source which offers clear local discretion, accountability and independence from central control.

Against the advantages of user charges are three main disadvantages. First, charges can be employed only for those services for which there is no redistributional intent (beneficial services), or otherwise access would be limited by ability to pay. For example, it is not possible to charge for services which are intended as income support. This severely restricts the amount of revenue that can be raised by charges. Second, charging policies need to be co-ordinated between local governments; otherwise they can stimulate migration and distort the basis of competition within the private sector. Third, it would be unusual if charges could reflect the entire cost of a given service since most local services are, in part at least, considered to be social or redistributive services and are hence subsidised. Where part of the costs is subsidised it often becomes very difficult to raise the level of charges in line with increasing costs and inflation because of political and other constraints.

Despite these disadvantages, user charges can certainly be employed to a much greater extent than the pattern obtaining in 1977/8 with the result that various attempts have been made to increase their use especially since the Conservative Government gained office in 1979. Particularly important areas in which increased use of service charges have been sought are refuse collection, public housing rents, school meals, school buses, planning application and building regulation consents, leisure facilities, libraries, and adult and special education. Moreover, the *Local Government, Planning and Land Act* (1980) transferred to local authorities a large number of (small-income) established charges which had previously been used by central government departments.

The recent impetus to the increased employment of user charges has come from two sources. First, there has been the desire by the Conservative Government to impose a 'more realistic market pricing' on services, and in some cases to allow 'privatisation' by selling off facilities or services to private contractors. Second, there is the suggestion that user charges encourage accountability, efficiency and autonomy, and reduce the need for grant support. The prime proponent of this approach has been Beasley (1980*a, b*) who suggests that fees and charges could yield an additional income of £29 m for cemeteries and crematoria, £54 m for other trading services, £91 m for baths, sports and recreation, and £137 m for town and country planning (1980 prices). Other workers have been less optimistic about the extent of potential for charges (see e.g. SCT, 1980; SMT, 1980; LESEWP, 1981) but it is clear that some scope for at least a 20–30% increase in charges does exist. On the basis of this evidence a recent independent research report for DoE (1981) concluded that considerable scope for user charges existed for refuse collection, waste disposal and leisure and recreation facilities; but this was dependent also upon the introduction of methods for closer monitoring of the performance of private contractors.

Because of the problems of charging for some redistributive services, attempts can also be made to integrate 'means testing' for payment of charges with the social security system better, especially employing similar criteria to those used by the National Health Service. However, the number of beneficial local services which are free from a significant degree of redistributional intent is relatively small. As a result it has usually been concluded that charges represent a useful source of additional revenue, but one which is inadequate to improve markedly the total balance of local revenues and expenditures (see for example G.B. Committee of Enquiry, 1976; G.B. Government, 1977; G.B. Consultative Council, 1979; G.B. DoE, 1981). Hence, they can and should be used to improve the balance of central–local revenues and expenditures, but it is not possible for them to dispense with the need for central grants and local taxes.

10.4 Conclusion

Within Britain a series of problems of balancing central–local expenditures and taxation have emerged over recent years. First, the tax base of local areas is extremely unequal giving variable capacity of local governments to support the same level of expenditure. Second, the expenditure requirements of local areas differ greatly depending upon their sparsity of population, infrastructure, economy and demographic structure. Third, the tax base of local government as a whole has proved increasingly inadequate to cater for the expenditure demands placed upon it. Relating to the inadequacy of the local tax base has been its relatively regressive character towards poorer income groups. Fourth, there has been increasing conflict between central and local governments as to the services that should be provided locally, at what level of quality, and at what cost. Fifth, central government has considered the behaviour of local government to have increasingly undermined its ability to manage the economy as a whole.

These problems are not specifically British, but have also characterised many other countries (see Hunter, 1977; Bennett, 1980). To overcome the problems of balancing central and local taxation and expenditure, various forms of co-operation and co-ordination between levels of government have evolved in different countries. The five major alternative means of co-ordination are new local taxes, grants, tax-base sharing, tax credits and deductions, and charges. This chapter has briefly reviewed each of these alternative means of allowing central–local tax and expenditure co-ordination and assessed their applicability to Britain. The conclusion of this discussion must be that it is a mixture of local taxation, tax-base sharing, tax credits, and equalisation grants which will provide the best means of co-ordination in the future. This conclusion extends to a great extent the proposals that have at present been implemented for reform of local finance through modified central grants (see G.B. Government, 1980*a, b*, 1981*b*). The details of the new financial system proposed here are discussed and its practical impacts appraised in the next chapter.

CHAPTER 11

AN ALTERNATIVE STRUCTURE FOR CENTRAL–LOCAL RELATIONS

11.1 Introduction

The discussion of the previous chapter has shown that the balance of revenue and expenditure functions at each level of government in Britain has given rise to a number of problems. Especially important have been: unequal local tax resources, unequal local expenditure needs; inadequate and unbuoyant total local tax base; the relatively regressive nature of local taxation; and the supposed conflict of local financial behaviour with central objectives of economic management. Each of the five main alternative means of intergovernmental financial co-operation discussed in the previous chapter permits some of these problems to be overcome, but each introduces further difficulties and complexities. As a result a combination of methods, rather than a single solution, must be sought.

In this chapter a *combined* method offering an alternative central–local financial structure is proposed. This combined method derives from considering simultaneous solution to each of the problems of local re-source-need imbalances, unbuoyant local revenues, income distribution effects of local finances, and central economic management in the face of local financial behaviour.

The alternative central–local financial structure proposed in this chapter consists of seven parts

(1)   Limited increased use of specific grants: termed *categorical specific grants*.
(2)   Introduction of an aggregate specific grant using a unitary grant structure for non-discretionary services: termed a *specific block grant*.
(3)   Modification and reduction in size of existing equalisation grants used under the Rate Support Grant for discretionary services: termed a *general block grant*.
(4)   Limited increased use of user charges.
(5)   Continued but slightly reduced use of local property tax, but employing capital valuation of property updated annually.
(6)   Provision of a shared local personal and corporate income tax from central government.
(7)   Introduction of tax credits against central tax bills of a proportion of local tax demanded, in the long term to be integrated with a system of *negative income tax* (NIT).

This alternative financial structure is considerably more complicated than the present structure but does offer major advantages for local governments, the central government, and the tax payer. Two principles underlying this proposal are the division of central and local expenditure functions, and the integration of central and local finances into a single tax and benefit system. Each of these principles is discussed in the next section before detailed discussion is given to the proposed central–local financial structure in sections 11.3 to 11.5.

11.2 Underlying principles

The first underlying principle for the proposed structure is the distinction of discretionary from non-discretionary local tax and spending decisions. The present balance, and confusion, of local discretionary and non-discretionary behaviour is neatly summarised by G.B. Government (1972*a*, p. 19)

'Local authorities cannot provide services unless empowered to do so by Act of Parliament; they are subject to statutory requirements to maintain specific minimum standards in some services; and much of their expenditure is in pursuance of (central) Government policies reflected, for example, in circulars giving departmental guidance. But there is no direct central control of their current expenditure either in total or by individual services. Local authorities, acting within their statutory powers, can raise money by levying rates and spend it for current purposes without seeking authority from any branch of central government. . . . Projections of local authority current expenditure are therefore forecasts of the aggregate amounts which individual local authorities, as separately elected bodies, will decide to spend. Nevertheless, the (central) government, through its relations with local authorities and the mechanism of the Rate Support Grant, is able to exercise a strong influence over the rate of growth of current expenditure'.

Hence local taxing decisions and central grant decision can be said to represent a complex convolution of conflicting aims at both levels of government.

If any attempt is made to improve central–local relations in Britain it must tackle the issue of separating discretionary from non-discretionary spending. It is this issue which has bedevilled the RSG, through the feedback effect, poverty traps, arbitrary definitions of GRE, and other factors recognised in earlier chapters. It is this issue which has also done much to cloud and complicate the issue of assessing economic efficiency, productivity and hence the extent of local over- (or under-) spending. Moreover, variation in non-discretionary spending still cannot be assessed sufficiently accurately to allow the most appropriate allocation of equalisation grants. Hence, if increased rigour is to be achieved in equalisation, and at the same time local discretion is not to be sacrificed, then it is necessary more clearly to differentiate, on the one hand, that expenditure directed at achieving a national equality in service levels, from, on the other hand, that expenditure which arises from the choice of local authorities to provide a given service at a level higher than that sought by central equalisation. The need for this clear distinction suggests that the local financial system should be divided into two parts: one responsible for funding what are essentially national services at national standards (where local government acts as a non-discretionary devolved agency of central government); and a second responsible for funding services provided as a response to the exercise of local preferences.

The aim of differentiating dicretionary from non-discretionary spending has been built into the RSG since 1981/2. On the one hand, the Grant-Related Expenditure (GRE) is intended to represent the non-discretionary component, whilst on the other hand, discretionary spending which produces 'feedback' effects on grant levels has been restricted by setting the threshold and tapering terms discussed in earlier chapters. Hence the proposal developed in this chapter, to differentiate more clearly the origins of expenditure decisions, is in line not only with the aims of increasing equity and efficiency in British local government, but also with recent thinking in Britain.

A second underlying principle for the central–local financial structure to be proposed here derives from considering the interrelations between central and local finances as a total tax and benefit structure. This determines the level of total tax burden and benefits received by each individual, household and family. The argument developed in this chapter is that, if full equalisation is to be achieved, it must be sought jointly at both the personal and geographical levels rather than merely at the local level (as in the present Rate Support Grant, between local authorities). For example, if it is sought, at central level, to equalise the access of different categories of client group to given services, to neutralise the effects on local tax rates of different numbers of clients per head of local population (and per head of local tax base), and then to provide the scope for local authorities to provide services of differing levels and qualities above the centrally set norm, then this can be fully achieved only through integrating the total tax system.

Take as an example the personal social services. A significant proportion of the local services is to some extent income-related; especially, for example, services to the old, public transport, and housing. Equalisation of tax rates between local authorities for these services, irrespective of local variations in the ability-to-pay of the clients concerned, and irrespective of the distribution of tax burdens between the local population, is highly perverse and inefficient. For example, consider local authorities with a large number of old people needing home helps, meals on wheels and other old people's services. In the case of one local authority these people may all be of high income or wealth quite able to pay the true costs of these services themselves, in which case public provision at free cost is inefficient; and taking account of these old people as 'needs' of that local authority, through an equalising grant, for example, will be unequalising. In the case of another local authority, there may again be a large number of old people, and these may be low-income old who can contribute nothing to the costs of service provision. But if these old people live in an area with a large number of rich people, and the old pay few or no taxes either to central or local government, then it may be well be the case that taking account in equalisation grants of these old people as 'needs' is again unequalising and inefficient, since the local rich can adequately support the services to the local old and needy. Hence there is the need to integrate the tax and benefit positions for both areas and people.

The implementation of central–local tax and benefit integration is motivated by the principle of achieving *fiscal equity* in the country as a whole. In loose terms, this means treating equals equally irrespective of where they live. As stated by Buchanan (1950, p. 596) 'The principle (of equity) establishes a firm basis for the claim that citizens of the low income states within a national economy possess the "right" that their states receive sums sufficient to enable these citizens to be placed in a position of fiscal equality with their equals in other states'. More precisely the principle of fiscal equity requires two aspects to be satisfied simultaneously: first, the resources of local authorities should be invariant to the need to spend, thus ensuring equal ability of areas to choose to provide discretionary services at the same marginal cost; second, the *fiscal residuum* for each individual for non-discretionary services should be equalised for similar individuals and be independent of geographical location of residence. The fiscal residuum, sometimes termed the 'fiscal wedge', is the difference between, on the one hand, contributions made through taxes and charges and, on the other hand, the value of public service benefits received.[1] This residuum will differ for different income groups and different areas since it is usually desired to redistribute incomes from rich to poor; hence, the residuum will be positive for low income groups and poor localities and negative for high income groups and rich localities. Thus there will be a designed feature of vertical inequity.[2] However, fiscal equity should not differ depending upon which local authority jurisdiction an individual happens to live in; every attempt should be made to eliminate horizontal inequity between areas.

The achievement of this aim of equalising the geographical incidence of fiscal residua is by no means simple. The procedure outlined in this chapter suggests that full client-need equalisation can be undertaken only by taking account of discretionary and non-discretionary spending, the interplay of central and local tax burdens, and integrating the central tax system with the local tax system, the Rate Support Grant, and other local financial procedures. This suggests the need for a unified system which integrates both local taxes and benefits with national taxes and benefits: the result is a proposal for a special form of negative income tax (NIT) which is discussed in section 11.5 of this chapter.

11.3 Reformed central grant structures

The primary aims of central grants are twofold: first, to support central services which are provided by local authorities as non-discretionary bodies; and second, to provide support for differences in the abilities of local authorities to provide services, both non-discretionary and discretionary, from their own tax bases.[3] There is no reform of the local revenue system which will not require these two functions of central grants to be maintained.

(1) Non-discretionary services

For the first function of central grants, support of non-discretionary services, the most appropriate form of grants is a specific grant for each non-discretionary service category. If central government can specify its objectives in terms of local service provision then it can fund each service as a specific hypothecated service. This places most emphasis on measuring local service needs and the cost of providing services to meet them. It also requires the full costs of these services to be borne by central government. The main areas in which specific grants are presently used (police, national parks, transport, magistrates courts, probation, Commonwealth immigrants, the urban programme and improvement grants) should certainly be continued. In addition there is scope for further development of specific grants to increase the support for education, administration of justice, transport, and some personal social services. For personal social services, provision has grown up in a rather haphazard fashion in Britain, and hence these are often deemed discretionary local services. However, since these services frequently involve a substantial degree of income support, and hence have important distributional effects, there is powerful argument that they should be taken much further into the area of direct central funding. Education is another service to which national minimum standards apply strongly and for which an aggregate specific grant has already been proposed (see G.B. Government, 1981*b*, chapter 9). Similarly transport, and most County level services, are also subject to standards which are largely national rather than local.

Where accurate need and cost assessment can be implemented, specific grants should be based as closely as possible on compensating for differences in the full costs of service provision by a unit cost or similar approach. This suggests a barrage of relatively small, service-by-service, categorical grants. However, for many services it is difficult or impossible to measure cost and need variation accurately enough to permit separate specific grants to be used for each service category. Hence, in these cases it may well be more appropriate to combine the specific grants into a single grant, but orientated towards achieving standard service levels in each service category. Thus a unitary grant, combined with rigorous guidelines, is perhaps the most appropriate means of funding many locally-managed services for which local authorities act as devolved agencies of central government. This proposal differs considerably in explicit motivation from the implementation of the unitary grant for the RSG since 1981/2, although implicitly it had many of these features.

This unitary specific grant would be most useful for many personal social services where measures of need of the total population are difficult to derive. However, it could most usefully be employed for education and many other personal social services where measures of need may be good, but where unit costing is greatly complicated by the differing quality of existing facilities, e.g. school infrastructure of variable age and design.

There is also a second motivation for implementing a unitary specific grant. This rests on the difficulties of determining the division between discretionary and non-discretionary spending. For many services these two components are very closely interlocked and the provision of one aspect represents either hidden costs or hidden economies for the other. For this reason it is not possible completely to define or to differentiate discretionary and non-discretionary spending. However, a unitary specific grant can be used in this case to support that proportion of local services which is deemed to be non-discretionary, and this can be enforced by statutory obligation or ministerial directive. That proportion of the local service which is discretionary then falls outside the area of direct central support, at least in the first instance, and is treated in a different way (as a problem of equalising the tax bases, to be discussed later below). This form of unitary specific grant therefore requires a method of unit costing for provision of minimum standard service levels. The hidden costs of additional discretionary spending are ignored, and any hidden economies to the discretionary sector are considered as a bonus to the local authority stimulating local participation and involvement. The form of specific unitary grant proposed is very like that used since 1981/2 for the RSG

$$G_i = \bar{E}_i - B_i \bar{t} - S_i. \tag{11.1}$$

This is in fact the unitary grant (equation 2.11) proposed by the Conservative Government in 1979 but not used. It differs from the actual RSG used since 1981/2 in that standard expenditure and poundages *only* are employed, there is no tapering or threshold, and there is a new term S_i denoting shared tax income. Local authorities have no effect on grant levels through discretionary spending since the standard is determined centrally by set standards of expenditure and poundage. These settings must reflect the true extent of minimum service standards where local authorities act as devolved agencies for central government. This grant requires that the standards set by central government are a fair reflection of local service requirements for the non-discretionary services. The shared revenue term relates to other central income received and could be omitted if grants alone were to be employed. Its inclusion, however, has the benefit of emphasising the central *responsibility* to guarantee revenue for the services which are undertaken locally on its behalf.

For non-discretionary services, therefore, the proposed grant structure consists of three parts: first, clear definition of service levels for national uniform minimum levels;[4] second, maintenance and expansion of categorical specific grants; and third, the implementation of a unitary specific grant, this latter motivated by the difficulties both of determining unit costs and of completely differentiating discretionary from non-discretionary functions. In each case it is suggested that the specific funding should be linked to firm central directives on minimum standards.

(2) Discretionary services

In addition to providing support for non-discretionary spending, central grants are also required for providing some support to discretionary services, but in this case only to the extent that tax bases and cost differences penalise authorities in their ability to provide such services. For discretionary services, therefore, we move from specific grants, linked to achievement of central directives on minimum service levels, to grants of a more general kind which have the clear aim of ensuring equal ability to provide services at a given discretionary level.

For discretionary services there should be no need for central involvement at all; the choice of enhanced service levels, resting with local councils and their electorate, should be a burden solely on those making the decision to provide enhanced services. But for such services it is, unfortunately, not possible for central government to remove itself completely from involvement. The problem that arises with discretionary services is that different local authorities have different tax bases and hence can fund the same level of discretionary services only at differing tax rates. As a consequence the marginal costs to each local taxpayer of each service provided (or consumed) differs in different local authorities, with the result that there is horizontal inequity and inefficiency between local authorities. These differing marginal costs for local discretionary services are the motivation for a form of central equalisation grant: to place each local authority in an equal position, with regard to marginal costs, to provide discretionary services.

If marginal costs differed only as a result of differing tax bases then a form of *resource* equalisation grant would be the most appropriate form of central support. However, costs also vary as a result of the varying level and costs of inputs which vary with the location, size and other characteristics of a local authority. This suggests that discretionary services require a combination of need and resource equalisation grant, where needs assessment relates to some form of unit cost assessment; but that the overall level of central support is as small as possible; just sufficient to reduce variation in marginal costs to an acceptable minimum.

In designing a form of central discretionary-service marginal-cost equalisation grant it would be quite inappropriate either to apply any criterion of 'standard expenditure' or to require any hypothecation to specific service categories. To permit local discretion for these services it should be required that each local authority is placed in an equal position to exercise discretion; i.e. an area should not be restricted in providing a given quality of service by its low tax base or high costs. Moreover, since local discretion should reflect local preferences as clearly as possible, central support should be restricted so that local tax rates relate directly to discretionary services provided; thus inducing clear lines of accountability. The proposal of this book is that the most appropriate grant structure is of a general block grant equalising local tax bases and local costs, and thence equalising the ability of local areas to exercise the same degree of discretion.

The form of grant equation proposed is defined as follows. Start with a unitary grant (such as unitary grant (2) in equation (2.16)) which compensates for all expenditure not raised from the local tax base (but without tapering and thresholds)

$$G_i = E_i - B_i t_i. \tag{11.2}$$

Now the proposed grant G_i' must apply only to discretionary expenditure, which is the excess of local expenditure over the standard expenditure of each authority defined for non-discretionary services. Hence, we must subtract from the above equation the grant allocated for standard expenditure equalisation $\bar{G}_i$. This is given by the unitary grant equation (11.1) above, i.e.

$$\bar{G}_i = \bar{E}_i - B_i \bar{t}.$$

To determine the discretionary expenditure equalisation grant subtract $\bar{G}_i$ from G_i to give

$$G_i - \bar{G}_i = E_i - B_i t_i - (\bar{E}_i - B_i \bar{t})$$
$$G_i' = E_i - \bar{E}_i - B_i (t_i - \bar{t}). \tag{11.3}$$

This gives the grant for discretionary spending only, G_i', as a full compensation for expenditure differences above standard minus the contribution from local tax resources as far as tax rates are above the standard.[5] Another way of writing equation (11.3) is

$$G_i' = (E_i - B_i t_i) - (\bar{E}_i - B_i \bar{t}) \tag{11.4}$$

which shows that the grant compensates for expenditure excess over local revenue resources, as far as this exceeds the standard expenditure excess over local resources at the standard tax rate.

There are a number of important features in this grant system. First, it places considerable significance on having accurate and fair settings for $\bar{E}_i$ and $\bar{t}$, otherwise there will be unintended bonuses and penalties in the system. Second, the grant gives the same marginal costs and hence marginal expenditure incentives to all local authorities. Third, the grant gives penalties on underspending as well as overspending: if an authority is spending less than the standard ($E_i < \bar{E}_i$), then negative grants will result. This is in accord with the objective that standard expenditure relates to non-discretionary services which local authorities perform as devolved agents of central government and hence $\bar{E}_i$ is a *minimum* standard which *should* obtain: the proposed grant system gives financial incentives to its achievement.

A further feature of this proposed grant is that it creates negative grants for areas with high tax base, i.e. there are horizontal transfers between authorities. For authorities with very high tax base these grants will present a strong disincentive to discretionary spending. This would be likely to apply to much of London and many other metropolitan areas. Since high rateable values reflect not only a large tax base but also the interaction of tax base with demand from the private economy for particular locations, this particularly inflates costs as well as rateable values in central cities. This suggests that there is a need to modify the grant by a weight measuring differences in local costs of service provision. Such a development of cost weights yields a grant equation which ensures the same marginal tax cost for expenditure in all areas at the same *service unit input*. This requires only a small modification of the grant equation to become

$$G'_i = E_i - \bar{E}_i - \frac{B_i}{C_i}(t_i - \bar{t})$$

or

$$G'_i = (E_i - B_i t_i/C_i) - (\bar{E}_i - B_i \bar{t}/C_i)$$

where C_i is an input cost weight. The costs weight the local tax rates and tax base such that areas with large base but high costs will receive larger grants. Hence rateable value is reduced to a scale of revenue base per £ of unit cost; i.e. revenue and expenditure are rendered in the same units of measurement. This equation retains all the features of the previous equation, but weights grants for differences in input costs.[6] Differences in output quality levels could also be included (see Le Grand, 1975) and have been discussed with respect to equation (2.20). This is a possibility for future developments which should be attempted, but present data sources preclude accurate estimation of outputs for most services; hence controls for output differences are not explored further here.[7]

It should be noted that this grant has the desired feature that it is at the lowest possible level to place the greatest possible burden of discretionary spending on to local taxes.

11.4 Reformed local revenue sources

The reform to the local revenue system proposed here consists of three elements: continued use of the property tax; increased employment of user charges; and the introduction of either local use or central–local tax-base sharing for personal income tax. This new balance of revenue sources has the advantages that it increases the diversity and buoyancy of local revenues available, provides additional sources, encourages more efficient provision, and reduces the need for employing grants to equalise between local tax bases.

(1) The property tax

The property tax is a major source of revenue to the public economy in Britain (accounting for 10% of all taxes in 1978/9). For this reason alone it is difficult to advocate its abolition. In addition, however, the property tax does possess the desirable features, outlined in chapter 10, that it is undeniably a local tax, amenable to use by overlapping local governments, and in some senses can be represented as a user charge on land. However, the present method of assessing liability to property tax is highly inadequate and the most important problem relates to inappropriate valuation procedures which are inadequately updated. The future role of the rates advocated here, therefore, is to base their revenue yield on their present levels, but to introduce two reforms into valuation practices which have been widely proposed elsewhere (see G.B. Committee of Enquiry, 1976, Appendix 9; G.B. Government, 1977, 1981*b*; Foster *et al.* 1980).

First, either a 'capital' or 'site-value' for property should be employed as a base for valuation, rather than 'nominal rental value'. Nominal rental value has certain desirable features if it can be reliably assessed, but in Britain it is now largely arbitrary. Either site-value or capital valuation have the advantages that they can be readily assessed from current sales or survey data, they can be readily updated, and they relate tax bases to wealth.[8] There may be certain regressive features that large families require large houses and may have low wealth or income, but this characteristic can be overcome by personal tax credits, as discussed below.

A second suggested reform of the rating system is to introduce annual updating of valuation lists. This is quite feasible from market price and survey data.[9] Alternatively, updating could be based on annual percentage increases (spatially variable, based upon allowances for inflation in prices or building costs), with less frequent full survey revaluations. Either method of maintaining up-to-date valuation lists has the advantage of eliminating the arbitrariness in present valuation practices. It also introduces buoyancy into local tax bases since local tax income will rise as a result of the increased size of the tax base without tax rates having to be adjusted. There are two main disadvantages of this reform. One is the increased administrative costs in preparing valuation lists, although these could be kept fairly small. The second disadvantage is the windfall gains in taxes to local governments derived from inflation in property values at rates higher than the general rate of price inflation. This latter problem was a major feature of the 1978 California tax revolt 'Proposition 13' against massive tax surpluses accumulating in the public sector. It is also the problem drawn attention to by Prest (1978) and Foster *et al.* (1980) of 'dubious political morality' which allows local authorities to gain higher real income from a fixed tax rate. In this case, however, it can be overcome either by (1) allowing continued increase of property tax valuations as part of a system of wealth tax, or (2) linking property tax valuations to some increase in average public service costs rather than general house value inflation rates.

(2) User charges

Increased utilisation of user charges is a reform of local finance which has been advocated a number of times in Britain, particularly by Conservative governments anxious to introduce a better form of market pricing into local service provision (e.g. G.B. Government, 1981*b*). The expanded use of charges is limited to the extent that services for which they are employed cannot have a major redistributional intent. However, in services such as transport, personal social services, and housing some expansion of charges is possible provided regressive effects are eliminated through the tax credit system discussed below.

(3) Tax-base sharing and LIT

The introduction of tax-based sharing of the personal income tax is a third reform of the local revenue system proposed here. The motivation for sharing of income tax is twofold. First, continued use of property tax and a small expansion of user charges, although helping to solve many problems, does not overcome the general difficulty of providing a more adequate total local revenue yield. This can only be provided by a new tax source. Since most possible tax sources for separate autonomous local use are low yielding, adequate yields can only be achieved by allowing local use of an existing central tax. Whilst this can introduce overlapping and competition for revenues between levels of government, any problem which arises can be considerably mitigated by use of an appropriate tax credit system.

The second motivation for sharing of the income tax is that it is this tax source which better than any other allows the integration of tax burdens and tax benefit assessment. This integration is essential for realisation of the principle of fiscal equity outlined in the introduction to this chapter; and can be achieved only through a complicated procedure of tax credits and negative income tax. It is also possible to integrate tax credits for the property tax into this procedure,[10] however, the use of LIT is preferable since it makes integration far more straightforward.

The form of tax-base sharing proposed here is heavily dependent upon the system of tax credits (NIT) developed in the next section. However, its five main characteristics can be sketched. First, the definition of taxable income, tax threshold, marginal rates and all other personal income tax criteria are *identical* for central and local government. Second, for the set of centrally-defined minimum levels of service benefits included in tax assessment (for NIT), the definitions of these, criteria of access to them, and any means testing for exemption from user charges are also *identical* for central and local government. Third, tax assessment must be undertaken by central agencies, i.e. the Inland Revenue. Fourth, benefit assessment can be undertaken by either central or local government, but for local services is probably most easily and satisfactorily undertaken by local authorities working to central guidelines within the national social security system. The local assessment allows more ready combination with perceptibility and accountability, but also permits integration of national minimum service levels with any local discretion to provide services at an enhanced level. Fifth, tax collection and benefit allocation of local services can be undertaken by local authorities; again this improves perceptibility and accountability in the overall financial system.

Within this broad structure it is proposed that two levels of local revenues should operate. First, personal income tax should be allocated at a level sufficient to fund non-discretionary services. It is not important, in revenue terms, whether this is an LIT, or is a shared income tax (as in Germany) allocated in line with measured non-discretionary service need. The choice between the two structures is, of course, important politically. The second level of revenue is a system of improved rates, as discussed above, for use in funding discretionary services. Moreover, the rates would be freed from excessive burdens and hence would be more available for funding discretionary service levels.

The proposed balance of local revenues which would be possible using the three reforms outlined above is shown in table 11.1. Whilst a very approximate assessment, the table does provide an indication of the impact and balance of the reforms suggested. A major characteristic of these is that they help to ensure revenue *adequacy* by making available to local

Table 11.1. *Impact of proposed reforms of local revenue system in comparison with revenue yields of local authorities in England and Wales using 1978/9 (at 1980 prices) as an example*

| | Yield £m | | | |
	Property tax	User charges	Shared income tax	Total
Existing use	7238	4358	—	11596
Proposed use	7238	4784	1585	13181
% change	0.0	+10.0	—	+13.7

authorities revenue resources of sufficient extent to permit them to fund discretionary and other services at a level, and in a manner, in accord with local preferences and political priorities. In the case of the present proposal, adequacy is assured by freeing the property tax of the burden of funding non-discretionary services. This also has the benefit that accountability and perceptibility are improved, as so much emphasised in the Layfield Report. As such the proposed reform of the finance system allows the provision of income sources which permit local authorities to be independent of central action using a *separate* income source for that significant proportion of their actions which is based on funding discretionary services.

It is often stated that the level of independent finance which will permit this level of independence is about 60% of total local expenditure. This seems to be a rule of thumb and no formal analysis is likely to clarify whether this is an appropriate level: indeed the level of local financial independence is largely a political decision. The level of truly independent revenue in the proposed system is relatively low: only the property tax of the rates. However, these are independent, separate and discretionary in their entire nature so that true independence should be fostered.

11.5 Co-ordination of central and local finance: tax credits and NIT

Increased use of specific instead of general grants, together with reform of the revenue system, should provide a means by which a significant shift can be made in the balance of revenues and expenditure between central and local government: the reduction in grant levels being made good by increased local tax yields from shared income taxes and user charges. However, this new pattern of revenue-raising does introduce two additional requirements. A first need is more closely to co-ordinate central and local revenues. When local authorities have separate tax sources from those of central government competition for the tax base is relatively restricted. However, when the two levels of government use the same tax

base, some method must be introduced to eliminate or restrict competition. A second result of the proposed reforms in local revenues is the
possible perverse effects on income distribution, resulting from both
site-value valuation of property and from use of LIT. This again requires a
co-ordination of local and central revenue policies to ensure that national
goals of income redistribution are not frustrated by local actions. Such a
co-ordination is already, to some extent, practised through the system of
rate rebates used since 1967/8, but the reforms suggested here would
require such methods of co-ordination to be greatly extended.

Tax credits are the method of central–local tax co-ordination proposed
here for British use to overcome these problems. The use of credits
proposed is based on a system of finance in which local taxes and benefits
are taken into account within both the local and central tax and benefit
systems. In its most equitable form this suggests, on the one hand, the use
of a method of tax credits for local taxes paid against central tax liability,
and on the other hand, a system of central taxation on local benefits
received. The combination of the two components represents the extension
of credits to become a special form of full negative income tax integrating
benefits and taxation at both central and local level, for both individual and
geographical variations.

Negative income tax (NIT)

The concept of NIT moves individual burden assessment into the area of
total income and total expenditure assessment: all fringe benefits, dis-
benefits, side payments, and expenditures are assessed in order to give an
equitable co-ordination of local geographical and personal variation in
costs, needs and burdens. The structure of NIT is shown in figure 11.1;
separate taxation and social security (income supplement) schemes shown
in figures 11.1(a) and (b) are combined into a unified system as shown in
figure 11.1(c).

In general terms, NIT is closely related to equalising grant payments
except that now redistribution is directed to the individual rather than the
geographical location. Thus the net tax paid or received, T, is given as $T =
t(Y - K)$, where t is the tax rate, Y the income level, and K the tax
threshold of the individual (or poverty level of income). The difference
between actual income and the tax threshold given by the term $(Y - K)$, is
frequently termed 'the poverty gap', but it can be either positive or
negative: in the former case a tax is exacted on the individual and in the
latter case a payment is made to the individual.

The major advantages of NIT are its simplicity and uniformity of
treatment, its capacity to give a smooth progression of marginal taxes or
benefits as individual income rises (overcoming the 'notch problem' or
'poverty trap' which can act as a strong disincentive factor), and its ability

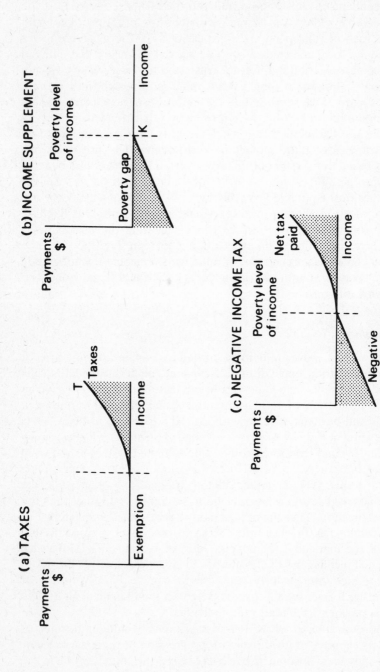

Fig. 11.1 Negative income tax as a composite of tax and income supplements at given income levels relative to the poverty level of income (tax threshold and limit for payment of welfare benefits combined, and assumed to be equal).

to integrate individual and geographically variable benefits and burdens into one composite and equalising format. Moreover, it is possible to combine NIT with other specific and general grant programmes aimed at targeting aid to specific client-groups, or to specific locations. The *demogrant* concept, for example, directs such transfers to age groups, and *wage rates subsidies*, based on rates of pay and hours worked, direct aid on the basis of productivity and incentives (Browning, 1975; Green, 1967). But any properly structured form of transfer of benefits by taxation meets the general aims of directing aid to the needs of an individual, and can be extended to a geographical basis.

NIT is much simpler than comparable grants and regulations. Since it is directly targeted to individuals it is far easier to use than the ever-burgeoning body of regulations which seek to direct individual and economic activity, since it emphasises the role of the market and places incentives on particular patterns of behaviour. It is also less odious than the role of social security and other regulations employed to ensure fairness in grant and transfer payment apportionment.

Despite its simplicity and other benefits, a number of technical problems arise in implementation of NIT (OECD, 1974): how is the family unit defined, what is the size of allowances and credits to be allowed, and how progressive should each be; how is income defined; finally, how are differences in costs and burdens between individuals and locations to be assessed? In addition, NIT can severely undermine national management of stabilisation and growth since flexibility in adjustment of taxation is more limited. Its usefulness is also affected by national incentives and attitudes to welfare payments in the national finances of the country.

NIT as a means of central–local financial co-ordination

The use of NIT as an instrument of central–local co-ordination and geographical equalisation, in addition to its purely social equalisation aspects, has been previously proposed by Bennett (1980, pp. 333–5), and the above discussion summarises that position. This present section aims to convert this novel suggestion into a programme which can be implemented in practice in Britain.

The starting point to describe this system, unlike the assessment of grants to local authorities in the past, is the individual or the household (whichever is the taxing unit). The level of income of the individual is assessed, both LIT and national income tax rates are applied to this income using national income definitions, and tax credits for the rate of LIT (or shared income tax) are applied to the national tax rate. This yields a level of total income tax paid, i.e.

$$T = (Y - K)t_1 + (Y - K)(t_2 - t) \tag{11.6}$$

where t_1 is the rate of LIT and t_2 the rate of national income tax. The local property tax is then taken into account. At the level of the individual taxpayer, if he is the head of the household, the local rates are then exacted (on a site-valuation basis), with local discretion in rate poundage set, after LIT. The sum of the rates levied, LIT and national income tax, then yields the total tax payments of the individual for NIT purposes. For a hypothetical set of data in a single local authority, this calculation is shown in the left-hand side of table 11.2. Note that rate payments, instead of LIT, could be credited against central income tax bills (a scheme for achieving this is discussed in footnote 10). Also other national taxes could be built into the NIT system, especially national insurance and other employee taxes.

The benefits side of NIT involves assessment, again on an individual or household basis, of the level of individual need for each social service. For those services which provide cash benefits (rather than in-kind benefits) the need level is then applied to the benefit structure to yield a level of cash benefits (or income support) for each individual for each service. As shown in table 11.2, the sum of these cash benefits over all categories of service, yields the total cash benefits received from the social security system. This assessment can include reimbursement of user charges for services at either national or local level; e.g. National Health Service charges, council house rents. It can also include reimbursement of rates as a direct rate rebate if crediting against the national tax bill is not employed.

The final negative income tax payment, or level of income support, is determined by subtracting assessed benefits from assessed tax payments for each individual or household. The result is a system, for the individual, which integrates tax burdens and tax benefits deriving from both the central and local levels of government. Such a scheme follows closely the lines of NIT suggested for Britain by G.B. Government (1972b) and also proposed more generally by OECD (1974). Its novel features are the integration of local and central government services. What has still to be discussed, however, is how the system of NIT integrates with the method of grants and revenue-sharing employed.[11]

The results of calculations of NIT discussed above, and sketched in table 11.2 for an hypothetical example, are an assessment of taxes or benefits for each individual resident within a given local authority. The sum of the various components of the NIT calculation for all individuals in a local authority provides the basis for the next stage of local financial integration: the assessment of requirements for support of the local authority (rather than the individual) by central government. This requires us to return to the treatment of discretionary and non-discretionary services, discussed earlier in this and the previous chapter.

It was suggested in the earlier discussion that non-discretionary services should be funded directly by central government by means of tax-base

Table 11.2. *An example tabulation of Negative Income Tax using hypothetical data and four example income levels, covering the assessment for a single local authority i*

	Negative Income Tax (NIT) (£)								Social Security and other services (include rate rebates)				Total NIT payment or receipt
	Personal Income Tax					Local Rates			Social service (k)				
									Level of need eligibility	Benefit level	Income benefits received	Total for all services (K)	
Individuals or households (j)	Personal income level	LIT rate	National income tax rate	National credit for LIT	Total income tax paid $[((1) - K) \times (2)] + [((3) - (2)) \times ((4) - (2))]$	RV of household	Rate poundate	Rates paid					
	(1)	(2)	(3)	(4)	(5)	(6)	(7)	(8) = (6) × (7)	(9)	(10)	(11) = (9) × (10)	(12) = Σ(9) × (10)	(13) = Σ(5) + (8) − (12)
	0	0	0	0	0	200	1.0	200	25	2.0	50	5000	+4800
	⋮			⋮			⋮			⋮			⋮
	2000 (assumed threshold for income tax K)	0	0	0	0	300	1.0	300	35	1.0	35	350	+50
	4000	0.05	0.30	0.05	600	400	1.0	400	30	0.5	15	150	−850
	⋮			⋮			⋮			⋮			⋮
	20000	0.10	0.60	0.10	10800	450	1.0	450	20	0.1	2	20	−11270
	⋮			⋮			⋮			⋮			⋮
Sum of all individuals in households in Local Authority (i)	$\Sigma_{(i,1)}$	$\Sigma_{(i,2)}$	$\Sigma_{(i,3)}$	$\Sigma_{(i,4)}$	$\Sigma_{(i,5)}$	$\Sigma_{(i,6)}$	$\Sigma_{(i,7)}$	$\Sigma_{(i,8)}$	$\Sigma_{(i,9)}$	$\Sigma_{(i,10)}$	$\Sigma_{(i,11)}$	$\Sigma_{(i,12)}$	$\Sigma_{(i,13)}$

sharing, categorical specific grants and a specific block grant. Once a decision has been made as to which services (or what proportion of which services) are non-discretionary, then it becomes possible to use the column totals for service costs in the NIT scheme of table 11.2 to provide assessment of the total spending requirements (need) of each local authority for this class of services as shown in table 11.3. In practice, the costs of providing non-discretionary services, in-kind benefits, capital costs, administrative and other fixed costs must be aggregated with the level of NIT payments within a local authority. With this level of revenue requirements calculated, there then follows a central decision as to the level of these requirements to be funded by shared LIT, specific grants and the specific block grant. The level of shared LIT allowed will then directly determine the local rate of LIT (column (3) in Table 11.3). It may well be that the level of shared LIT can be kept sufficiently high that the specific block grant is not required.

A similar path of calculation is followed also for discretionary services as shown in table 11.3. The number of clients in each service group, when multiplied by the level of discretionary benefits chosen in a given local authority yields the total service costs in that area. When this is multiplied by the proportion allowable for equalisation purposes, we obtain the level of block grant in each service category which, when summed for all services, yields the level of total central grant. The remainder of these discretionary service costs then falls on the rates and appears in the setting of the rate poundage as column (7) of table 11.3.

The system of NIT proposed above allows enormous advantages to be obtained in integrating the central and local financial systems in Britain. Moreover, it is the only system to be proposed to date which allows complete equity in the overall system whilst keeping local government free to decide on discretionary service levels. Its main advantages can be only briefly summarised here. First, it maintains the principles of fiscal equity introduced in the Introduction to this chapter: equals are treated equally, and unequals are treated equally unequally, *wherever* they live, except for discretionary services. This should provide geographical variety consistent with preferences. As a result the financial system is not only horizontally equitable, it is also horizontally efficient in a Pareto sense: national (non-discretionary) services are nationally funded in accordance with national preferences, and local (discretionary) services are locally funded (except for equalisation grants to overcome differences in costs and size of the tax base) in accordance with local preferences. Services are hence in accordance with preferences, as far as practically possible.

A second advantage of the proposed system is that it is vertically equitable: unequals are treated in precisely the same way for tax and benefit purposes irrespective of where they live (again except for discretionary services). As a result, the proposal is also vertically efficient in a Pareto

Table 11.3. *Hypothetical display of information required in calculating grants and shared income for a scheme of NIT as developed in table 11.2*

	Central grant levels	Local authorities (i) 1 2 . . . . . . N	Total for all local authorities
(A) *Non-discretionary Services*			
(1)	Sum of costs of non-discretionary services provided by local government: level of need × benefit level summed for all individuals $\Sigma(i, 11)$		
(2)	Percentage to be funded by shared LIT		
(3)	Rate of LIT (1)/(2)		
(4)	Percentage to be funded by specific grants		
(5)	Percentage to be funded by specific block grants		
(B) *Discretionary Services*			
(6)	Sum of clients in service group k		
(7)	Level of benefits chosen by local authority		
(8)	Total costs for service K (6) × (7)		
(9)	Level of equalising general block grant for service k to be applied to discretionary costs		
(10)	Total level of central grant for service (k) (8) × (9)		
(11)	Total level of central grants for all services (sum of (10) for all services k)		

sense. Since the proposed system is both horizontally and vertically equitable and efficient there should be no significant stimulus to fiscal migration, except insofar as local discretionary services do not accord with individual preferences. Since the scale of these is relatively small, migration is unlikely to be of significance as a result of fiscal stimuli.

The third main advantage is the improved constitutional/structural properties of the proposed framework. The large level of specific grants proposed recognises the preference in Britain for a high degree of uniformity in service levels which derives from both the smallness of the State and its historical tradition. However, the use of a shared income tax, equalisation grants, and freedom to set the rates (now unfettered from the need to fund a large proportion of central non-discretionary services), should permit improvements in local independence, accountability and perceptibility of both local and central services. At the same time, the responsibility for central services will be more visible locally.

The fourth main advantage of this system is its smoothness of progression and freedom from both the personal and geographical problems of poverty traps (the 'notch problem'). Geographical and individual costs and benefit incidence can be smoothly integrated into a total tax structure.

11.6 Assessment of alternative grant and revenue system

It is not possible in the space available here to give more than a broad indication of the way in which the suggestions proposed above can be implemented in practice. Clearly a large number of detailed and practical decisions require to be made in order to permit 'fine tuning' of the proposed system. Nevertheless, some aspects of the proposed grant and revenue system are already sufficiently well-defined, and are also sufficiently robust to underlying practical assumptions, that a good indication of the likely financial patterns can be estimated. This is undertaken below.

The simulation of the proposed grant and tax system follows five stages in the analysis below

(1)   Specific grants, although it is proposed that these should be increased, are assumed constant.

(2)   User charges are assumed to produce the same income per head as the upper quartile per head used in the past. This allows an approximately 10% increase on existing use.

(3)   Shared income from the local income tax is allocated to each local authority on a per capita basis. An initial level of sharing of 5% of national income tax is employed.

(4)   Grants for non-discretionary spending are allocated using equation (11.1), after shared revenue has been allocated. Standard expenditure is defined by the representative index employed in chapter 6 and standard tax rates are set at the England and Wales average (as in RSG since 1981).

(5)   Grants for discretionary spending are allocated using both equations (11.2) and (11.5); the expenditure and tax rate standards and costs are derived from those used in constructing the needs index (table 6.3) revalued to the relevant year. Expenditures are actuals, estimated actuals, and relevant expenditure after fees, specific grants, etc., have been deducted.

These assumptions ignore the need to estimate local effects as the sum of individuals living within an area, which is what would be required for a full system of NIT. However, they do allow a broad assessment of the spatial characteristics of the proposed grant and revenue system at the level of the local authorities. To permit comparative results calculations are made for two years, 1977/8 and 1980/1.

The distribution of the resulting grants and shared income is shown in table 11.4 for the main administrative areas. Shared income derived from the income tax is large for the English Counties and metropolitan Districts because of the larger populations living there. Similarly fee income from user charges, and specific grants, are also highest for these areas, although in both cases Wales, despite its small population, is also a very large recipient. In each of these cases, however, the allocation produced in the table totally reflects that used in the past.

The non-discretionary grant income received by the different areas

Table 11.4. *Means and standard deviations (in parentheses) of proposed shared income and grants for 1977/8 and 1980/1 (£m. at 1980 prices)*

	Non-met. Cos. England	Non-met. Cos. Wales	Met. Districts	Inner London	Outer London	Total
1. Specific grants						
1977/78	24.9(12.1)	17.6(5.4)	13.7(8.5)	8.3(1.8)	5.4(1.9)	15.8(11.3)
1980/81						
2. User charges						
1977/78	12.5(5.3)	6.7(1.9)	3.3(1.9)	2.9(0.5)	2.2(0.6)	6.4(5.5)
1980/81	13.7(5.8)	7.3(2.0)	3.6(2.1)	3.2(0.6)	2.4(0.7)	7.0(5.9)
3. Shared income						
1977/78	22.8(10.3)	11.1(4.2)	10.3(5.6)	6.6(1.4)	7.2(1.7)	13.6(9.5)
1980/81	23.7(10.7)	11.5(4.4)	10.7(6.9)	5.9(1.5)	7.5(1.8)	14.2(9.9)
4. Non-discretionary services grant						
1977/78	60.3(26.6)	93.9(30.6)	57.4(25.8)	7.8(103.1)	21.3(9.4)	63.3(57.4)
1980/81	61.2(26.9)	88.4(28.5)	47.8(20.4)	0.0(112.0)	14.8(10.3)	52.1(57.6)
5. Deviation: shared income and non-dis. grant minus actual RSG						
1977/78	−17.4(19.5)	3.7(9.2)	−13.8(16.9)	−38.2(98.9)	−20.4(8.7)	−17.5(35.5)
1980/81	11.9(12.5)	26.8(7.5)	−7.1(18.4)	−50.1(108.0)	−18.0(8.9)	−4.6(41.4)
6. Discretionary services grant						
1977/78	0.0(45.9)	0.0(20.0)	0.0(31.9)	0.0(87.9)	0.0(10.5)	0.0(51.7)
1980/81	44.4(45.4)	0.0(15.4)	16.4(24.7)	95.7(197.4)	25.7(14.5)	33.9(73.3)
7. Deviation: shared income and both grants minus actual RSG						
1977/78	0.0(59.9)	−73.8(24.2)	−70.1(45.6)	−67.1(15.7)	−59.1(15.6)	−86.2(51.7)
1980/81	56.3(38.4)	16.8(13.4)	9.2(13.7)	45.4(90.0)	7.8(9.1)	29.2(43.1)
8. Standard expenditure need						
1977/78	245.0(106.0)	148.5(48.1)	118.3(63.1)	110.6(10.7)	82.7(19.8)	156.3(97.5)
1980/81	245.6(106.1)	148.3(47.3)	115.6(59.4)	102.0(10.7)	82.0(21.0)	154.7(97.9)

Discretionary grants are calculated using equation (11.3) in row 6. Average service costs for calculation of row 8 are derived from the representative needs index developed in chapter 6.

displays an interesting pattern. The cities and London are all substantial net losers, when the grant is added to shared income, compared with their actual allocations of RSG. Wales makes small net gains on actual allocations, whilst the English non-metropolitan Counties make a small net loss in 1977/8 and a small net gain in 1980/1 in comparison with actual RSG allocations. 1977/8 represents a time when all areas were favourably treated, especially the cities, whilst 1980/1 represents a period when substantial grant cuts had been made.

The discretionary grant presents interesting contrasts since it is calculated using the deviation of actual rate poundage above average rate poundage for the relevant administrative area, after shared income and non-discretionary grants have been taken into account; its magnitude reflects variance of local poundage relative to the total variance of poundages for any administrative category. For 1977/8 there would have been a substantial number of net losers of grant as many poundages were set relatively low, i.e. the distribution of poundages was skewed downwards from the mean for each administrative area resulting in potential loss of grant. In each case these potential grant losers receive no grant: a deficiency payments principle similar to the pre-1980 resources element is assumed to operate. For 1980/1 the distribution of poundages is skewed above the mean with many areas levying relatively high poundages. This was stimulated by more rigorous grant control at this date. The result is a substantially higher level of mean discretionary grant payments.

It is also interesting to compare standard expenditure, given in the final row of table 11.4, with actual expenditure levels. For all areas, actual expenditure is exceeded by assessed need to spend. The major deficit in the English Counties results from their small extent of discretionary service provision. This becomes reflected in the level of discretionary service grants which is generally higher in London. The spatial distribution of non-discretionary and discretionary grants per head of local population is given in figures 11.2 and 11.3. These figures should be compared with the results from allocation of the RSG over the period up to 1981 shown in figures 8.1 to 8.5, especially those for total grant levels. From figure 11.2 it is clear that the non-discretionary grant, like total RSG and RSG needs element strongly favours inner cities and suburban Counties which have high concentrations of needy client groups; although in addition the poor tax base of some rural areas leads to substantial per capita receipts in these areas as well. London, with its high tax base, fares worst from non-discretionary grant. The distribution of the discretionary grant reflects the extent to which discretion in service provision is employed and the inequality in tax bases which is present. The grant will be most favourable in its distribution to those areas exercising a high degree of discretion on a low tax base, and hence using a high rate poundage relative to other areas of the same administrative type. Figure 11.3 shows these areas to be

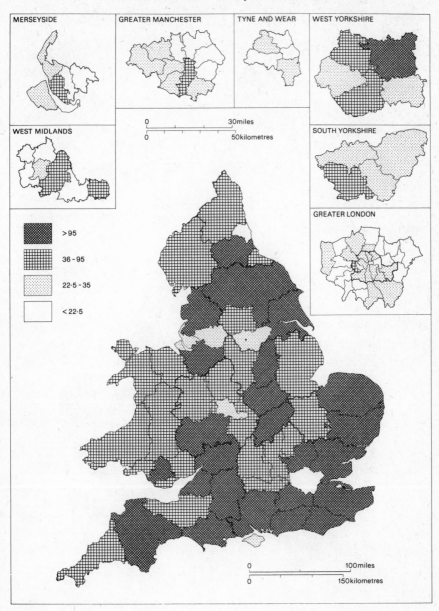

Fig. 11.2 Level of non-discretionary grant per head relative to the median of £35 based on data for 1980.

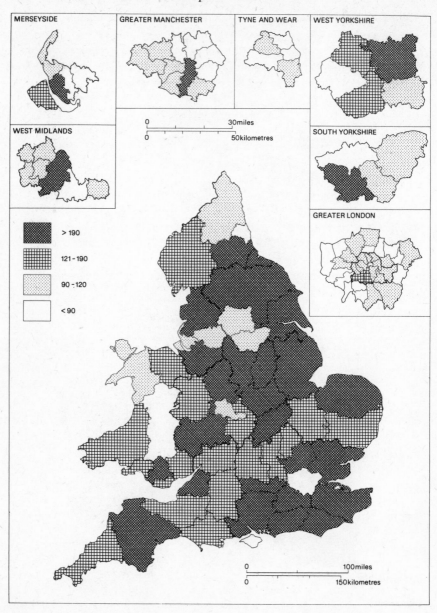

Fig. 11.3 Level of discretionary grant per head relative to the median of £120 based on data for 1980.

concentrated in many urban locations, especially inner cities, but also in many rural areas. It is the areas controlled by local Labour councils (as well as some other party-controlled councils) which are generally believed to exercise the greater degree of discretion. An ANOVA of the level of per capita discretionary grant with administative area and local party colour as factors, and tax base and need as covariate controls, demonstrates that local party colour is indeed a significant correlate of discretionary grant receipt (at the 99% level). Hence, implementation of such a grant implies a commitment by central government, in an open-ended way, to subsidise local discretion to provide services. Hence, the proposals given here, whilst emphasising the importance of minimum standards and central control, nevertheless also accentuate local autonomy through both tax base sharing and the discretionary grant.

Clearly the results displayed in table 11.4, and figures 11.2 and 11.3 depend on a specific set of assumptions, and another set of assumptions would produce a different conclusion. Hence, the results must be treated as an approximate and heuristic device, to be used to gain insight into an alternative local financial system, rather than to give diagnostic figures for any area. Nevertheless, these results do give an indication of the orders of magnitude involved in switching to a more rational financial structure. Three features should be noted. First, the total level of grant support is slightly reduced, the burden being shifted to shared revenue. This fosters a better relation of local expenditure to local discretion and accountability, and also encourages more efficiency. Second, the size of equalisation grants required is greatly reduced. This depends in part upon greater use of the more adequate and less regressive local revenue system using income tax sharing. A third feature is the size of the unitary specific grant. This compares most closely to the old RSG needs element (existing from 1967/8 to 1980/1; compare table 11.4 with table 8.3), but in comparison with that, it is a considerably smaller component of local revenues. Again this is accounted for, in part, by reform of the local revenue system. As a result of these features, the total level of grant support is reduced from a mean of 63% to less than 40% of total local current expenditure. Within this reduced proportion of central grants, however, the most significant change is the marked shift in emphasis towards separation of discretionary from non-discretionary support.

The proposed form of central grants shown in table 11.4 is not without its problems, and four in particular stand out. First, there is the difficulty of defining for what services, and at what level, minimum central standards should be set. Second, there is the difficulty of differentiating discretionary from non-discretionary spending. A better and more detailed assessment than that provided in table 11.4 will require greatly improved data on local service provision. In particular, additional statistics are required on the levels of services provided in different categories of need. This applies particularly to local public transport, services for the old, and special schools, which also require that output quality should be taken into account. This in turn raises a third difficulty, that the level of service provision must be measured against the background of the local eligible population. As a result improved Census indicators, or indicators of latent need, may be required. Providing them is an exceedingly difficult task, but as has been previously suggested, it can in many cases be approached through use of data presently assembled by the Social Security and Inland Revenue offices. The use of these data creates new problems of confidentiality, but provided these can be overcome the use of these data draws us towards the national integration of central and local revenue and benefits systems which is discussed more fully earlier in this chapter. The fourth major difficulty of the proposed system is presented by the fact that in those local authorities with the lowest tax base nearly all discretionary spending will be funded by central government. This will minimise the impact on local tax rates, giving essentially free services and thus is likely both to induce inefficiency, and to undermine accountability. It is for this reason, among many others, that the unitary block grant form of the RSG which has operated since 1981/2 has combined together the discretionary and non-discretionary service sectors, and has set thresholds and tapering to the levels of central involvement. Despite this problem, however, the proposed system of grant and shared revenue support can do much to ameliorate the difficulties in the present local financial system and, by the addition of a better integration of central and local taxation and benefits through NIT, can do much to overcome the geographical inequities of public finance in Britain.

11.7 Conclusion

The aim of this chapter has been to introduce discussion of a new central-local financial structure including grants, shared revenues and a total tax/benefit framework. It has been possible here only to sketch the outlines of this proposed scheme. However, sufficient detail has been given to show that major advantages can be achieved by use of such a new structure.

The proposals evaluated offer the potential for a greatly improved structure for central–local relations in Britain. The increased emphasis on specific grant support more truly recognises the constitutional position of British local authorities, the smallness of the country, and the need for minimum standards in non-discretionary services. Increased use of user charges, although of limited potential, should encourage efficiency and accountability. The new block grant for discretionary services will increase local autonomy for these services by ensuring that all authorities experience the same marginal tax rate in providing them. The shift in the valuation basis of the rates will overcome present arbitrariness; and the provision of a shared income tax plus NIT will ensure a high level of central–local co-ordination, which the use of the rates alone makes difficult, and which can fully achieve fiscal equity both between individuals and between local authorities.

This proposal will find its critics. Local authorities will no doubt dislike the increased restrictions of minimum standards and specific aid for non-discretionary services. Central government will no doubt resist attempts to share the income tax or introduce a tax credit system. Many of these changes have been, to some extent, held back by the method of Inland Revenue processing of tax returns. However, as this becomes progressively more automated through use of computers, the NIT system, and its integration with local taxes and benefits, becomes fully possible.

The proposals here are, of course, preliminary and full implementation in practice will require very considerable developments and refinements in detail. In making such refinements account must be taken of a number of particular difficulties such as the definition of the services and the level to which minimum central standards should apply; the differentiation of discretionary from non-discretionary spending and the problem of identifying the local eligible population; and the development of adequate measures of output and service quality variation. Whilst unit costs, and variations in these costs, have been included in the proposed grant programmes, equalisation of these service inputs does not guarantee equalisation of service outputs. The measurement of output levels and qualities is still one of the most complex and under-researched fields, but clearly attempts must be made to include this factor in grant formulae.

Despite these difficulties, and despite the need for many practicalities and definitions to be determined, it is clear that many major problems with the existing local financial structure, and other proposals that have been made, are either overcome or are by-passed by the proposals developed in this chapter. Certainly there is the need for much greater research and appraisal of these proposals before they can be implemented; however, by sketching its properties at this time, it is hoped that it will be possible to open up a new set of possibilities for discussion and also to stimulate constructive criticism and appraisal of the present structure of local finance.

CHAPTER 12

THE BRITISH EXPERIENCE OF CENTRAL GRANTS – A WAY FORWARD?

Central grants to local governments are now a major part of the economic and social fabric of most western countries. Grants have major impacts on local economies, factor prices in the private sector, and hence on the basis of competition affecting regional patterns of economic growth; grants also have major implications for social distribution by modifying service levels and the cost of locally-provided public goods; and they also modify the constitutional balance between central control and local autonomy in the organisation of the State. This book set out by discussing the general characteristics of central grants to local governments and has then charted the evolution of the British history of central grants, especially the Rate Support Grant, as one of the major examples of such programmes to be developed in the western world.

Of their various aspects, it is the effect of grants on social equity which has been the major emphasis of this book and this has been used as the basis for evaluation of both the RSG and proposals for its reform. This in turn has drawn attention to the need more closely to integrate central and local taxes into a total tax system. This would yield the benefits of eliminating, as far as possible, individual and spatial patterns of inequity (except insofar as these reflect the overall distributional aims). The *principle of fiscal equity* proposed in chapter 11 has as its aim the equal treatment of individuals irrespective of where they live. This has two major implications for central grants: first, it requires a level of resource equalisation such that all local authorities are placed in an equal position to provide discretionary services at the same marginal cost; second, a level of tax and expenditure equalisation which permits the difference between benefits received and taxes paid (the fiscal residuum) for non-discretionary services to be equal for similar individuals irrespective of locations.

The proposition of a principle of fiscal equity has profound implications for the system of local finances which results and, in particular, the most appropriate form which central grants should take. For Britain, the implications of the principle which has been proposed are that discretionary and non-discretionary services, and the levels of these services, have to be separated in their treatment by both central and local government. This in turn leads to profound implications for the structure of central control and local autonomy. For discretionary services, there is no objection, from the overall equity point of view, to variation in service levels and costs, provided these variations reflect true local preferences. For non-discretionary services, however, variation in service levels and costs should be eliminated as far as possible. Thus autonomy in local decisions is restricted to discretionary services alone. In contrast, non-discretionary services must be organised as a form of central-local partnership which is subject to a high degree of central control. This reflects the legal and constitutional position of local authorities in Britain. Of course a shift to a federal structure would lead to a different definition of discretionary and non-discretionary services and hence to a different balance of central control and local autonomy, but under the present constitutional structure the extent of local autonomy is limited.

These observations serve to provide the basis for evaluation of central grants in this book: they have underlain evaluation of the systems of RSG which operated from 1967 until 1980; they have permitted assessment of the form of RSG implemented since 1981; and they have permitted the evaluation of a set of changes proposed to the system of local finance in chapter 11. The conclusions of this analysis are briefly summarised below.

12.1 The RSG up to 1980/1

It is clear from the discussion of earlier chapters that the evolution of the RSG has been exceedingly complex and at various stages has sought to satisfy different objectives.

The major *explicit* aim of the RSG has always been concerned with forms of equalisation, but the precise concept of equalisation employed has varied over time. In early years equalisation tended to favour needs and potential need more highly than resources. However, in more recent years resource equalisation and per capita allocations were favoured to a greater extent; for example, the needs element reduced from 82 to 61% of the Aggregate Exchequer Grant, whilst the resources element increased from 16 to 30% from 1967 to 1980. This shift from need to resource equalisation resulted mainly from the effect of raising the standard rateable value per head. Moreover, increased emphasis on the domestic element, made the grant more of a direct per capita subsidy and hence reduced the need-equalising effect of transfers. In addition, the effect of closed-ending, the damping of the needs element, the variable priorities of area-biased needs, and lack of full needs equalisation payments until 1979/80, also reduced the equalising effect of needs compensation. On the resources side, closed-ending, the 'clawback' from London, and the immunity of the London equalisation scheme to national redistribution also reduced the degree of equalisation possible. It is clear, therefore, that although need and re-source equalisation were major explicit aims, subsidiary factors considerably diminished the equalisation that was achieved, and produced a number of effects which were often perverse to the original stated goals of the RSG programme.

As well as equalisation, the RSG has had major impacts in other areas of central-local relations. Lack of equalisation between areas has also undermined the neutrality of the grant or income distribution in the country as a whole. Thus, for example, individuals living in areas with large tax bases in relation to need, and also in areas with low tax bases and low need, received much larger allocations of RSG than required for equalisation, which was often passed on in reduced local tax bills. Conversely, people living in areas with small tax bases and large need were penalised by the deficiency payments effect of the resources element, and people living in London Boroughs with large tax base but high needs were penalised by 'clawback'. Moreover, where specific aid through the RSG was directed towards low income or high cost areas, the effects of this aid were often diffused to other areas by using need indicators which prevented specific targeting to the areas of highest need. Although the deleterious effects of the RSG on income distribution are minimised at the lower end of the income profile by the effect of rate rebates, these do not guarantee equity between areas over the lower-middle and upper income levels.

The fiscal balance between central and local government has been continually modified over the period up to 1980. Whilst local authorities perform a large level of expenditure functions, their own revenue has not kept up in real terms. In the early years a major aspect of the RSG was to overcome this imbalance between revenues and expenditures. This was clearly stated by the G.B. Government (1966, p. 6) White Paper on local finance which set up the initial RSG 'the paramount need is not so much to encourage or assist the development of particular services as to ensure that the total cost of all services does not place an impossible burden on ratepayers and in particular on households'. Even for the 1974/5 settlement of the RSG it was still stated as a major aim of the grant that 'where under the new distribution formula authorities receive more grant . . . it is intended that ratepayers should benefit and that whenever possible rates should be reduced' (G.B. Government DoE Circular 19/74, 1974, p. 6). By the end of the 1970s, however, the emphasis had shifted to using the RSG as a means of encouraging greater efficiency in local spending. For example M. Heseltine, Secretary of State for the Environment (reported in *The Times*, 17 January 1980) suggested that the RSG was the 'main means of influencing local authorities' rating and spending decisions'; in particular the 1980/81 RSG 'was based on a logical extension of (national) economic policy . . . (requiring) careful planning and the co-operation of local government'. Hence, the late 1970s saw a marked shift away from using central support as a means of tax-base sharing, and towards the employment of central support as a means of both encouraging local efficiency and of gaining greater central control of the level, and also to some extent, the distribution of local expenditures.

As well as shifts in the concepts of equalisation and fiscal balance employed in the RSG up to 1980 it was also inadequate in its independence of local behaviour, in the data employed and in its certainty and stability. Inadequacy of independence resulted from the use of allocation procedures which could, to a significant extent, be manipulated by the local authorities to achieve higher grants. The use of local authority estimates of expenditure, when combined with the use of past expenditure levels in the regression formula, and damping of needs, gave the RSG a prodigious bias in favour of those authorities with a history of high expenditure levels in comparison with those with low expenditure levels. Inadequacy in the data base derived from two factors: first, the heavy use of out of date information (e.g. the 1971 Census data even in 1980/1) for distribution of the RSG; second, inadequate surveys of pay and price changes included in RSG Increase Orders. Inadequacies in the certainty and stability of the RSG arose from a range of factors: the unknown magnitude of Increase Orders and other central government legislation; the frequently late or delayed amendment of the level of RSG each year (especially the effect of subsequent cuts due to exigencies of national macro-economic management); the change in form of distribution from one year to the next (due mainly to changes in need indicators, level of needs damping used, and the level of standard rateable value per head adopted); and differences in the form and impact of central directives over the fiscal year on manning levels, priority areas for spending, and allowable levels of local tax increases. Central government frequently imposed constraints, shifted macro-economic policies or changed directives to authorities without considering or compensating for the resulting expenditure consequences.

12.2 The RSG since 1981/2

The reform of the RSG which took place for 1981 combined the previous need and resource elements into a new block grant and maintained the previous domestic element. The analysis of previous chapters has shown that some of the results of this change have been less radical than have been claimed. Nevertheless over the period during which the block grant is used it is likely to progressively modify central–local relations and local service characteristics.

As we have seen, the latter years of the previous RSG showed a decline in the use of area weights in the needs element and an increasing rigour of central control. The post-1981 RSG block grant was very much an extension of this thinking and was concerned less explicitly with equalisation and providing for fiscal balance than with promoting local economic efficiency and controlling local spending: its prime expressed aim was 'to deal with the problems of the major overspending authorities and the pre-emption of grant that flows from high spending . . . It is critically important to provide a grant system which encourages economy and efficient management of resources by all authorities and which provides them with an incentive to keep their expenditure within reasonable limits'. (M. Heseltine, Secretary of State for the Environment; reported in *The Financial Times*, 13 March 1980).

The new form of RSG has conferred some undoubted benefits on the structure of local government finance. It has combined resource and needs equalisation into a unitary grant thus permitting full equalisation to be achieved (see chapter 2) and has hence overcome many of the imperfections in the previous grant system. In addition a first synoptic attempt has been made to differentiate national services as merit goods for which non-discretionary minimum standards should apply. This has had the benefit of allowing improvements in equalisation and made an initial step towards the financial system proposed in chapter 11. Moreover the consequential publication of GREs, manpower levels and other comparative local authority statistics should be welcomed as a means for better monitoring of local behaviour and efficiency and improving local participation, awareness and accountability. Technically many aspects of the new RSG are also useful steps forward: the grant is now allocated to all levels of relevant local authorities, thus overcoming many of the difficulties resulting from precepting and allocations within Counties; feedback from past expenditure levels has been reduced, although not eliminated; and there is a more direct relation of allocations to actual costs and client numbers. In these ways the new grant is a more rational, comprehensible and equitable system than the previous RSG.

Despite these benefits, the new grant system has introduced a large number of new difficulties and has ignored other pressing problems. It does not equalise the marginal costs for discretionary services above GRE; the definition of needs (i.e. GRE) is still influenced by a large number of political as well as objective characteristics; major inequities are introduced by the contradictions between thresholds, abatements, and tapering penalties from differing volume and GRE target levels; uncertainty in final grant levels has greatly increased rather than diminished; complexity deriving from the number of factors used in allocation has increased and GRE is not in all cases rational; there are still many weak and questionable sources of data employed particularly for Districts in both metropolitan and non-metropolitan areas; regression feedback from past expenditure levels is still present in some GREs; there is not enough allowance for cost differences between authorities, especially between urban and rural areas in salary levels; multipliers are still the major and complicated method of making grant adjustments; and closed-ending of the grant still introduces major conflicts with equalisation goals.

These are only a few of the major difficulties outlined in previous chapters for the block grant. The domestic rate relief grant, as it is identical in structure with the previous domestic element, possesses the same drawbacks as it had up to 1981: it arbitrarily subsidises domestic as opposed to commercial ratepayers; the level of spatial differentiation is largely arbitrary and, in the case of London, is still confused with the independent issue of resource equalisation; and there is still no clear relation to local income measures of ability to pay rates.

Hence, although the new RSG has severely modified the pattern of equalisation, central–local fiscal balance and incentives to efficiency, it has not reduced the perverse effects on income distribution. Moreover it is still not entirely neutral to local behaviour, it has not been fully supported by collection of adequate data, it is still too complicated and sophisticated, it has suppressed local discretion and undermined local accountability, it has increased rather than diminished uncertainty, and it is still largely unintelligible to both the public and to the politicians.

A further problem, which interrelates with the form of the RSG, is the structure of relations between the Counties and Districts (or in the case of London, between the GLC and the Boroughs). A major difficulty present in the pre-1981 RSG was that the method of grant allocation to the upper level government in the non-metropolitan areas, and the lower level government in the metropolitan areas considerably confused both accountability and grant allocation. This has been overcome in the post-1981 RSG. The difficulty that remains, however, is that the income of the higher level governments deriving from local sources is still derived from the use of precepts on the lower level governments. This has diffused the lines of accountability and in 1981 and 1982 has been the occasion for a series of legal battles between the two levels of government; mainly as a result of Labour-controlled Counties making heavy precepts and supplementary rate levies on Conservative-controlled Districts and London Boroughs. The clear implications of these difficulties is that as well as reform of RSG to give allocation directly to each tier of government, the revenue system also requires reform to permit direct access to local tax base for each tier of government. This could be achieved by allowing both tiers direct access to the rates. However, in the context of the discussions of implementing alternative local tax sources to that of the rates, the allocation of a new tax source to the County level would seem a more attractive possibility. Either sales tax or local income tax would fit this objective. In the context of the suggestions for reform discussed in chapters 10 and 11, the LIT may well suit this purpose best.

Despite this problem and despite major flaws over the whole period of evolution since 1967, however, the RSG in Britain in the context of international comparisons can, in many ways, be judged to have been a most successful programme. To a greater extent than in many countries it has involved the central support by national taxpayers of local services. The level of central support of about 60% of current account relevant spending and 50% of total local spending is higher than almost all other western countries. Moreover, the extent of equalisation criteria built into distribution of the grant has allowed more specific targeting to areas of need than is possible in many other countries and this has permitted a higher degree of equalisation.

Perhaps international comparison shows greatest contrast with the USA. Here the support by higher level government of local authorities is rather less: 19% of local spending from federal support and 40% from federal and State levels combined in 1978. Also in the USA the level of targeting of aid is much less: the Congressional 'pork-barrel' for funds necessary in the enactment of most grant and intergovernmental transfer programmes has considerably diffused aid and hence reduced the overall levels of equalisation possible (see, e.g., Dommel, 1974; Nathan *et al.*, 1976, 1977). However, the major contrast with the USA is not just in equalisation grants, but also in tax base and local expenditure. In Britain the classic features arising in the local governments of USA urban areas have not arisen. In Britain, as shown in previous chapters, the tax base of urban areas is *not* small, nor is it declining; instead it is relatively large and in many cases is increasing at a more rapid rate than most other areas. Moreover, expenditure levels, although high and increasing rapidly, are lower than in other urban areas. Again, rate poundages, although high, have increased more slowly in urban stress areas than in all other areas. Although there are undoubtedly 'pockets of poverty' within these areas, the overall result is that rate yields are relatively equal and up to 1980 have contributed a decreasing proportion of local finance, such that dwellers in urban stress areas have been able to shift to national tax payers an increasing burden of local finance (through the Rate Support Grant).

If, therefore, there are a number of major flaws in the RSG, it must be remembered that in comparison with many countries, especially the USA, it possesses a number of beneficial characteristics. Moreover, the experience with RSG suggests that some of its characteristics may usefully be applied to other countries. Nevertheless major difficulties remain and in overcoming these it is the experience of other countries with other financial arrangements, particularly tax-base sharing, which is suggestive of reforms.

12.3 The proposed system of central–local relations

Although this book has been primarily concerned with central grants it is not possible to exclude from consideration the interrelations of central grants with local revenues and, hence, with the total tax system. As a result, chapters 10 and 11 above have been used to evaluate various alternative financial arrangements in Britain and from this a complex set of proposals has been developed. These proposals are underlain by two principles: of clearly differentiating discretionary from non-discretionary services, and of increasing the overall fiscal equity of the central–local financial system. These two principles lead to a proposed structure of local finance composed of seven main elements

(1)   increased use of specific grants;
(2)   replacement of RSG block grant by a similar 'unitary specific grant' for non-discretionary services (perhaps primarily for education);
(3)   introduction of a new block grant to equalise the marginal cost of discretionary services;
(4)   increased application of user charges where possible;
(5)   continued, but reduced, use of local rates with a shift to a capital or site valuation basis;
(6)   the introduction of a shared local income tax; and
(7)   integration of central and local taxes and benefits, in the long term, into a total tax system based on negative income tax.

Not all of these proposals are novel. The increased use of specific grants and user charges has already been advocated by a number of workers and has been implemented, to some extent, since 1979. Additionally, the RSG block grant used since 1981/2 already possesses many of the features proposed for the 'unitary specific grant'. The shift to capital and site valuation for rating has been considered by various governments. It was a specific proposal of the Layfield Report may soon be implemented However, the proposal for a division of discretionary from non-discretionary support is relatively novel (although also advocated by Davies, 1975); and the proposals for a shared income tax and system of central–local negative income tax are highly novel (although presaged in rather different forms by a number of authors: e.g., Ilersic, 1973; G.B. Committee of Enquiry, 1976; Prest, 1978; Beasley, 1980*a*, *b*).

The proposed structure makes several significant improvements in local finance and central–local relations. It does clearly differentiate discretionary from non-discretionary spending. This allows linkage of the revenue for non-discretionary services to a fully shared set of sources which are both buoyant and guaranteed. This should overcome the problems arising from a lack of central support for services mandated onto local authorities, which has often been characteristic of the past. For discretionary services the new block grant allows full equalisation of the marginal costs of discretionary spending whilst reducing feedback, thus fostering both equity and also autonomy. The proposed structure should also improve economic efficiency. It has, in common with the block grant used since 1981, a form of GRE which has the effect of pressuring both overspenders and underspenders to perform at national levels of efficiency; this feature also pressures authorities to provide non-discretionary services at least at minimum standards (which many did not previously do). The proposals also give a more rational basis for local taxation which can be more easily maintained and updated. Most important, however, the proposal recognises the full ramifications of attempting to achieve fiscal equity in a decentralised government system. Although this has been discussed before (see, e.g., Oates, 1972, 1977; Foster *et al.*, 1980), it is only by a full system of negative income tax that complete equity can be achieved.

Of course there are various practical problems in implementing the proposals suggested here which have yet to be overcome. The major difficulties are outlined in chapter 11, but particularly important amongst these is the need for: improvements in data availability, especially on local incomes; better measures of service output levels; and improved measures of local costs. It has certainly not been possible in the space available here to do more than sketch the general characteristics of the proposed system and evaluate the broad structure of the likely repercussions as far as either spatial or social distribution are concerned. What is intended, therefore, is that these proposals, and the brief assessment of them which has been possible, will be stimulative of further research in this area; but most of all it is hoped that practical assessment and implementation of these proposals will follow in due course. If this book has facilitated some of the first cautious steps in this direction, then it will have achieved its main objective.

DATA SOURCES AND METHODS OF ANALYSIS

1962–73

The data for the period from 1962 to 1973 were derived from a variety of sources for various years:

> *Rates and Rate Products* (DoE, London) provide rateable values and rate poundages.
>
> *Return of Rates* (Institute of Municipal Treasurers and Accountants, London) and other IMTA sources provide expenditure estimates, population, needs element and resources element.
>
> *General Grant Orders* (HMSO, London), general grants.
>
> *Rate Support Grant Orders* (HMSO, London), rate support needs elements.
>
> *Rate Deficiency Grant* estimates (DoE, London), rate deficiency grant.
>
> *The Times*, local election results.
>
> *Municipal Year Book* (London), local election results.

For various reasons it proved impossible to obtain a comparable data set for all local authorities for all years. For this reason some authorities are omitted from the analysis: for the period 1962–73 an authority is omitted if statistics are unavailable for one variable for one year. As a result the following local authorities were omitted from the analysis: Hartlepool, Luton, Solihull, Teeside, Torbay and Warley. All are County Boroughs.

In addition it was necessary to reconstruct some later-period authorities from the constituent parts for earlier years. This affected Cambridgeshire and Isle of Ely, and Huntingdon and Peterborough which merged together to give, respectively, Cambridgeshire and Huntingdonshire after 1966. The major problem, however, is the reconstruction of London Boroughs prior to reorganisation of London government in 1965. From these direct aggregations from old County Boroughs, Metropolitan Boroughs (M.B.), Boroughs (B.) and Urban Districts (U.D.) are possible in most cases. However, for some new London Boroughs approximations were necessary where new boundary lines did not coincide with old. In the following cases the following multipliers were applied to convert old to new London Boroughs:

New London Borough	*Previous local authority*
Lambeth	Wandsworth M.B. × 0.333
Wandsworth	⎰ Wandsworth M.B. × 0.666 ⎱ Battersea M.B.
Bexley	⎰ Bexley B. ⎮ Erith B. ⎮ Crayford U.D. ⎱ Chislehurst and Sidcup U.D. × 0.5
Bromley	⎰ Beckenham B. ⎮ Bromley B. ⎨ Chislehurst and Sidcup U.D. × 0.5 ⎮ Orpington U.D. ⎱ Penge U.D.

For a further four Boroughs the boundaries of old and new areas coincide closely but not exactly. No adjustments were made and hence these areas have only estimated data: Greenwich, Barking, Newham, and Redbridge. Hence, approximations are required for gathering data for eight out of the 33 London Boroughs. In addition to these problems, it proved impossible to reconstruct any reliable information on the City of London (for expenditure per head and General Grant 1962/3) and for London as a whole (for the General Grant 1962/3).

As a result of these various limitations and difficulties with data sources, the analysis of the years 1962/3–1973/4 is undertaken with the following set of local authorities:

English Counties	44
Welsh Counties	13
County Boroughs	72
Inner London Boroughs	13
Outer London Boroughs	20
City of London	1
Total	**163**

For comparison purposes with later years it should be noted that two changes occur with reorganisation of local government in 1974: first, changes in boundaries; second, changes in responsibilities for services. With respect to boundaries the old and new London Boroughs coincide exactly, many counties also coincide exactly, but only three Metropolitan Districts can be derived from previous County Boroughs. With respect to changes in service responsibilities, the London Boroughs again show little change, but the Counties assume much greater importance after 1974. As a result of these changes only comparisons within London can be treated with great confidence; comparisons between non-London authorities pre-

and post-1974 have to be treated with extreme caution and can yield only broad indicia.

To prepare a second set of comparisons of pre- and post-1974 data, some analyses were undertaken of pre-1974 data reduced where possible, to post-1974 boundaries. This was possible for 20 Counties in England and four Counties in Wales where boundary changes were small or relatively unimportant. It was also possible for all London Boroughs, but for only four Metropolitan Districts, most of which are inner city areas (Manchester, Liverpool, Wolverhampton, and Sandwell (as an aggregation of Warley and West Bromwich)). This reorganised data set does allow some additional comparisons to be drawn, but all conclusions must be treated with extreme caution.

1974–81

The data for this period were derived for the most part directly from the Department of Environment. They are those used for Rate Support Grant purposes and are for the most part unpublished, although in some cases near approximations are published. These sources provide information on rateable value, rate poundage, relevant expenditure total, grant levels at final Increase Order and indicators of need used in the needs element up to 1980. In addition unpublished expenditure outturn data for 1976/7, 1977/8 and 1978/9 compiled by the Department of Environment was also used, disaggregated into service categories. Source of income as between specific grants and fees, and levels of debt costs were also derived from the outturn sources. The outturn data are at outturn prices, whereas the Rate Support Grant data are at RSG prices for November of the year preceding the relevant grant year.

Price inflators and deflators for these data were derived from the Retail Price Index. Although recognised as not being entirely relevant to local government services, there is no satisfactory local government price index for the entire period covered and hence RPI was thought to be a satisfactory substitute.

In addition to these data, the need index required a further set of client group data which was derived either direct from the RSG needs indicator statistics supplied by the Department of Environment, or from other sources. Other sources were unpublished statistics of school pupils and education client groups supplied by the Department of Education and Science, personal social service clients information supplied from unpublished Department of Health and Social Security Statistics and CIPFA *Community Indicators*, serious crime from *Criminal Statistics*, transport services from *Transport Statistics* and a number of other sources noted in chapter 6. *The Census, National Dwellings and Housing Survey*, and *New Earnings Survey* were also used.

After 1981

For this period sources of data were the same as over 1974–81, except that for grant levels, expenditure estimates and poundage estimates, the CIPFA *Financial General and Rating Statistics* were employed as the analysis was undertaken before 1981/2 actual expenditure and final poundages were known. Hence, the results exclude supplementary rate changes occurring after June 1981. These give rise to only a small level of changes from CIPFA estimates. In other respects the book includes legislative changes up to May 1982.

NEED INDEX TABULATIONS PER CAPITA 1974–81

Local authority	74/5	75/6	76/7	77/8	78/9	79/80	80/1
Avon	86.4	87.6	87.3	86.8	87.1	86.7	86.9
Beds.	87.3	89.1	89.2	90.4	90.6	92.6	93.0
Berks.	86.3	84.9	85.6	86.6	86.8	86.2	86.4
Bucks.	81.2	82.3	83.4	85.0	85.2	90.4	91.8
Cambs.	84.3	85.4	86.5	87.1	87.4	88.1	89.1
Cheshire	84.7	85.5	86.5	87.6	87.8	91.8	92.2
Cleveland	99.4	100.3	99.6	99.0	99.2	98.8	98.6
Cornwall	91.6	92.7	93.3	94.6	95.0	97.0	98.1
Cumbria	96.3	96.1	95.9	96.2	96.5	97.5	97.7
Derby	85.7	86.6	86.2	86.1	86.3	88.7	89.0
Devon	88.8	89.6	90.1	90.9	91.3	91.9	92.5
Dorset	82.2	82.7	83.4	84.4	84.8	85.2	85.6
Durham	90.1	89.5	89.5	87.8	88.1	88.7	88.5
E.Sussex	85.6	85.1	85.6	86.1	86.6	87.2	87.8
Essex	82.0	82.8	84.0	84.9	85.2	87.0	87.6
Glouc.	89.3	89.5	91.2	91.4	91.7	92.7	92.9
Hants.	84.4	85.2	86.7	87.6	87.8	88.4	88.9
Hereford & Worcs.	87.9	88.2	89.4	91.8	92.1	94.1	94.9
Herts.	90.2	91.2	91.4	91.1	91.3	91.7	91.9
Humberside	92.5	92.8	93.6	93.3	93.5	94.1	94.1
IoW	80.9	83.9	84.9	85.4	85.8	89.2	90.5
Kent	82.7	83.3	83.9	85.5	85.8	87.5	88.1
Lancs.	88.4	91.1	91.5	91.4	91.8	93.2	94.3
Leics.	88.6	89.9	90.4	91.2	91.4	91.3	91.6
Lincs.	93.2	94.4	95.5	96.9	97.3	100.0	100.7
Norfolk	88.9	89.2	90.1	90.7	91.0	92.6	93.5
Northants.	97.9	101.4	101.3	102.0	102.3	104.7	105.6
Northumberland	94.4	93.5	95.0	94.1	94.5	97.6	98.1
N. Yorks.	80.9	80.7	84.4	83.0	85.3	85.3	86.1
Notts.	89.3	91.5	93.2	93.3	93.6	93.9	94.1
Oxon.	85.2	85.4	86.0	85.4	85.7	85.7	85.9
Salop.	98.0	98.4	98.5	100.6	101.0	103.5	104.5
Smst.	90.5	89.4	95.0	95.5	95.9	98.3	99.2
Staffs.	84.3	85.6	87.1	87.0	87.2	88.5	89.0
Suffolk	83.3	83.7	84.0	84.8	85.1	86.8	87.6
Surrey	75.9	76.4	76.8	76.5	76.7	78.6	78.7
Wrks.	85.4	87.5	88.7	89.4	89.6	91.5	91.8
W. Suss.	83.5	83.9	84.4	85.1	85.5	87.3	88.0
Wilts.	90.2	90.0	91.9	92.7	92.9	92.1	92.4
Clwyd	110.5	109.2	111.0	110.4	110.7	114.0	114.8
Dyfed	118.7	118.9	120.5	118.2	118.5	120.8	121.4

315

Local authority	74/5	75/6	76/7	77/8	78/9	79/80	80/1
Gwent	110.2	108.8	110.0	108.2	108.3	109.8	109.9
Gwynedd	115.9	119.5	120.9	120.0	120.5	123.1	123.7
M. Glamorgan	105.9	105.7	107.0	104.0	104.1	105.8	105.8
Powys	149.4	149.8	151.3	151.2	151.7	155.7	157.1
S. Glamorgan	115.5	113.4	116.0	114.1	114.2	112.2	112.3
W. Glamorgan	103.5	105.3	104.4	101.8	102.0	103.2	103.2
Bolton	99.7	102.8	102.5	101.9	101.6	98.6	99.4
Bury	84.9	85.4	86.5	87.2	87.1	94.1	94.2
Manchester	105.4	104.7	106.5	100.6	100.1	94.5	94.1
Oldham	90.9	94.0	94.2	93.7	93.5	98.3	98.6
Rochdale	100.5	101.5	102.1	100.0	99.6	102.3	102.3
Salford	101.9	101.8	99.9	97.4	96.9	92.2	91.5
Stockport	88.9	90.3	90.7	91.2	90.9	93.7	93.6
Thameside	93.1	94.2	93.8	93.2	92.8	93.2	93.4
Trafford	97.2	100.5	103.2	103.9	103.5	102.6	102.3
Wigan	94.1	89.1	88.9	96.3	95.8	96.4	97.2
Knowsley	103.2	101.9	102.5	100.1	98.9	100.7	98.8
Liverpool	101.5	100.4	101.3	97.1	96.4	90.2	89.0
St Helens	97.7	99.5	100.2	98.8	98.4	100.3	100.6
Sefton	92.7	94.9	96.1	94.2	93.8	95.5	95.0
Wirral	100.1	96.2	99.0	97.9	97.6	94.6	94.3
Barnsley	94.4	95.8	96.2	96.0	95.7	98.9	99.1
Doncaster	111.5	116.4	114.0	109.8	109.2	109.2	109.3
Rotherham	93.9	96.5	97.4	96.7	96.1	97.0	97.5
Sheffield	90.6	90.0	91.0	90.9	90.7	89.6	89.4
Gateshead	95.4	96.7	97.3	94.9	94.6	96.2	95.9
Newcastle	97.9	95.9	96.2	93.4	93.1	88.6	88.0
N.Tyneside	93.3	92.6	93.4	89.8	89.4	91.4	91.0
S.Tyneside	106.4	108.0	107.1	102.7	102.1	104.0	103.4
Sunderland	93.5	93.5	93.6	93.1	92.5	93.5	94.0
Birmingham	96.9	95.9	96.0	94.3	93.7	91.7	91.2
Coventry	94.8	93.8	93.8	92.9	92.3	91.3	90.9
Dudley	80.7	81.7	82.8	83.6	83.2	82.5	82.8
Sandwell	91.6	92.5	94.1	92.1	91.7	91.3	91.2
Solihull	105.0	111.1	110.3	111.4	110.9	114.8	114.9
Walsall	91.6	93.4	94.8	93.4	92.8	93.8	93.5
Wolverhampton	99.6	98.7	98.7	97.1	96.5	95.2	94.6
Bradford	97.3	97.2	97.0	95.5	95.1	96.3	96.7
Calderdale	107.7	105.5	107.3	107.5	107.4	110.8	111.1
Kirklees	97.3	98.6	98.1	99.4	99.1	99.0	99.5
Leeds	93.1	91.9	92.1	92.3	92.0	91.3	91.8
Wakefield	93.7	94.1	95.0	94.5	94.0	96.1	96.6
Camden	142.7	137.6	131.8	146.3	146.5	138.4	137.1
Greenwich	142.5	137.3	131.5	145.1	145.4	138.1	137.0
Hackney	152.0	146.3	140.3	141.2	141.5	133.6	132.3
Hammersmith	167.9	160.2	155.1	156.2	156.4	146.9	145.3
Islington	166.1	160.0	153.2	154.3	154.4	145.3	143.7
Kensington	151.0	163.1	156.0	157.3	157.4	147.6	146.0
Lambeth	121.0	107.8	103.5	113.1	113.5	108.3	107.5
Lewisham	134.7	129.5	124.6	125.9	126.3	120.1	119.2
Southwark	143.9	138.4	133.0	134.0	134.3	127.6	126.6
Tower Hamlets	182.4	175.9	168.5	170.3	170.4	160.1	158.3
Wandsworth	124.7	120.2	115.8	116.6	117.0	111.7	111.0
Westminster	147.5	142.3	136.8	137.7	138.1	130.9	129.7

Local authority	74/5	75/6	76/7	77/8	78/9	79/80	80/1
Barking	100.8	101.1	99.9	98.4	98.7	91.2	90.5
Barnet	92.3	97.9	97.1	96.8	97.1	95.5	95.5
Bexley	90.4	90.8	91.7	92.9	93.1	95.6	95.9
Brent	97.5	98.6	99.9	99.2	99.5	105.7	105.1
Bromley	96.9	96.1	96.3	96.6	96.9	100.2	99.9
Croydon	96.2	96.4	97.2	95.3	95.5	96.2	95.8
Ealing	90.5	90.3	89.8	88.4	88.7	92.5	92.1
Enfield	94.8	94.3	95.2	94.4	94.7	96.7	96.6
Harringey	92.9	92.8	92.8	92.5	92.8	95.7	94.7
Harrow	96.3	93.6	96.4	93.6	93.9	92.0	91.9
Havering	99.9	100.7	100.8	99.4	99.6	103.8	103.7
Hillingdon	96.7	96.1	95.8	95.2	95.4	100.0	99,7
Hounslow	103.3	101.7	101.8	101.1	101.4	96.7	96.7
Kingston	97.4	96.2	98.4	98.5	98.8	86.8	86.9
Merton	90.4	91.9	92.2	89.4	89.8	90.4	90.3
Newham	101.6	102.8	104.7	102.8	103.0	97.0	96.6
Redbridge	80.9	81.8	80.9	82.2	82.4	90.7	90.7
Richmond	95.1	97.5	98.5	91.0	91.2	90.7	90.3
Sutton	90.9	93.8	92.8	92.9	93.2	94.2	94.4
Waltham Forest	96.1	93.7	92.7	99.9	93.5	98.9	97.3

Notes

1. Central grants to local governments

1 See, for example, Smith (1977), Johnston (1979), Bennett (1980).
2 Bradford, Malt and Oates (1969), Gramlich and Galper (1973), Musgrave and Musgrave (1980).
3 Musgrave (1959), Musgrave and Musgrave (1980), Oates (1972, 1977), Head (1973).
4 Davies (1964, 1968), Black (1958), Buchanan (1965).
5 Wilson and Banfield (1964), Arnold (1977), Johnston (1979, 1980).
6 Nathan *et al.* (1976, 1977), Caputo and Cole (1976), Caputo (1975), Juster *et al.* (1976).
7 Smith (1977, 1979), Johnston (1980).
8 Bennett (1980).
9 Boyle (1966), Cripps and Godley (1976), Foster *et al.* (1980), Burgess and Travers (1980).
10 See Layfield appendices (G.B. Government Committee of Enquiry, 1976) and the *CES Review*.
11 See, e.g., Birch (1959), Bulpitt (1967), Bealey *et al.* (1965), Newton (1976), Jones (1969), Sharpe (1967) and Burgess and Travers (1980).
12 See, e.g., Smith (1974, 1977), Johnston (1980), Bennett (1980, 1981c).
13 See Foster *et al.* (1980), Burgess and Travers (1980), Heald (1980).
14 See, e.g., Jackman and Sellars (1977a,b), Alt (1971), Ashford (1975), Nicholson and Topham (1971), Boaden (1971) and chapter 9 below.
15 Needs assessment is discussed in full in chapter 6 below where references to the various approaches are given.
16 This is the approach suggested by Page (1980) for Britain. It has also been used by Jackman and Sellars (1977c).
17 This is the approach used by Oliver & Stanyer (1969), Jackman and Sellars (1977a), Gibson (1980).
18 This approach is favoured by many political scientists and has been used by Alt (1971), Ashford *et al.* (1976), Strauss (1974), Boaden (1971), Gibson (1980) and many others.
19 This approach has been little but was an important addition to the models of Ashford *et al.* (1976) and Strauss (1974) suggested by Gibson (1980) and found to be a significant explanatory variable for total expenditure and total expenditure minus grants for English Counties and metropolitan Districts in 1978/9.
20 For example, urban stress indexes are given by Nathan *et al.* (1977), Bradbury (1978).

318

21 The *New Earnings Survey* for 1977 is used throughout the subsequent analyses as this is the most complete. Even this, however, needs supplementing for data on Richmond on Thames, Sutton, Bromley and Powys from the 1976 and 1975 surveys; data is unavailable for metropolitan Districts in all years, although it is provided for metropolitan Counties: see Appendix 1.

2. Categories of central grants

1 This classification of grants is based upon data given by Musgrave & Musgrave (1980), Maxwell (1946) and Mathews (1974). Other introductory discussions to general forms of grants are Break (1980), Bennett (1980).

2 Break (1967) argues the case for open-ended matching and suggests that such grants must be categorical, with size related to spillover, with national level contributions equal to the ratio of external to total benefits and open-ended to allow local flexibility (since, as Oates (1972) notes, spillovers do not stop at a particular level of service). However, the experience with such grants in US Medicaid and Aid to Families with Dependent Children (AFDC) demonstrates that they have underpriced local services and this has stimulated rapid expenditure growth. Moreover, the matching requirement has favoured high-income states, resulting in violent inequities in levels of service provision. They have also been extremely fiscally perverse, have been exploited by local politicians by substituting for other local expenditures, and have been characterised by extreme fraud and abuse (see Derthick, 1975; Beam, 1978).

3 This has been suggested in the UK by Hicks and Hicks (1945) Boyle (1966), and Smith (1979).

4 See the discussion in Musgrave (1961), Boyle (1966), Mathews (1974), Cuciti (1976), Reischauer (1974), Heald and Jones (1980), Le Grand (1975), CES (1978), G.B. Government (1977).

5 See speech of M. Heseltine, Secretary of State for the Environment, 16 November 1979 (DoE *Press Release* 522, 1979) and DoE evidence to Layfield G.B. Government Committee of Enquiry (1976, Appendix 7, pp. 21–50).

6 *Local Government, Planning and Land Act* (1980).

7 This is also noted by Le Grand (1975).

8 Regression analysis of costs, e.g. Schmid *et al.* (1975).

9 Such sectional unit costing has been particularly favoured in the UK by the DHSS and DES, and their proposals are discussed further in chapter 6.

10 The importance of output levels of services is particularly emphasised by Le Grand (1975).

4. The Rate Support Grant

1 Further details of the allocation of RSG are given by Burgess and Travers (1980), Bennett (1981*b*), Foster *et al.* (1980) and various papers in the CES *Review* (especially Harrison *et al.* (1979) and Griffin *et al.* (1980)).

2 The Local Authority Associations up to 1974 consisted of: the Association of County Councils, Association of County Boroughs, Association of Rural Districts, Greater London Council, London Boroughs Association, and the National Association of Local Councils. After the 1974 reorganisation of government, they consisted of: The Association of County Councils, Association of District Councils, Association of Metropolitan Authorities, Greater

London Council, London Boroughs Association, and the National Association of Local Councils (for town, parish and community councils).

CIPFA, the Chartered Institute of Public Finance, was formed in 1974 following local government reorganisation. Prior to 1974 the corresponding bodies were the Institute of Municipal Treasurers and Accountants (IMTA) and the Association of County Councils (ACC).

3 This derived from the theoretical work of Fabricant (1952), Woodham (1953) and Boyle (1966). However, it was a fundamentally flawed argument since, apart from the difficulties noted in this chapter there was no reason to believe that increases or decreases in the weight placed on one factor in one area would be offset by compensating effects in other factors. It was purely a *statistical* fit which could have arbitrary effects between areas and was likely to be very unstable over time (see e.g. Godley and Rhodes, 1972; G.B. Committee of Enquiry, 1976).

4 Detailed discussion of the London RSG is given by Jackman (1979), Rose (1978) and Jenkins and Rose (1976a,b).

5 For example, CDP (1976) and Burgess and Travers (1980) point out that targeting to the inner areas of London gives aid to Boroughs with high needs, but also to resource-rich areas such as Westminster and Kensington and Chelsea.

5. The distribution of the RSG since 1981

1 Schedules of marginal incentive to spend are given by Burgess and Travers (1980), Harrison and Jackman (1980) and Heald and Jones (1980).

2 See *The Guardian* (1980) where it was suggested that the block grant should be used to prevent subsidies of local council house rents. See also G.B. DoE (1979c) who state that the grant is aimed 'to enable the local authorities to make the adjustment to lower levels of expenditure', e.g. by 'charging for school meals, milk and transport'.

3 See Martin (1980), ACC (1980a), Jacobs (1980) and Heseltine (1979).

6. Local expenditure need

1 Examples of the use of such indicators are given by Davies (1964, 1968, 1974), Davies *et al.* (1971a), Packman (1968), Imber (1977), Hakin (1977) and Pinch (1979) for social services, and Culyer (1974, 1976) and G.B. DHSS (1976) for health services, and Little and Mabey (1972) for education.

2 See Schmid *et al.* (1975), Nathan and Adams (1976), Downs (1978), Dommel *et al.* (1978), Bradbury (1978), Isserman and Brown (1980), Neuber *et al.* (1980). Other composite need studies in Britain are those of Davies (1968), Davies *et al.* (1971a,b), SPNW (1973), Roweth (1980) and Craig and Driver (1972).

3 Brazer (1959), Wood (1961), Adams (1965), G.B. Royal Commission on Local Government in England (1968a,b), Oliver and Stanyer (1969), Nicholson and Topham (1971, 1972), Boaden (1971), Danziger (1974), Newton and Sharpe (1976).

4 Examples of the use of this method are given by Horowitz (1968), Auten (1971, 1972, 1974), Ohls and Wales (1972) and Reischauer (1974, chapter 4).

5 Further discussion of the unit cost approach is given by Le Grand (1975), ACC *et al.* (1979), CIPFA (1979), Harrison and Jackman (1978).

6 See CIPFA (1979), Newton and Sharpe (1976), Harrison and Jackman (1978).
7 See Davies (1968), Edwards (1975), Webber and Craig (1978).
8 Further discussion of the representative index is given by Reischauer (1974), Bennett (1982*b*), G.B. Treasury (1979) and, in relation to GRE, in the relevant RSG *Reports* since 1980.

9. Modelling expenditure and grants

1 See, e.g., Musgrave and Musgrave (1980), Groves (1964), Rechtenwald (1971). Attempts to include the private sector in expenditure models have also been made by Barr and Davis (1966), Dahlby (1979) and Schott (1978).
2 See the British examples of Dye (1960), Birch (1959), Bulpitt (1967), G.B. Royal Commission (1968*a,b*), Boaden (1971), Davies (1968), Oliver and Stanyer (1969); and the USA examples of Brazer (1959), Henderson (1968), Bahl (1968), Stern (1973), Fabricant (1952), Hawley (1973), Hansen (1965), Wood (1964).
3 See for example Gupta and Hutton (1968), Barr and Davis (1966), Alt (1971), Ashford (1975), Ashford *et al.* (1976), Aiken and Alford (1970), Hawley (1973), Sullivan (1973), Sharkansky (1967), Dawson (1976), O'Cleireacain (1977), Frey (1978), Frey and Schneider (1978), Page (1978, 1980), Nicholson, Topham and Watt (1975), Foster *et al.* (1980).
4 See for example Boaden (1971), Davies (1968*a,b*), Oliver and Stanyer (1969), Nicholson and Topham (1971, 1972), Fried (1975), Danziger (1974), Foster *et al.* (1980), Ashford *et al.* (1976). It is also the case that the local rate poundage does not seem to be major determinant of electoral results (see e.g., Birch *et al.* (1959), Bealey *et al.* (1965), Bulpitt (1967), Sharpe (1967), Jones (1969), Dearlove (1973)).
5 See individual local authority analyses of Hinckley (1966) for Coventry, Alt (1971), Newton (1976) for Birmingham, Weir (1979) and Page (1978, 1980).
6 See Oliver and Stanyer (1969), Ashford (1974), Jackman and Sellars (1977*b,c*), Gibson (1980), Page (1978, 1980), Rhodes and Bailey (1980) and chapters 6–8 above.
7 See Bhangava (1953), Buchanan (1950), Scott (1952), Musgrave (1959), Gramlich and Galper (1973).
8 See Breton (1965), Tiebout (1956), Pigou (1947) and Oates (1972, 1977).
9 See Smith (1968), Thurow (1966), Buchanan (1952), Scott (1952), Osman (1968), Morss (1966), Pogue and Sgortz (1968), Oates (1968), Musgrave and Musgrave (1980), Oates (1972, 1977), Ashford (1974), Strauss (1974).
10 See especially the work on US Revenue-Sharing and Community Development Block Grant by Nathan and Adams (1977), Nathan *et al.* (1977), Nathan and Dommel (1977).
11 See e.g., Morss (1966), Oates (1968), Pogue and Sgortz (1968).
12 See for example Osman (1968), Strauss (1974), Pogue and Sgortz (1968).
13 See Ohls and Wales (1972), Bergstrom and Goodman (1976), Ashford *et al.* (1976), Strauss (1974), Jackson (1972), Dahlby (1979). An earlier approach was also made by Barr and Davis (1966).
14 The Strauss/Ashford approach has been also extended for Britain by Gibson (1980) to include a political interaction variable. This is discussed further in section 9.6.

15 An alternative approach by Foster *et al.* (1980) has recently been used in Britain in which local rateable value has been used as an inverse measure of price of public goods whilst controlling for the private economy. This definition is not followed here since tax base is assumed to be a variable separate from that of price (tax rate), although related to it. Moreover, the results of estimation with this assumption provide positive elasticities of expenditure with rateable value (in contrast to the negative elasticities of Foster *et al.* which bear no simple explanation as price effects!).

16 Analyses using the Welfare basis for expenditure assessment have been made, for example, by Greene, Neenan and Scott (1974), Peterson (1979), Le Grand and Winter (1977), and are often combined with the outputs approach in assessments of local authority services (cf. Newton and Sharpe, 1976); e.g., the analysis by Peterson (1979) for the USA, found considerable differences in spending patterns between redistributional and other categories.

17 See Chapters 4 and 5.

18 A more complex model for local debt has been given by Schott (1978).

19 For expenditures this derivation is given as follows
    Starting from

$$E_i = c_0 + c_1 f(B_i) + c_2(E_i - d_1 G_i^N)/d_1 \bar{B} + c_3 G_i^N + c_4 N_i$$

$$E_i - \frac{c_2}{d_1 \bar{B}} E_i = E_i \left(1 - \frac{c_2}{d_1 \bar{B}} = E_i \frac{(d_1 \bar{B} - c_2)}{d_1 \bar{B}} \right)$$

$$= c_0 + c_1 f(\bar{B}_i) - \frac{c_2 d_1}{d_1 \bar{B}} G_i^N + c_4 N_i$$

Therefore,

$$E_i = \frac{d_1 \bar{B}}{d_1 \bar{B} - c_2} \left[c_0 + c_1 f(B_i) + \left(c_3 - \frac{c_2}{\bar{B}} \right) G_i^N + c_4 N_i \right]$$

gives the final expenditure patterns.

20 For tax rates starting from

$$t_i = d_1(t_i B_i + t_i(\bar{B} - B_i) + G_i^N)$$

rearrange

$$t_i = d_1(t_i \bar{B} + G_i^N).$$

Substitute the right-hand side of (9.22)

$$t_i = d_1 \left\{ \frac{d_1 \bar{B}}{d_1 \bar{B} - c_2} \left[c_0 + c_1 f(B_i) + \left(c_3 - \frac{c_2}{\bar{B}} \right) G_i^N + c_4 N_i \right] + G_i^N \right\}$$

Therefore,

$$t_i = d_1 \left\{ \frac{d_1 \bar{B}}{d_1 \bar{B} - c_2} \left[c_0 + c_1 f(B_i) + \left(c_3 - \frac{c_2}{\bar{B}} + 1 - \frac{d_1 \bar{B}}{d_1 \bar{B} - c_2} \right) G_i^N + c_4 N_i \right] \right\}$$

10. An appraisal of alternatives

1 The issue of buoyancy is a contentious one. Foster *et al.* (1980) argue that a buoyant revenue source has a doubtful political morality: it allows increases in revenue yield without a conscious (or a perceptible) political decision to raise

more revenue through increased tax rates. This has most frequently occurred with the effect of inflation on the income tax pushing taxpayers into successively higher tax brackets of marginal rates. This is a feature, of so-called fiscal drag, which arises by not linking the tax threshold, and threshold for changes in marginal rates, to an index of inflation: for example, as in the Rooker-Wise Amendment in Britain. Such windfall increases in tax yields also characterise the property tax when it is linked for market values and was an important stimulus to the success of the State of California Proposition 13 which forced the government to repay much of its windfall gains in 1978. Bennett (1981c) suggests that although the absence of indexation of tax thresholds gives the income tax an elasticity far in excess of unity (estimates for the USA and Britain for the recent past put the elasticity of personal income tax at between 1.5 and 1.7) with indexation of the tax thresholds the elasticity of personal income tax can be reduced to equal one. In this case the revenue system guarantees a *stable* revenue yield against changes in income (in real terms), increases in the level of *real* yields can be achieved only by increasing tax rates, and the objection to the doubtful political morality of buoyant revenues disappears.

A second issue relating to buoyancy is also raised by Foster *et al.* (1980). It is disputed by these authors that the rates are not buoyant or are less buoyant than an LIT would be. This is based on taking a long-term view: that after periodic revaluation of property the tax base rises very markedly. The weakness in this argument (Bennett, 1981a) is that in the short term local authorities are not helped by a revaluation that may be ten or fifteen years away in the future. As a consequence local authorities experience, between revaluations, a continual erosion of their tax base: it is not buoyant or stable with inflation and income changes. Hence, their revenue yield becomes progressively more inadequate if they hold to an unchanging tax rate. Local authorities in Britain are, therefore, forced into raising tax rates merely to keep real income stable. Ferry (1978, 1979) also suggests that local government is less able to raise tax rates than central government. In addition rate increases are generally lower in an election year than at other times.

2   Note that in the USA 'revenue-sharing' is a form of block grant.

11. An alternative structure for central–local relations

1   The criterion of the fiscal residuum as an equalisation norm is due mainly to Buchanan (1950) and Scott (1952). However, there is dispute as to the validity of this concept for different local authorities: see Musgrave (1969), Foster *et al.* (1980). The recent development of the concept is discussed by Oates (1972, 1977) and Bennett (1980).

2   Further discussion of the principles of fiscal equity involves consideration of spillover and tapering of access to public goods with distance: see Tiebout (1961), Smith (1977) and Bennett (1980).

3   This distinction and method of approach are similar to that proposed by Davies (1975). It can probably be argued that at least 65% of relevant expenditure is non-discretionary and has identifiable client groups. Davies (1975) notes that it is not important whether or not the groups are the sole or indeed most important 'beneficiaries' of the services so long as the definition of client group makes all those counted as members of a group similar with respect to the social costs and benefits of the service interventions implied by

the expenditure. In this case the representative need index approach of chapter 6, which is similar to Davies' (1968) territorial justice principle, can be applied.

4 What Davies (1975) terms 'inescapable expenditure'.

5 This grant structure is closely related to the proposals by Hicks and Hicks (1943), Musgrave (1961), Boyle (1966) and Cripps and Godley (1976).

6 An alternative output weighting is suggested by Cripps and Godley (1976).

7 This issue is discussed further by Le Grand (1975) and Newton and Sharpe (1976).

8 In a study of a 20% sample of house sales, Evans (1975) found that although ratios of capital to rateable values were very variable between properties, they could be described by a logarithmic function of price

$$\frac{\text{capital value}}{\text{rateable value}} = \frac{(\text{price})^{0.33}}{(9\,750)}$$

Nationally about 55% of houses would have a decreased, and 45% an increased liability, assuming that total values for both systems were equal. However, the pattern is spatially very variable with most increases concentrated in London, the south east, east anglia, Wales and the south west; and most decreases in the north and midlands. However, these spatial differences are compounded with the price effect on any individual dwelling and would be difficult to predict.

9 Most local authority associations favour this; see. e.g. ACC(1977, 1980).

10 A scheme of tax credits from central tax bills for the local rates is an alternative to LIT tax credits, if LIT is not introduced. This could take the form of equation (10.2) and allow a credit to central income tax bills for the equivalent central income tax rate on the local rate bill. This credit is equivalent to the yield from tax on the ratepayer's income at the basic income tax rate in respect of the actual rate bill paid and is given by the following equation

$$T = (Y - K)\,(t_1 - R/(Y - K))$$

where R is the local rate bill, and the term $R/(Y - K)$ is equivalent to the local income tax rate t_2 of the equation (10.2). This allows credit at only the basic rate of income taxation, and leaves the total tax burden on different income groups unchanged (unless there is the unlikely circumstance that the equivalent tax rate $R/(Y - K)$ exceeds the total basic rate tax percentage). If a local authority levies rates at a level equivalent to a rate of basic income tax of, say, 7%, then this is subtracted from the central income tax rate. For a basic rate of central government income tax at 30%, this means that the rate receipt by central government is reduced to 23%, and the difference of 7% equivalent rate for that taxpayer is paid by the Inland Revenue to the local authority. Administratively this could be accomplished by including a separate section relating to rate bills on each taxpayer's income tax return, but this would be expensive. The alternative would be for each tax office to code out local ratebills on some average or flat rate basis, with the Inland Revenue receiving information (already available) on local average domestic rate bills direct from the Department of Environment. The first method is clearly the best, but it is very expensive and increases the total number of tax returns which the Inland Revenue (IR) have to process. The second alternative is cheap and represents an improvement over the flat rate credits evaluated by the IR (in G.B.

Committee of Enquiry, 1976, Appendix 8), but is still given at a flat rate within each local authority irrespective of total tax bills. This could, however, have a beneficial distribution effect since it would benefit households with low rateable values (usually low income) at the expense of high rateable value households.

11 The problems of individual and spatial differences in benefit and tax incidence are now a major focus for research in many countries. Bennett (1980) gives a recent review of these and the issue is discussed further by Boulding and Wilson (1978). In Britain attention on this problem has been particularly focused by the difficulties of inner cities, and small areas within inner cities. Thus Bristow (1972), Webster and Stewart (1974), Holman and Hamilton (1973), Travers (1978), CDP(1976) and Roweth (1980) have drawn attention to the lack of attack on these difficulties in the RSG and other programmes which treat local authorities as a whole.

BIBLIOGRAPHY

Accountancy Research, Research Working Party (1949–53). The Effects of the
Local Government Act 1948, and other recent legislation on the finances of
Local Authorities. *Accountancy Research* **1**, 85–105; 281–302; **2**, 189–251; **3**,
266–304; **4**, 1–29.
ACC (1977). *Green paper on Local Government Finance (Cmnd. 6813). Comments
of the Association.* London: Association of County Councils.
ACC (1980a). *Local Government, Planning and Land (No. 2) Bill: Memorandum
of the Association's Views* London: Association of County Councils.
ACC (1980b). *Rate Support Grant: Needs Element Statistics 1980–81* London:
Association of County Councils.
ACC et al. (1979). *Rate Support Grant 1979–80* London: Association of County
Councils, Association of Metropolitan Councils, Association of District Coun-
cils, London Boroughs Association, Greater London Council.
Adams, R.F. (1965). On the variation in the consumption of public services.
Review of Economics and Statistics, **47**, 400–5.
Aiken, M. and Alford, R. (1970). Community structure and innovation: the case of
public housing. *American Political Science Review*, **64**, 843–64.
Alt, J.E. (1971). Some political and social correlates of County Borough Expendi-
tures. *British Journal of Political Science*, **1**, 49–62.
Alt, J.E. (1977). Politics and expenditure models, *Policy and Politics*, **5**, 83–92.
AMA (1979). *Value for Money: Local Authorities and Cost Effectiveness*, London:
Association of Metropolitan Authorities.
Arnold, R.D. (1977). *Congressmen, Bureaucrats and Constituency Benefits: the
Politics of Geographic Allocation.* Unpublished Ph.D. Thesis, Faculty of the
Graduate School, Yale.
Ashford, D.E. (1974). The effects of central finance on the British local govern-
ment system. *British Journal of Political Science*, **4**, 305–22.
Ashford, D.E. (1975). Resources, spending and party politics in British local
government. *Administration and Society*, **7**, 286–311.
Ashford, D.E., Berne, R. and Schramm, R. (1976). The Expenditure-financing
decision in British local government. *Policy and Politics*, **5**, 5–24.
Auten, G.E. (1972a). *The Measurement of Local Public Expenditure Needs.*
Unpublished Ph.D. Thesis, University of Michigan.
Auten, G.E. (1972b). *An approach to measuring local public expenditure needs.*
National Tax Association – Tax Institute of America Proceedings,
pp. 283–302.
Auten, G.E. (1974). The distribution of revenue-sharing funds and local public
expenditure needs. *Public Finance Quarterly*, **2**, 352–75.
Bahl, R.W. (1968). Studies on determinants of local expenditure: a review. In

Functional Federalism: Grants in aid and PPB systems, ed. S.J. Mushkin and J. F. Cotton. Washington, D.C.: Public Services Laboratory.

Barr, J. and Davis, O. (1966). An elementary political and economic theory of expenditure of local governments. *Southern Economic Journal*, **33**, 149–63.

Barr, N.A., James, S.R. and Prest, A.R. (1977). *Self-assessment for Income Tax*. London: Heinemann.

Basmann, R.L. (1957). A generalized classical method of linear estimation of coefficients in a structural equation. *Econometrica*, **25**, 77–83.

Bealey, F., Blondel, J. and McCann, W.P. (1965). *Constituency Politics*. London: Faber.

Beam, D.R. (1978). *Economic theory as policy prescription: pessimistic findings on 'optimising' governments*. Presented at Conference of American Political Science Association, New York.

Beasley, M. (1980*a*). How rates can be abolished. *Local Government Chronicle*, 4 January, 13–17.

Beasley, M. (1980*b*). The light of the charge brigade. *Local Government Chronicle*, 21 March 1980, 334–5.

Bennett, R.J. (1979). *Spatial Time Series: Analysis – forecasting – control*. London: Pion.

Bennett, R.J. (1980). *The Geography of Public Finance: Welfare under fiscal federalism and local government finance*. London: Methuen.

Bennett, R.J. (1981*a*). The costs of government. *Geographical Magazine*, **53**, 658–62.

Bennett, R.J. (1981*b*). The rate support grant in England and Wales 1967–8 to 1980–1: A review of changing emphases and objectives. In *Geography and the Urban Environment*, ed. D. Herbert and R.J. Johnston. London: Wiley.

Bennett, R.J. (1981*c*). The Local Income Tax in Britain: A critique of recent arguments against its use. *Public Administration*, 59, 295–311.

Bennett, R.J. (1982*a*). The financial health of local authorities in England and Wales (1974/5–1980/1: resource and expenditure position, *Environment and Planning, A.* **14**.

Bennett, R. J. (1982*b*). A representative index of local authority expenditure need in England and Wales. *Environment and Planning, A.* **14**.

Bennett, R.J. (1982*c*). The financial health of local authorities in England and Wales 1974/5–1980/1: The role of the rate support grant. *Environment and Planning, A.* **14**.

Bennett, R.J. (1982*d*). Modelling local authority expenditure in England and Wales. *Papers, Regional Science Association*.

Bentham, C.A. (1980). The classification of local authorities in the UK Inner Urban Areas Act 1978. *Environment and Planning A* **12**, 703–12.

Bergstrom, T.C. and Goodman, R.P. (1973). Private demands for public goods. *American Economic Review*, **63**, 280–96.

Bhangava, R.R. (1953). Theory of federal finance. *Economic Journal*, **63**, 84–97.

Binder, B. (1978). The importance of political decisions. *Local Government Chronicle*, **11**, 872–4.

Birch, A.H. (1959). *Small town Politics*. London: Oxford University Press.

Black, D. (1958). On the rationale of group decision making. *Journal of Political Economy*, **56**, 23–34.

Boaden, N. (1971). *Urban Policy Making*. Cambridge: Cambridge University Press.

Boulding, K.E. and Wilson, T.F. (ed.) (1978). *Redistribution through the Financial System: The Grants Economics of Money and Credit.* Eastbourne: Praeger.

Boyle, L. (1966). *Equalisation and the Future of Local Government Finance.* Edinburgh: Oliver and Boyd.

Bradbury, K.E. (1978). *The Fiscal Distress of Cities.* Washington, D.C.: Brookings Institution, mimeo.

Bradford, D.F., Malt, R.A., and Oates, W.E. (1969). The rising cost of local public services: some evidence and reflections. *National Tax Journal*, **22**, 185–202.

Bramley, G. and Evans, A. (1981). GREat complexity causes controversy. *Municipal Journal*, 6 February, 80–81.

Brazer, H.E. (1959). *City Expenditures in the United States.* New York: National Bureau of Economic Research.

Break, G. (1967). *Intergovernmental Fiscal Relations in the United States.* Washington D.C.: Brookings Institution.

Break, G. (1980). *Financing Government in a Federal System.* Washington D.C.: Brookings Institution.

Breton, A. (1965). A theory of government grants. *Canadian J. of Economics and Political Science*, **31**, 175–87.

Bristow, S.L. (1972). The criteria for local government reorganisation and local authority autonomy. *Policy and Politics*, **1**, 143–62.

Browning, E.K. (1975). *Redistribution and the Welfare System.* Washington D.C.: American Enterprise Institute.

Buchanan, J.M. (1950). Federalism and fiscal equity. *American Economic Review*, **40**, 583–97.

Buchanan, J.M. (1952). Federal grants and resource allocation. *Journal of Political Economy*, **60**, 208–17.

Buchanan, J.M. (1965). An economic theory of clubs. *Economica*, **32**, 1–14.

Buchanan, J.M. (1966). *The Demand and Supply of Public Goods.* Chicago: Rand McNally.

Bucknall, B. (1978). LGCs financial experts assess the 1979/80 RSG. *Local Government Chronicle*, **12**, 1316–17.

Bulpitt, J.G. (1967). *Party Politics in English Local Government.* London: Longmans.

Burgess, T. and Travers, T. (1980). *Ten Billion Pounds: Whitehall's Takeover of the Town Halls.* London: Grant McIntyre.

Caputo, D.A. (ed.) (1975). General revenue sharing and federalism. Special Issue of the *Annals of the American Academy of Political and Social Science.* Philadelphia.

Caputo, D.A. and Cole, R.L. (ed.) (1976). *Revenue Sharing: Methodological Approaches and Problems.* Lexington, Mass: D.C. Heath.

Caulcott, T. and Hale, R. (1978). The Metropolitan view on Rate Support Grant 1978/79. *Local Government Chronicle*, **11**, 607–9.

CBI (1979). *Value for Money*: Report on Cheshire County Council by PA Management Consultants on behalf of the Confederation of British Industry. Manchester: CBI.

CDP (1976). *Rates of decline: an unacceptable base of Public Finance.* Community Development Project Submission to Layfield Report.

CES (1978). Unitary grant: Bad arguments in a good cause. *CES Review*, **2**, 96–100.

Chester, D.N. (1951). *Central and Local Government: Financial and Administrative Relations*. London: Macmillan.

CIPFA (1979). *Community Indicators*. London: Chartered Institute of Public Finance and Accountancy.

CIPFA (1981a). *Local Government Comparative Statistics*. London: Chartered Institute of Public Finance and Accountancy.

CIPFA (1981b) *Financial, General and Rating Statistics 1981–82*. London: Chartered Institute of Public Finance and Accountancy.

Craig, J. and Driver, A. (1972). The identification and comparison of small areas of adverse social conditions. *Applied Statistics*, **21**, 25–35.

Cripps, F. and Godley, W. (1976). *Local Government Finance and Its Reform: a Critique of the Layfield Report*. Cambridge: Department of Applied Economics.

Cuciti, P.L. (1976). *The distribution of grants to local governments: equalisation in the American Policy*. Presented at the American Political Science Conference, Chicago.

Culyer, A.J. (ed.) (1974). *Economic Policies and Social Goals: Aspects of Public Choice*. London: Martin Robertson.

Culyer, A.J. (1976). *Need and the National Health Service: Economics and Social Choice*. London: Martin Robertson.

Culyer, A.J., Lavene, R.J., and Williams, A. (1972). Health indicators. In *Social Indicators and Social Policy*, ed. A. Shonfield and S. Shaw. London: Heinemann.

Dahlby, B.G. (1979). *The Demand for Labour by Local Governments: Cross-section and Time Series Evidence Since World War Two*. Unpublished Ph.D. Thesis London School of Economics.

Danziger, J.N. (1974). *Budget-making and Expenditure Variation in English County Boroughs*, unpublished Ph.D. Thesis, Stanford University, California.

Davies, B.D. (1964). An index of variation in 'Need' of County Boroughs for Old People's Homes. *Sociological Review*, **12**, 5–38.

Davies, B.D. (1968). *Social Needs and Resources in Local Services*. London: Joseph.

Davies, B.D. (1974). Personal Social Services. In *Review of UK Statistical Sources, vol. 1*, ed. W.F. Maunder. London: Heinemann.

Davies, B.D. (1975). *On local expenditure and a 'standard level of service'*, Evidence submitted to G.B. Committee of Enquiry on Local Government Finance; Discussion paper No. 43 (Personal Social Services Research Unit, University of Kent).

Davies, B.D., Barton, A.J., McMillan, I.S. and Williams, A. (1971a). *Variations in Services for the Aged*. London: Bell.

Davies, B.D., Barton, A.J. and McMillan, I.S. (1971b). *Variations in Children's Services among British Urban Authorities*. London: Bell.

Dawson, D.A. (1976). *Determinants of local authority expenditure*, p. 1–20 in Appendix 7 to G.B. Committee of Enquiry into Local Government Finance. London: HMSO.

Dearlove, J. (1973). *The Politics of Policy in Local Government*. Cambridge: Cambridge University Press.

Derthick, M. (1975). *Uncontrollable Spending for Social Service Grants*. Washington D.C.: Brookings Institution.

Dommel, P. (1974). *The Politics of Revenue Sharing*. Bloomington: Indiana University Press.
Dommel, P., Nathan, R. and associates (1978). *Decentralising Community Development*. Washington D.C.: Brookings Institution.
Downs, A. (1978). Urban policy. In *Setting National priorities: The 1979 Budget*, ed. J.A. Pechman, Washington, D.C.: Brookings Institution.
Drewett, R., Goddard, J. and Spence, N. (1976). Urban Britain: Beyond Containment. In *Urbanization and Counter-urbanization*, ed. B.J.L. Berry. London: Sage.
Dye, T.R. (1966). *Politics, Economics and the Public Policy: Policy Outcomes in the American States*. Chicago: Rand McNally.
Edwards, J. (1975). Social indicators, urban deprivation and positive discrimination, *Journal of Social Policy*, **4**, 275–87.
Evans, A. (1975). *Commentary on the effect of changing the basis of rating domestic property from rateable values to capital values, based on data from the 20 per cent sample of sales of dwellings reported to the Inland Revenue*. (Unpublished report, School of Advanced Urban Studies, Bristol.)
Fabricant, S. (1952). *Trends of Government Activity 1900–1950*. (NBER and Princeton University Press.)
Ferry, J. (1978). Politics and the rates. *CES Review*, **4**, 56–9.
Ferry, J. (1979). Rates and elections. *CES Review*, **5**, 5–7.
Foster, C.D., Jackman, R.A. and Perlman, M. (1980). *Local Government Finance in a Unitary State*. London: Allen and Unwin.
Frey, B.S. (1978). Politico-economic models and cycles. *Journal of Public Economics*, **9**, 203–20.
Frey, B.S. and Schneider, F. (1978). A politico-economic model of the United Kingdom. *Economic Journal*, **88**, 243–53.
Fried, R.C. (1972). *Comparative Urban Performance*. Los Angeles: UCLA, European Urban Research Paper 1.
G.B. Advisory Committee on Local Government Audit (1980) *First Report*, Chairman B.A. Maynard. London: HMSO.
G.B. Committee of Enquiry (1976). *Local Government Finance*. Report of Committee of Enquiry, Chairman: F.A. Layfield. London: HMSO with 10 Appendices.
G.B. Consultative Council on Local Government Finance (1979). *Report of the Joint Working Group on Fees and Charges* CCLFG (79) 2. London: Dept. of Environment.
G.B. DoE (1979a). *Publication of financial and other information by local authorities: A consultative document*. London: Dept. of Environment.
G.B. DoE (1979b). *Publication of financial and other information by local authorities: Draft Code of Practice*. London: Dept. of Environment.
G.B. DoE (1979c). Press Notice 522. *Rate Support Grant 1980–81*, statement by M. Heseltine, Secretary of State for the Environment, to the Consultative Council on Local Government Finance, 16 November. London: Dept. of Environment.
G.B. DoE (1980a). *Joint Manpower Watch – September 1980 Return*. Press release No. 534. 16 December. London: Dept. of Environment.
G.B. DoE (1980b). *Joint Manpower Watch – June 1980 Return*. Press release No. 377, 18 September. London: Dept. of Environment.
G.B. DoE (1981). *Service Provision and Pricing in Local Government: Studies in Local Environmental Service*. London: HMSO.

G.B. DoE (1982). *Local Authority Expenditure targets for 1982–83*. Memorandum. Dept. of Environment: London.

G.B. DHSS (1976). *Sharing resources for health in England*. London: HMSO, Report of Resource Allocation Working Party.

G.B. Government (1914). *Final Report of the Departmental Committee on Local Taxation*. Cmnd. 7315, 7316. London: HMSO.

G.B. Government (1953). *Report of the Committee appointed to investigate the operation of the Exchequer Equalisation Grant in England and Wales*, Edwards Committee. London: HMSO.

G.B. Government (1957). *Local government finance*. Cmnd. 209. London: HMSO.

G.B. Government (1959). *Report of the committee on the working of the monetary system*, chairman C.J. Radcliffe. London: HMSO.

G.B. Government (1961). *The Control of Public Expenditure*, Cmnd. 1432. London: HMSO, Report of Plowden Committee.

G.B. Government (1962). *Report of the working party appointed to investigate the working of the Rate-Deficiency Grants in England and Wales*. London: HMSO.

G.B. Government (1966). *Local Government Finance England and Wales*, Cmnd. 2923. London: HMSO.

G.B. Government (1969). *Public Expenditure: a new presentation*, Cmnd. 4017. London: HMSO, Green Paper.

G.B. Government (1971). *The future shape of local government finances*, Green paper, Cmnd. 4741. London: HMSO.

G.B. Government (1972*a*). *Public Expenditure White Papers – a handbook on methodology*. London: HMSO.

G.B. Government (1972*b*). *Proposals for a tax-credit system*, Cmnd. 5116. London: HMSO.

G.B. Government (1974). *Local Government Act*. London: HMSO.

G.B. Government (1977). *Local government finances*. Green paper, Cmnd. 6813. London: HMSO.

G.B. Government (1978). *Inner Urban Areas Act*, Circular 68/78. London: Dept. of Environment.

G.B. Government (1979). *Organic change in local government*, Cmnd. 7457. London: HMSO.

G.B. Government (1980*a*). *Local Government, Planning and Land Act*. London: HMSO.

G.B. Government (1980*b*). *Explaining the Local Authority Rate bill*, London. HMSO.

G.B. Government (1981*a*). *Local Authority Annual Reports: Code of Practice*. London: HMSO.

G.B. Government (1981*b*). *Alternative to Domestic Rates*, Cmnd, 8449. London: HMSO.

G.B. Government (various dates). *DOE Circulars,* Department of the Environment circulars, London: HMSO, see Appendix to Bibliography.

G.B. Government (various dates). *House of Commons Papers,* London: HMSO, see Appendix to Bibliography.

G.B. Government (various dates). *MHLG Circulars*, Ministry of Housing and Local Government Circulars. London: HMSO, see Appendix to Bibliography.

G.B. Government (various dates). *Statutory Instruments,* London: HMSO, see Appendix to Bibliography.

G.B. Royal Commission on Local Government in England (1968*a*). *Performance and Size of Local Education Authorities*, Research Study No. 4. London: HMSO.

G.B. Royal Commission on Local Government in England (1968*b*). *Local Authority Services and the Characteristics of Administrative Areas*. Research Study No. 5. London: HMSO.

G.B. Royal Commission on the Constitution (1973). *Report*, 2 vols, Chairman: Lord Kilbrandon. London: HMSO.

G.B. Treasury (1979). *Needs Assessment Study: Report*. London: HM Treasury.

Gibson, J.G. (1980). The effect of matching grant on local authority user charges. *Public Finance*, **25**, 372–9.

Godley, W. and Rhodes, J. (1972). *The Rate Support Grant System*. Cambridge, England: Department of Applied Economics.

Goschen, G.J. (1872). *Reports and speeches on local taxation*. London: Macmillan.

Gramlich, E.M. and Galper, H. (1973). State and local fiscal behaviour and federal grant policy. *Brookings Papers on Economic Activity*, **1**, 15–65.

Green, C. (1967). *Negative Taxes and the Poverty Problem*. Washington D.C.: Brookings Institution.

Greene, K.V., Neenan, W.B. and Scott, C.D. (1974). *Fiscal Interactions in a Metropolitan Area*. Lexington, Mass: D.C. Heath.

Griffin, C., Harrison, A. and Lansley, S. (1980). Rate Support Grant: the 1979 Settlement. *CES Review*, **8**, 41–50.

Groves, H.M. (1964). *Financing Government*, 7th edn. New York: Holt, Rinehart and Winston.

Gupta, S.P. and Hutton, J.P. (1968). *Economics of Scale in Local Government Services*. London: HMSO, Royal Commission on Local Government in England, Research Study No. 3.

Hale, R. (1981). Betwixt and between on penalties. *Local Government Chronicle*, 3 July, 694–5.

Hall, P., Gracey, H., Drewett, R. and Thomas, R. (1975). *The Containment of Urban England*, 2 vols. London: Allen and Unwin.

Hakin, C. (1977). *Social and Community Indicators from the Census*. Occasional Paper No. 5. London: OPCS.

Halstead, K. (1978). *Tax Wealth in Fifty States*. Washington D.C.: National Institute of Education.

Hansen, N.M. (1965). The structure and determinants of local public investment expenditures. *Review of Economics and Statistics*, **47**, 150–62.

Harrison, A. and Jackman, R. (1978). Rate Support Grant. *CES Review*, **4**, 22–6.

Harrison, A. and Jackman, R. (1980). Changing the incentive to spend: a critical look at the new block grant system. *CES Review*, **9**, 22–4.

Harrison, A., Jackman, R. and Papadachi, J. (1979). Needs grant: which way now? *CES Review*, **6**, 18–31.

Harrison, A. and Smith, M. (1978). Reacting to crisis: the impact of economic fluctuations on local authority finance. *CES Review*, **4**, 40–4.

Hawley, W.D. (1973). *Non-Partisan Elections and the Case for Party Politics*. New York: Wiley.

Head, J.G. (1973). Public goods and multi-level government. In *Public Finance, Planning and Economic Development: Essays in Honour of Ursula Hicks*, ed. W.L. David. London: Macmillan.

Heald, D. (1980). *Financing Devolution within the United Kingdom: A Study of Lessons from Failure*. Canberra: Australian National University.

Heald, D. and Jones, C.A. (1980). *The nature of resources equalisation in British Local Government.* Discussion paper No. 33, Centre for Urban and Regional Research, University of Glasgow.

Henderson, J.M. (1968). Local government expenditures: a social welfare analysis. *Review of Economics and Statistics*, **50**, 156–63.

Heseltine, M. (1979). *Rate Support Grant 1980–81*: Statement to Consultative Council on Local Government Finance, DoE Press Notice 522. London: Dept. of Environment.

Hicks, J.R. And Hicks, U. (1943). *Standards of Local Expenditure: a Problem of the Inequality of Incomes.* Cambridge: Cambridge University Press and NIESR.

Hicks, J.R. and Hicks, U. (1944*a*). *The Problem of Valuation for Rating.* Cambridge: Cambridge University Press and NIESR.

Hicks, J.R. and Hicks, U. (1944*b*). The Beveridge Plan and local government finance. *Review of Economic Studies*, **11**, 1–19.

Hicks, J.R. and Hicks, U. (1945). *The Incidence of Local Rates in Great Britain.* Cambridge: Cambridge University Press and NIESR.

Hinckley, D. (1966). Factors influencing Local Government Elections. In *The Political Structure of Local Government in Coventry*, ed. P. Spencer, pp. 102–12. Coventry: Institute for Operational Research.

Holman, R. and Hamilton, L. (1973). The British Urban Programme *Policy and Politics*, **2**, 97–112.

Horowitz, A.R. (1968). A simultaneous-equation approach to the problem of explaining inter-state differences in State and local government expenditures. *Southern Economic Journal*, **34**, 459–76.

Hunter, J.S.H. (1977). *Federalism and Fiscal Balance: A Comparative Study.* Canberra: Australian National University.

Ilersic, A. (1973). Grant determination and its distribution. *Proceedings of Conference on Local Government Finance.* London: Institute of Fiscal Studies.

Imber, V. (1977). *A Classification of English Personal Social Service Authorities.* DHSS Statistical and Research Report Series No. 16, London: HMSO.

Isserman, A.M. and Brown, M.A. (1980). Community need: its measurement and incidence. *Papers, Regional Science Association*, **45**, 139–58.

Jackman, R. (1979). London's Needs Grant. *CES Review*, **5**, 28–34.

Jackman, R. and Sellars, M.B. (1977*a*). The distribution of Rate Support Grant: the hows and whys of the new needs formula. *CES Review*, **1**, 19–30.

Jackman, R. and Sellars, M.B. (1977*b*). Why rate poundages differ: the case of the Metropolitan Districts. *CES Review*, **2**, 26–32.

Jackman, R. and Sellars, M.B. (1977*c*). Local expenditure and local discretion. *CES Review*, **3**, 63–73.

Jackson, K.T. (1972). Metropolitan government versus political autonomy: politics on the Crabgrass frontier. In *Cities in American History*, ed. K.T. Jackson and S.K. Schultz. New York: A. Knopf.

Jacobs, A. (1980). Mr. Heseltine presses on with Block Grant despite criticism. *Local Government Chronicle*, **13**, 813.

Jenkins, J. and Rose, M. (1976*a*). *Rate Support Grants: the system and its application to London.* London: Greater London Council, Research Memorandum No. RM 479.

Jenkins, J. and Rose, M. (1976*b*). Rate support grants and their application to London. *Greater London Intelligence Quarterly*, **35**, 9–25.

Johnston, J. (1972). *Economic Methods.* 2nd edn. New York: McGraw-Hill.

Johnston, R.J. (1979). *Political, Electoral and Spatial Systems*. London: Oxford University Press.

Johnston, R.J. (1980). *The Geography of Federal Spending in the United States of America*. London: John Wiley.

Jones, G.W. (1969). *Borough Politics*. London: Macmillan.

Jones, G. (1977). *Responsibility and Government: an Inaugural Lecture*. London: LSE.

Juster, F.T. (ed.) (1976). *The Economic and Political Impact of General Revenue Sharing*. Washington D.C.: US Printing Office for NSF, Survey Research Center, University of Michigan, Ann Arbor.

Key, V.O. (1951). *Southern Politics in State and Nation*. Chicago: Rand McNally.

Lees, D.S. (1957). Reform of Local Finance; The implications of block grants. *Local Government Finance*, **61,** 180–85, 209–14.

Lees, D.S. *et al*. (1956). *Local Expenditures and Exchequer Grants*. London: Institute Municipal Treasurers and Actuaries.

Le Grand, J. (1975). Fiscal equity and central government grants to local authorities. *Economic Journal*, **85,** 531–47.

Le Grand, J. and Winter, D. (1977). Towards an economic model of local government behaviour. *Policy and Politics*, **5,** 23–39.

LESEWP (1981). Research notes, unpublished; quoted in G.B. DoE (1981) London: Local Expenditure Survey, Expenditure Working Group.

Little, A. and Mabey, C. (1972). An index for the designation of educational priority areas. In *Social Indicators and Social Policy*, ed. A. Shonfield and S. Shaw. London: Heinemann.

Lynch, B. and Perlman, M. (1977). Local Authority predictions of expenditure and income. *CES Review*, **3,** 12–24.

Margolis, J. (1961). Metropolitan finance problems: territories, functions and growth. In National Bureau of Economic Research, *Public Finances: Needs, Sources and Utilisation*. New Jersey: Princeton University Press.

Martin, P. (1980). Block grant is a complex irrelevance, *Local Government Chronicle*, **13,** 218–19.

Mathews, P. (ed.) (1974). *Intergovernmental Relations in Australia*. Sydney: Angus and Robertson.

Maxwell, J.A. (1946). *The Fiscal Impact of Federalism in the United States*. Cambridge, Mass.: Harvard University Press.

Maxwell, J.A. (1962). *Tax Credits and Intergovernmental Fiscal Relations*. Washington, D.C.: Brookings Institution.

Mills, E.S. and Oates, W.E. (1975). *Fiscal Zoning and Land Use Controls*. Farnborough: Saxon House.

Miner, J. (1972). British local expenditure analysis: an American evaluation. *Policy and Politics*, **1,** 357–61.

Moore, B. and Rhodes, J. (1971). *The relative needs of local authorities*. In Appendix 7 to G.B. Committee of Enquiry into Local Government Finance, pp. 86–145. London: HMSO.

Morss, E.R. (1966). Some thoughts on the determinants of state and local expenditure. *National Tax Journal*, **19,** 97–104.

Musgrave, R.A. (1959). *Theory of Public Finance*. New York: McGraw-Hill.

Musgrave, R.A. (1961). Approaches to a fiscal theory of political federalism. In National Bureau of Economic Research, *Public Finances: Needs, Resources and Utilisation*. New Jersey: Princeton University Press.

Musgrave, R.A. and Musgrave, P.B. (1980). *Public Finance in Theory and Practice*, 3rd edn. New York: McGraw-Hill.

Nathan, R.P. and Adams, C.F. (1976), Understanding central city hardship. *Political Science Quarterly*, **91**, 47–62.

Nathan, R.P. and Dommel, P. (1977). The cities. In *Setting National Priorities: the 1978 Budget*, ed. J.A. Pechman. Washington D.C.: Brookings Institution.

Nathan, R.P. and associates (1976). *Monitoring Revenue Sharing*. Washington D.C.: Brookings Institution.

Nathan, R.P. and associates (1977). *Revenue Sharing: the Second Round*. Washington D.C.: Brookings Institution.

Neild, R. and Ward, T. (1976). Evidence to G.B. *Committee of Enquiry*, Appendix 6, pp. 81–97. London: HMSO.

Neuber, K.A., Atkins, W.T., Jacobson, J.A. and Reuterman, N.A. (1980). *Needs Assessment: model for Community Planning*. London: Sage.

Nevitt, D.A. (1973). The 'burden' of domestic rates. *Policy and Politics*, **2**, 1–25.

Newcomer, M. (1917). Separation of state and local revenues in the United States. *Columbia University Studies in History, Economics and Public Law*, **76**, 295 pp.

Newton, K. (1976). *Second City Politics: Democratic Processes and Decision Making in Birmingham*. Oxford: Clarendon Press.

Newton, K. (1981). *Balancing the Books*. London: Sage.

Newton, K. and Sharpe, L.J. (1976). *Service Outputs in Local Government: Some Reflections and Proposals*. Mimeo, Oxford.

Newton, K. and Sharpe, L.J. (1977). Local outputs research: some reflections and proposals. *Policy and Politics*, **5**, 61–82.

Nicholson, R.J. and Topham, N. (1971). The determinants of investments in housing by local authorities: an econometric approach. *Journal of the Royal Statistical Society A*, **134**, 273–303.

Nicholson, R.J. and Topham, N. (1972). Investment decisions and the size of local authorities. *Policy and Politics*, **1**, 23–44.

Nicholson, R.J., Topham, N. and Watt, P.A. (1975). Housing investment by different types of local authorities. *Bulletin of Economic Research*, **27**, 65–86.

Oates, W.E. (1968). The dual impact of federal aid on state and local government expenditures: a comment. *National Tax Journal*, **21**, 220–3.

Oates, W.E. (1969). The effects of property taxes and local public spending on property values: An empirical study of tax capitalization and the Tiebout hypothesis. *Journal of Political Economy*, **77**, 957–71.

Oates, W.E. (1972). *Fiscal Federalism*. New York: Harcourt, Brace Jovanovich.

Oates, W.E. (1977). *The Political Economy of Fiscal Federalism*. Lexington, Mass.: D.C. Heath.

O'Cleireacain, C.C. (1977). *The Determinants of Local Government Expenditure in English and Welsh County Boroughs, 1971*, unpublished Ph.D. Thesis, University of London.

OECD (1974). *Negative Income Tax: an Approach to the Coordination of Taxation and Welfare Policies*. Paris: OECD.

Ohls, J.C. and Wales, T.J. (1972). Supply and demand for state and local services. *Review of Economics and Statistics*, **54**, 424–30.

Oliver, F.R. and Stanyer, J. (1969). Some aspects of the financial behaviour of County Boroughs. *Public Administration*, **47**, 169–84.

Osman, J. (1968). On the use of intergovernmental aid as an expenditure determinant. *National Tax Journal*, **21**, 437–47.

Owens, H. (1980). Social Indicators. In *Public Expenditure and Social Policy in Australia, Volume II: The First Fraser Years 1976–8*, ed. R.B. Scotton and H. Ferber, chapter 7. London: Longmans.

Packman, J. (1968). *Child Care: Needs and Numbers*. London: Allen and Unwin.

Page, E. (1978). Why should central–local relations in Scotland be different to those in England? *Public Administration*, **28**, 51–72.

Page, E. (1980). *Comparing local expenditure: lessons from a multinational State*. Studies in Social Policy No. 60, Centre for Study of Public Policy, University of Strathclyde.

Peterson, P.E. (1979). A unitary model of local taxation and expenditure policies in the United States. *British Journal of Political Science*, **9**, 281–314.

Pigou, A.C. (1947). *A Study in Public Finance*, 3rd edn. London: Macmillan.

Pinch, S. (1979). Territorial justice in the City: a case study of the social services for the elderly in Greater London. In *Social Problems and the City*, ed. D.T. Herbert and D.M. Smith. London: Oxford University Press.

Pogue, T.F. (1974). Deductions versus credits. *National Tax Journal*, **27**, 659–62.

Pogue, T.F. and Sgortz, L.G. (1968). The effects of grants-in-aid on local spending. *National Tax Journal*, **21**, 190–9.

Prest, A.R. (1978). *Intergovernmental Financial Relations in the United Kingdom*. Research Monograph No. 23, Centre for Research on Federal Financial Relations, Australian National University, Canberra.

Rechtenwald, H.C. (1971). *Tax Incidence and Income Redistribution*. Detroit: Wayne State University Press.

Reischauer, R.D. (1974). *Rich Governments – Poor Governments: Determining the Fiscal Capacity and Revenue Requirements of State and Local Governments*. Washington D.C.: Brookings Institution.

Rhodes, G. (1976). *Local Government Finance 1918–1966*. In Appendix 6, G.B. Committee of Enquiry into Local Government Finance, pp. 102–73. London: HMSO.

Rhodes, T. and Bailey, S.J. (1979). Equity, Statistics and the distribution of the rate support grant. *Policy and Politics*, **7**, 83–97.

Rose, M. (1978). *Rate Support Grant: Needs element, rate equalisation and London's inner urban areas*. London: Greater London Council, Research Memorandum No. RM 547.

Roweth, B. (1980). Statistics for policy: needs assessment in the Rate Support Grant. *Public Administration*, **59**, 173–86.

Schmid, G., Lipinski, H. and Palmer, M. (1975). *An Alternative to Revenue Sharing: a Needs Based Allocation Formula*. Menlo Park, California: Institute for the Future.

Schott, K. (1978). *An analysis of local authorities*. University of Southampton, Department of Economics, Discussion paper.

Scott, A.D. (1952). The evaluation of federal grants. *Economica*, **19**, 377–94.

SCT (1980). *Comments on Joint Working Group on Fees and Charges*. Society of County Treasurers, Beverley, N. Humberside.

SCT (1981). *Block Grant Indicator 1981–82*. Society of County Treasurer: Beverley, N. Humberside.

SMT (1980). *Report of Study Group on Comparative Fees and Charges*. London: Society of Metropolitan Treasurers.

Sharkansky, I. (1967). Economic and political correlates of state government expenditures: general tendencies and deviant cases. *Mid-West Journal of Political Science*, **2**, 1973–92.

Sharpe, L.J. (ed.) (1967). *Voting in Cities*. London: Macmillan.

Smith, D.J. (1983). Local Government debt in England and Wales: an inter-authority comparison. (*Unpublished paper*.)

Smith, D.L. (1968). The response of state and local governments to federal grants. *National Tax Journal*, **21**, 349–57.

Smith, D.M. (1974). Who gets what where and how: a welfare focus for human geography. *Geography*, **59**, 289–97.

Smith, D.M. (1977). *Human Geography: A Welfare Approach*. London: Arnold.

Smith, J. (1979). Equalising local authority resources: what are the options? *CES Review*, **5**, 69–74.

Smith, S. de (1971). *Constitutional and Administrative Law*. Harmondsworth: Penguin.

SPNW (1973). *Strategic Plan for the North West Region*. London: HMSO.

Stern, D. (1973). Effects of alternate state aid formulae on the distribution of public school expenditure in Massachusetts. *Review of Economics and Statistics*, **55**, 91–7.

Strauss, R.P. (1974). The impact of block grants on local expenditures and property tax rate. *Journal of Public Economics*, **3**, 269–84.

Sullivan, J.L. (1973). Political correlates of social, economic and religious diversity in the American states. *Journal of Politics*, **35**, 70–84.

Sunley, E.M. (1977). The choice between deductions and credits. *National Tax Journal*, **30**, 243–7.

Taylor, G. (1980). Sir Godfrey Taylor and the Local Government Bill. *Local Government Chronicle*, 7 March, 271–2.

Theil, H. (1953). *Estimation and Simultaneous Correlation in Complete Equation Systems*. The Hague: Central Planning Bureau.

Thurow, L.C. (1966). The theory of grants-in-aid. *National Tax Journal*, **19**, 373–7.

Tiebout, C.M. (1956). A pure theory of local expenditures. *Journal of Political Economy*, **64**, 416–24.

Travers, T. (1978). *Rate Support Grant: Changes and consequences 1974/5 – 1977/8*. Centre for Institutional Studies, N.E. London Polytechnic.

US ACIR (1962). *Measures of State and Local Fiscal Capacity and Tax Effort*. Washington D.C.: Report M-16, US Advisory Commission on Inter-governmental Relations.

US ACIR (1971). *Measuring the Fiscal Capacity and Effort of State and Local Areas*. Washington D.C.: Report M-58, US Advisory Commission on Intergovernmental Relations.

US ACIR (1977). *Measuring the Fiscal 'Blood Pressure' of the States – 1964–1975*. Washington D.C.: Report M-111, US Advisory Commission on Intergovernmental Relations.

Webber, R. and Craig, J. (1978). *A Socio-Economic Classification of Local Authorities in Great Britain*. OPCS Studies in Medical and Population Subjects No. 35. London: HMSO.

Webster, B. and Stewart, J. (1974). The area analysis of resources. *Policy and Politics*, **3**, 5–16.

Weir, S. (1979). How Islington set the rates. *New Society*, vol. **47**, (22.3.79).

Welsh Office (1980). *Welsh Rate Support Grant. A Methodology for the Assessment of Standard Expenditure.* Report of the Distribution Sub-group to the Welsh Consultative Council on Local Government Finance. Cardiff: Welsh Office.

Wilde, J.A. (1968). The expenditure effects of grant-in-aid programs. *National Tax Journal*, **21**, 340–8.

Wilensky, H. (1975). *The Welfare State and Equality.* Berkeley: University of California Press.

Williams, A. (1958). The finance of local government in England and Wales since 1948, part 1. *National Tax Journal*, **11**, 302–13.

Williams, A. (1959). The finance of local government in England and Wales since 1948, parts 2 and 3. *National Tax Journal*, **12**, 1–21, 127–50.

Wilson, J.Q. and Banfield, E.C. (1964). *City Politics.* Cambridge, Mass.: Harvard University Press.

Wonnacott, R.J. and Wonnacott, T.H. (1970). *Econometrics.* New York: John Wiley.

Wood, R.C. (1961). *1400 Governments.* Cambridge, Mass.: Harvard University Press.

Woodham, J.B. (1953). *Education Rates and the Education and Equalisation Grants.* London: Institute of Municipal Treasurers and Actuaries.

Zellner, A. and Theil, H. (1962). Three-stage least-squares: simultaneous estimation of simultaneous equations. *Econometrica*, **30**, 54–78.

Bibliography Appendix: publications of G.B. Government relevant to Rate Support Grant

(1) Public Expenditure White Papers

Session	Date	Title
1967–68	1968	Public Expenditure in 1968–69 and 1969–70, Cmnd. 3515.
1968–69	1969	Public Expenditure in 1968–69 to 1970–71, Cmnd. 3936.
1968–69		Public Expenditure: a new presentation, Cmnd. 4017.
1969–70		Public Expenditure in 1968–69 to 1973–74, Cmnd. 4234.
1970–71	1971	Public Expenditure to 1974–5, Cmnd. 4578.
1971–72		Public Expenditure to 1975–76, Cmnd. 4829.
1972–73	1972	Public Expenditure to 1976–77, Cmnd. 5178.
1973–74	1973	Public Expenditure to 1977–78, Cmnd. 5519.
1974–75	1975	Public Expenditure to 1978–79, Cmnd. 5879.
1975–76	1976	Public Expenditure to 1979–80, Cmnd. 6393.
1976–77	1977	The Government's Expenditure Plans, 2 vols. Cmnd. 6721–I, 6721–II.
1977–78	1978	The Government's Expenditure Plans, 2 vols, Cmnd. 7049–I, 7049–II.
1978–79	1979	The Government's Expenditure Plans, Cmnd. 7439.
1979–80	1980	The Government's Expenditure Plans 1980–81 to 1983–84, Cmnd. 7841.
1980–81	1981	The Government's Expenditure Plans 1981–1982 to 1983–1984, Cmnd. 8175.
1981–82	1982	The Government's Expenditure Plans 1982–1983 to 1984–1985, Cmnd. 8494–I, 8494–II.

(2) House of Commons Papers

Session	Paper	Title
1966–67	HC-252	Rate Support Grant Order 1966.
1967–68	HC-19	Rate Support Grant (Increase) Order 1967.
1968–69	HC-24	Rate Support Grant Order 1968.
1969–70	HC-21	Rate Support Grant (Increase) Order 1969.
1970–71	HC-172	Rate Support Grant Order 1970.
	HC-173	Rate Support Grant (Increase) Order 1970.
1971–72	HC-21	Rate Support Grant (Increase) (No. 1) Order 1971.
	HC-22	Rate Support Grant (Increase) (No. 2) Order 1971.
1972–73	HC-26	Rate Support Grant (Increase) Order 1972.
	HC-27	Rate Support Grant Order 1972.
1973–74	HC-47	Rate Support Grant (Increase) (No. 1) Order 1973.
	HC-48	Rate Support Grant (Increase) (No. 2) Order 1973.
		Rate Support Grant 1974–75, 5532.
1974–75	HC-74	Rate Support Grant (Increase) Order 1974.
	HC-75	Rate Support Grant (Increase) Order 1975.
	HC-460	Rate Support Grant (No. 2) Order 1974.
1975–76	HC-31	Rate Support Grant Order 1975.
	HC-32	Rate Support Grant (Increase) (No. 2) Order 1975.
1976–77	HC-26	Rate Support Grant Order 1976.
	HC-27	Rate Support Grant (Increase) Order 1976.
	HC-28	Rate Support Grant (Increase) (No. 2) Order 1976.
1977–78	HC-57	Rate Support Grant Order 1977.
	HC-58	Rate Support Grant (Increase) Order 1977.
	HC-59	Rate Support Grant (Increase) (No. 2) Order 1977.
1978–79	HC-63	Rate Support Grant Order 1978.
	HC-64	Rate Support Grant (Increase) Order 1978.
	HC-65	Rate Support Grant (Increase) (No. 2) Order 1978.
1979–80	HC-280	Rate Support Grant Order 1979.
	HC-281	Rate Support Grant (Increase) Order 1979.
	HC-282	Rate Support Grant (Increase) (No. 2) Order 1979.
1980–81	HC-52	Rate Support Grant Report (England) 1980.
	HC-53	Rate Support Grant Report (Wales) 1980.
	HC-57	Rate Support Grant (Increase) Order 1980.
	HC-58	Rate Support Grant (Increase) (No. 2) Order 1980.
1981–82	HC-141	Rate Support Grant Report (England) 1982/3.
	HC-145	The Welsh Rate Support Grant Report 1982.
	HC-139	Rate Support Grant (Increase) Order 1982.
	HC-140	Rate Support Grant Supplementary Report (England) 1982.
	HC-146	The Welsh Rate Support Grant Supplementary Report 1982.

(3) Department Circulars

(a) *Ministry of Housing and Local Government (MHLG)*

Year	No.	Title
1966	42/66	Public Expenditure 1966.

1967	9/67	International obligation and local authority purchasing.
	12/67	Local Government Act 1966. Financial Provisions.
	18/67	Rate demands 1967–68. Information about reductions in poundage under section 6 of the Local Government Act 1966.
	21/67	The Rate Support Grant Regulations.
	26/67	General Rate Act 1967.
	28/67	The Rate Product (Amendment of Enactments) Regulations 1967.
1968	6/68	General Rate Act 1967. Rate Demands and Rate Rebates.
	7/68	International Obligations and Local Authority Purchasing.
	9/68	Local Expenditure.
	21/68	General Rate Act 1967. The Rate Product Rule 1968.
	39/68	Rate Rebates from Autumn 1968.
1969	11/69	Local Government Act 1966. The Rate Support Grant (Amendment) Regulations 1969.
	18/69	Part 1. Loan consent for small amounts. Part 2. Loan period for all loans.
1970	43/70	Local Government Statistics and Central Forecasting for Rate Support Grant.
	56/70	The Rate Support Grant (Amendment) Regulations 1970.

(b) *Department of the Environment*

Year	No.	Title
1970	2/70	Capital programmes.
1971	65/71	Capital programmes: Classification of detail of Circular 2/70
	66/71	Capital programmes: Arrangements for 1972/3.
1973	77/73	Public Expenditure in 1974–75.
	157/73	Public Expenditure in 1974–75
1974	19/74	Rate Fund expenditure and rate calls in 1974–75.
	171/74	Rate Fund expenditure and rate calls in 1975–76
1975	51/75	Local Authority expenditure.
	88/75	Local Authority expenditure in 1976–77 – Formal planning.
	129/75	Rate Support Grant settlement 1976–77.
1976	45/76	Local Authority current expenditure 1976–77.
	84/76	Local Authority expenditure 1976–78.
	120/76	Rate Support Grant settlement 1977–78.
	123/76	Reductions in Public Expenditure in 1977–78 and 1978–79.
1977	37/77	The Government's expenditure plans (Cmnd. 6721): Implications for Local Authority expenditure 1976–79.
1978	8/78	Rate Support Grant settlement 1978–79.
	28/78	The Government's expenditure plans (Cmnd. 7049): Implications for Local Authority expenditure 1978–82.
	68/78	Inner Urban Areas Act 1978.

1979	6/79	Rate Support Grant settlement 1979–80.
	15/79	The Government's expenditure plans (Cmnd. 7439): Implications for Local Authority expenditure 1979–83.
	21/79	Local Authority expenditure in 1979-80.
1980	14/80	Publication of rate demands and supplementary information by local authorities.
1981	3/81	Publication of Annual Reports and financial statements by local authorities.
	24/81	Publication by local authorities of information about manpower and about employment of disabled people.

(4) Statutory Instruments

(a) *Orders and Increase Orders*

Year	No.	Title	Volume
1966	1612	Rate Support Grant Order	1966 **III**, p. 5053
1967	1877	Rate Support Grant (Licence) Order	1967 **III**, p. 5098
1968	1956	Rate Support Grant Order	1968 **III**, p. 5356
1969	1806	Rate Support Grant (Increase) Order	1969 **III**, p. 5618
1970	1875	Rate Support Grant (Increase) Order	1970 **III**, p. 6159
1970	1876	Rate Support Grant Order	1970 **III**, p. 6163
1971	2031	Rate Support Grant (Increase) (No. 1) Order	1971 **III**, p. 5821
1971	2032	Rate Support Grant (Increase) (No. 2) Order	1971 **III**, p. 5825
1972	2033	Rate Support Grant (Increase) Order	1972 **III**, p. 6016
1972	2034	Rate Support Grant Order	1972 **III**, p. 6020
1973	2180	Rate Support Grant (Increase) (No. 1) Order	1973 **III**, p. 7714
1973	2187	Rate Support Grant (Increase) (No. 2) Order	1973 **III**, p. 7734
1974	550	Rate Support Grant Order	1974 **I**, p. 2251
1974	2108	Rate Support Grant (Increase) Order	1974 **III**, p. 8192
1974	2109	Rate Support Grant (No. 2) Order	1974 **III**, p. 8198
1975	2148	Rate Support Grant (Increase) Order	1975 **II**, p. 4259
1975	2149	Rate Support Grant Order	1975 **III**, p. 7976
1975	2150	Rate Support Grant (Increase) (No. 2) Order	1975 **III**, p. 7983
1976	2201	Rate Support Grant (Increase) Order	1976 **III**, p. 6151
1976	2202	Rate Support Grant (Increase) (No. 2) Order	1976 **III**, p. 6155
1976	2203	Rate Support Grant Order	1976 **III**, p. 6161
1977	2113	Rate Support Grant Order	1977 **III**, p. 5772
1977	2114	Rate Support Grant (Increase) Order	1977 **III**, p. 5789
1977	2115	Rate Support Grant (Increase) (No. 2) Order	1977 **III**, p. 5796
1978	1867	Rate Support Grant Order	1978 **III**, p. 5431
1978	1868	Rate Support Grant (Increase) Order	1978 **III**, p. 5447
1978	1869	Rate Support Grant (Increase) (No. 2) Order	1978 **III**, p. 5449
1980	57	Rate Support Grant Order	1980 **I**, p. 281
1980	58	Rate Support Grant (Increase) Order	1980 **I**, p. 297
1980	59	Rate Support Grant (Increase) (No. 2) Order	1980 **I**, p. 314
1982	186	Rate Support Grant (Increase) Order 1982	1982

(b) *Regulations*

Year	No.	Title	Volume
1974	428	Rate Support Grant Regulations	1974 **I**, p. 1384
1974	788	Rate Support Grant (Specified Bodies) Regulations	1974 **II**, p. 3036
1974	1987	Rate Support Grant (No. 2) Regulations	1974 **III**, p. 6964
1975	5	Rate Support Grant (Specified Bodies) Regulations	1975 **I**, p. 4
1975	1950	Rate Support Grant Regulations	1975 **III**, p. 7246
1976	214	Rate Support Grant (Specified Bodies) Regulations	1976 **I**, p. 529
1976	1939	Rate Support Grants (Adjustments of Needs Element) Regulations	1976 **III**, p. 5200
1976	2071	Rate Support Grants Regulations	1976 **III**, p. 5681
1977	1342	Rate Support Grants (Adjustment of Needs Element) (Amdt.) Regulations	1977 **II**, p. 3955
1977	1941	Rate Support Grant Regulations	1977 **III**, p. 5420
1977	2002	Rate Support Grants (Adjustment of Needs Element) (Amdt. No. 2) Regulations	1977 **III**, p. 5506
1978	171	Rate Support Grant (Specified Bodies) Regs.	1978 **I**, p. 377
1978	1701	Rate Support Grant Regulations	1978 **III**, p. 5068
1979	337	Rate Support Grant (Adjustment of Needs Element) (Amdt.) Regulations	1979 **I**, p. 791
1979	1514	Rate Support Grant Regulations	1979 **III**, p. 3690

INDEX

abatement of grants, 121–5, 306; *see also* penalties
ability to pay, 82, 271, 276–7
accountability, 5, 6, 19, 23, 43, 63, 85, 95, 102, 109, 124, 126, 129, 254–60, 262–3, 267, 271–3, 285–6, 293, 299, 306–7
adequacy of revenues, 15, 22, 254, 257, 262, 284, 285–6; *see also* buoyancy
administrative class of authority, 7–12, 225–7, 228–49
advances, internal, 220–1
adversary model, 211–13
Africa, 49
ageing of population, 141
Aggregate Exchequer Grant, 72, 75, 102–3
agriculture, 47–9
aid, overseas, 47–8
Allen Committee on rates, 63
analysis of variance (ANOVA), 10, 158–9, 170, 176, 299
Annual Reports of local authorities, 65
arbitrariness of grants, 125, 130
area weights of grants, 87–90, 95–7, 108, 303, 305
assigned revenues, 49, 262; *see also* revenue sharing
Association of County Boroughs, 319
Association of County Councils, 319
Association of District Councils, 319
Association of Metropolitan Authorities, 92, 319
Association of Rural District Councils, 319
Associations of Local Authorities, definitions, 93, 107, 126, 319–20, 324
Audit Commission, 66
auditing of local accounts, 66, 84
Australia, 1, 3, 20, 22, 49, 54, 262
Australian tax sharing, 1
Austria, 3, 262
autonomy, local, 5, 6, 63, 67, 127–8, 271; *see also* discretion
Avon, 123–4

balance, fiscal, 22, 24, 43, 70, 74–5, 304–5
balances, revenue, 220

Balfour, Lord, 51–2, 67
Barking, 312
Barnsley, 164, 243
base authority, 228
base year, 116
Basic Grant, 52, 59–60
basic payment, 90–2
Battersea, 312
Beasley, Michael, 272
Beckenham, 312
Bedfordshire, 139, 247
beneficial services, 42, 215, 271–2
benefits, 2, 5, 24, 213, 249, 276, 285–6, 287–93, 300–2, 309, 323, 325; access to, 5; in-kind, 290–2; tapering of, 323; utility of, 217
Berkshire, 247
Bexley, 240, 243, 245, 312
bias, statistical, 221
Birmingham, 149, 154, 240
birthrate, 141
Blackburn, 102
Blaenau-Gwent, 162
block grant (RSG), 64–7, 103–30, 205, 217, 223–4, 238–9, 244–9, 268–9, 305–9; *see also* grants
bonds, 220
Boundary Commission, 55
Bradford, 240, 247
Brent, 247
Bromley, 319
Buckinghamshire, 241, 243, 245–6
Budget, The, 72
budgets: local authority, 209, 217, 248; scaled minimum volume, 121–3
buoyancy of revenue, 2, 15, 43, 55, 64, 74–5, 164, 254–5, 262–5, 267, 283–4, 284–5, 309, 322–3
burden of taxes, 249, 276, 285, 287, 300–2, 304, 309, 325
Bury, 240, 241, 243, 245–7

Calderdale, 243, 245–6
California (Proposition 13), 284, 323
Cambridgeshire, 169, 241, 311

343

General Exchequer Contribution, 51–3, 67
General Grant, 6, 55, 57–60, 61, 67, 86,
    90–1, 97–8, 208, 311–12
general grants, *see* grants
geographical effects, 248, 276–7, 287, 289,
    293–4, 300
Germany, 1, 3, 22, 262, 264, 285; Bund, 4
GLC, *see* Greater London Council
GLIM, 226; estimates, 229–34
Gloucester, 240
Goschen, G. J., 15, 49, 50, 54, 63, 68–9, 262
grants, 219–21, 262–3, 267–9, 283, 294–300,
    308; block, 19–20, 51–3, 55, 61, 269, 274,
    278–83, 292, 323 (*see also* block grant);
    categorical, 19, 319; closed-ended, 20, 22,
    84, 85, 109–10, 117, 122, 127–9, 222, 303,
    306; cost equalising, 39–42; elasticity to
    expenditure, 209, 232–48; general, 19, 23,
    213–14, 262, 274, 286 (*see also* block
    grant); marginal rates of, 102; matching,
    19–21, 24, 213–15; model of, 221–49;
    need-equalising, 19–21, 30–3, 281, 306;
    non-matching, 19–20; objectives of, 21–5;
    open-ended, 20, 22, 84, 127, 319; pooling
    of, 39; proposed new structure of, 259,
    267–70, 274, 294–301; proportional, 29,
    33–4, 139; resource-equalising, 19–21,
    26–30, 280–1 (*see also* resources element);
    specific, 3, 19–20, 23–4, 49, 55, 58–9, 60–1,
    72, 74–8, 84, 105, 107, 122, 130, 173,
    213–14, 254–5, 262, 269, 274, 278–83, 286,
    292–6, 299–300, 309; typology of, 19–21;
    unitary, 12, 34–9, 84, 103–30, 203–8, 279,
    306; *see also* block grant
Grant Related Expenditure (GRE), 104–9,
    111, 116–28, 136, 138, 141, 148, 150,
    153–4, 159, 218–19, 239, 256, 275–6, 306
Grant Related Poundage (GRP), 104,
    109–11, 112–16, 118, 121, 127–9, 256
Grants Working Group, 104
GRE, *see* Grant Related Expenditure
Greater London Council (GLC), 103, 105,
    119, 138–9, 176, 307, 319–20
Greater Manchester, 195, 199
Greenwich, 243, 245–7, 312
Green Paper, (1971), 4, 63; (1977), 63, 125;
    (1981), 66
growth, economic, 2, 4, 256, 302
GRP, *see* Grant Related Poundage
Gwent, 154
Gwynedd, 154

Hackney, 243, 245, 247
Hammersmith, 162, 247
Hampshire, 176
handicapped, services for, 143
Harringey, 243, 245
Harrogate, 129
Harrow, 240, 241

Hartlepool, 311
Havering, 243, 245
health service, 19, 47–8, 54, 58–9, 62–3, 67,
    74, 137, 211, 320; Regional Health
    Authorities, 54
Hereford and Worcestershire, 240
Heseltine, Michael, Secretary of State for
    the Environment, 102, 105, 121, 319
Hicks, J. R. and Hicks, U., 53–4, 68, 174–6
Hillingdon, 243, 247
holdback, 66, 117
hold-harmless, *see* damping
home helps, 134, 257, 276
hospitals, 46, 54, 215
housing: publicly provided, 10, 46, 49, 54–5,
    58, 62–3, 77, 106, 126, 134, 148, 203,
    211–12, 256, 270–1, 276, 284; starts, 203
Housing Improvement Grants, 278;
    Revenue Account, 77, 108, 270–1
Hounslow, 247
HRA, *see* Housing Revenue Account
Humberside, 199, 240, 247
Huntingdonshire, 311

ILEA, see Inner London Education
    Authority
IMTA, see Institute of Municipal Treasurers
    and Actuaries
incentives to spend, 20, 289
incidence, fiscal, 5, 210; *see also* benefits,
    burdens
income, data, 310, 313; effects of grants, 213;
    groups, 3; levels of, 93, 216, 264–7; tax, *see*
    tax, local income tax; *see also* salaries,
    wages
independence of local authorities, 56, 58,
    134, 138–9, 153, 243, 254–8, 262–7, 271,
    284, 285–6, 293, 299, 303, 305, 307; *see*
    also autonomy, discretion
India, 49
inflation, 73–5, 99, 117, 122, 271, 284, 323;
    elasticity of taxes, 2, 63, 84; *see also* prices
inflexibility of revenues, 43
infrastructure, 19, 40, 215, 272, 279
Inland Revenue, 55, 259, 265–6, 285, 300–1,
    324–5
inner cities, 148–9, 155, 158–9, 161, 164, 169,
    176, 190, 195, 206–8, 296, 299, 325
Inner City, Partnerships, 10–11, 226;
    Programmes, 190, 226; Special Powers,
    226
Inner London Education Authority, 103,
    105, 120–1, 138–9
inputs to services, 6, 31, 40–2, 135–6, 213,
    219, 282
Institute of Municipal Treasurers and
    Actuaries, 320
intelligibility of grants, 23–4, 95, 126
intensity of services, 74, 213, 218–19

privatisation, 272
probation service, 49, 278
productivity, 24–5, 216, 275, 289; *see also* labour/output ratio
professional judgment, 109
progression, 21–2, 287–8, 293; *see also* taxes
property tax, *see* the rates
protection from penalties of grants, 118–21, 128–9
provision, 229
pss, *see* personal social services
Public Expenditure Survey Committee, 70–2, 77
public finance, economic theory, 229, 248
public goods, 23, 92, 126, 271, 302, 322; local, 23; private/public, 126, 267, 302
pupil/teacher ratio, 40–1, 65, 219

quality of services, 40–2, 75, 87, 126, 176, 218–19, 282; *see also* extensiveness, intensiveness

Radcliffe Report, 55
Rate Deficiency Grant, 6, 55, 57–60, 82–6, 192–3, 205, 208
rate rebates, 60, 73, 75, 90, 92, 106, 164, 287, 304
Rate Support Grant (RSG), *see under* separate elements
rateable value, 26–30, 156, 158, 162–73, 174, 195–6, 206–7, 282, 311–14, 322; commercial, 162, 165, 168–70; credited, 82, 170; industrial, 165, 169–70; residential, 162, 167, 169–70; standard, 27–30, 170–2, 228, 232, 235, 237–8, 303, 305; *see also* tax capacity, tax base
ratepayers, 55, 57, 61, 307
rates, the (property tax), 43–6, 49, 63–4, 162–73, 195, 201, 206–7, 258–60, 263, 283–4, 304, 307, 323; commercial, 78–81, 86; domestic, 66, 78–81, 116, 130, 266; industrial, 78–81, 86; reductions of, 50; reform of, 60; regressivity of, 63–4, 75; supplementary, 65, 66, 117, 119, 121, 128–9; *see also* tax base, tax rate, valuation, yield
rational choice, 216
rebates, 73, 263–5; *see also* rate rebates
recreation services, 58, 137, 271–2
Redbridge, 312
reduced form, 215, 217, 221, 225–34
refuse collection and disposal, 41, 62–3, 105, 137, 270–2
regional bodies, 64; *see also* health
Registrar General's Population Estimates, 99
regression: categorical, 40, 90, 92–9, 106–7, 109, 127, 134, 153, 182, 197, 201, 206–7, 257, 262, 306, 319; coefficients, 95–8; formula, 208, 305; indicator, 197

regulations, central government, 289; *see also* Circulars
relevant expenditure, *see* expenditure
rents, public housing, 55, 73, 108, 270–1, 320
Rent Acts, 60
Reorganisation of Local Government (1974), 6, 8, 54, 60, 61, 63, 67, 69, 73–4, 81, 161, 312
Reorganisation of London Government (1963), 6, 8, 19, 65, 311
Reports (RSG), 69, 72–3, 255
residential location, 259; *see also* migration
residuals, 244–7
resources: clawback (*see* clawback); element, 60–1, 78, 82–6, 103, 114, 116, 123, 125, 130, 192–7, 201, 203–7, 217, 222–3, 228–44, 280–1, 303–5, 311–14; expenditure, 84; *see also* grants
retailing, 165
revenue: estimates, 51–2, 224–5; of British local authorities, 43–6, 254–8; sharing, 14, 19, 20, 22, 49, 57, 254, 258, 262–7, 269, 284–7, 294–7, 299, 304, 308–10 (*see also* assigned revenues); sharing of LIT, 262–7, 279, 285–98, 309–10
Richmond-on-Thames, 319
roads: expenditure, 40–1, 47–8, 54, 66, 211, 220, 313 (*see also* transport services); mileage, 52, 56, 57–8, 59, 61, 88–90, 148–9, 193; safety, 58; vehicles, 137
Rochdale, 243, 245
Rooker-Wise amendment, 323
Rotherham, 247
rural areas, 6, 50, 53, 56, 57, 61, 67, 99, 107, 108, 143, 148, 199, 212, 217, 245–9, 296, 306

safety net, 98, 100, 110, 115–16, 127
Saint Helens, 164, 170, 247
salaries: public, 5, 9, 185, 257, 259, 305; teachers, 99; *see also* income
sales tax, *see* taxes
Salford, 243, 247
Sandwell, 241, 243, 245–6, 313
scale economies, 108; *see also* size
school children, 137; meals and milk, 65, 270–1, 320; *see also* education
Scotland, 51, 61, 64, 68, 72, 86, 87, 121, 128, 254
Scunthorpe, 162
Secretary of State for the Environment, 69, 70; *see also* Heseltine, M.
Seebohm Committee, 63
self-assessment for taxes, 260–1
service maintenance, 213–14
services, *see* under separate headings
sewerage, 74
shape of local authorities, 40
Sheffield, 154, 170